T0174716

Ionizing Radiation Effects in Electronics

From Memories to Imagers

EDITED BY
MARTA BAGATIN
University of Padova, Padova, Italy

SIMONE GERARDIN
University of Padova, Padova, Italy

Devices, Circuits, and Systems

Series Editor
Krzysztof Iniewski
ET CMOS Inc.
Vancouver, British Columbia, Canada

PUBLISHED TITLES:

PUBLISHED TITLES:

FORTHCOMING TITLES:

Silicon on Insulator System Design
Bastien Giraud

Semiconductor Devices in Harsh Conditions
Kirsten Weide-Zaage and Malgorzata Chrzanowska-Jeske

Smart eHealth and eCare Technologies Handbook
Sari Merilampi, Lars T. Berger, and Andrew Sirkka

Structural Health Monitoring of Composite Structures Using Fiber Optic Methods
Ginu Rajan and Gangadhara Prusty

Tunable RF Components and Circuits: Applications in Mobile Handsets
Jeffrey L. Hilbert

Wireless Medical Systems and Algorithms: Design and Applications
Pietro Salvo and Miguel Hernandez-Silveira

Ionizing Radiation Effects in Electronics

From Memories to Imagers

EDITED BY
MARTA BAGATIN
University of Padova, Padova, Italy

SIMONE GERARDIN
University of Padova, Padova, Italy

KRZYSZTOF INIEWSKI
MANAGING EDITOR
Emerging Technologies CMOS Inc.
Vancouver, British Columbia, Canada

CRC Press
Taylor & Francis Group
Boca Raton London New York

CRC Press is an imprint of the
Taylor & Francis Group, an **informa** business

MATLAB® is a trademark of The MathWorks, Inc. and is used with permission. The MathWorks does not warrant the accuracy of the text or exercises in this book. This book's use or discussion of MATLAB® software or related products does not constitute endorsement or sponsorship by The MathWorks of a particular pedagogical approach or particular use of the MATLAB® software.

CRC Press
Taylor & Francis Group
6000 Broken Sound Parkway NW, Suite 300
Boca Raton, FL 33487-2742

First issued in paperback 2020

© 2016 by Taylor & Francis Group, LLC
CRC Press is an imprint of Taylor & Francis Group, an Informa business

No claim to original U.S. Government works

ISBN-13: 978-1-4987-2260-5 (hbk)
ISBN-13: 978-0-367-65595-2 (pbk)

This book contains information obtained from authentic and highly regarded sources. Reasonable efforts have been made to publish reliable data and information, but the author and publisher cannot assume responsibility for the validity of all materials or the consequences of their use. The authors and publishers have attempted to trace the copyright holders of all material reproduced in this publication and apologize to copyright holders if permission to publish in this form has not been obtained. If any copyright material has not been acknowledged please write and let us know so we may rectify in any future reprint.

Except as permitted under U.S. Copyright Law, no part of this book may be reprinted, reproduced, transmitted, or utilized in any form by any electronic, mechanical, or other means, now known or hereafter invented, including photocopying, microfilming, and recording, or in any information storage or retrieval system, without written permission from the publishers.

For permission to photocopy or use material electronically from this work, please access www.copyright.com (http://www.copyright.com/) or contact the Copyright Clearance Center, Inc. (CCC), 222 Rosewood Drive, Danvers, MA 01923, 978-750-8400. CCC is a not-for-profit organization that provides licenses and registration for a variety of users. For organizations that have been granted a photocopy license by the CCC, a separate system of payment has been arranged.

Trademark Notice: Product or corporate names may be trademarks or registered trademarks, and are used only for identification and explanation without intent to infringe.

Visit the Taylor & Francis Web site at
http://www.taylorandfrancis.com

and the CRC Press Web site at
http://www.crcpress.com

Contents

Preface

There is an invisible enemy that constantly threatens the operation of electronics: ionizing radiation. From sea level to outer space, ionizing radiation is virtually everywhere. At sea level and even more at aircraft altitudes, atmospheric neutrons, originating from the interaction of cosmic rays with the atmosphere, constantly bombard electronic devices. Alphas, emitted by radioactive contaminants in the chip materials, are another major threat in terrestrial applications. In space, satellites and spacecraft can be hit by highly energetic particles such as protons, electrons, and heavier particles, due to radiation belts, solar activity, and galactic cosmic rays. Finally, man-made radiation in environments such as nuclear power plants or high-energy physics experiments can expose devices to extreme amounts of radiation.

Depending on the type and characteristics of the impinging radiation and struck device, different effects, both irreversible (hard) and reversible (soft), may arise. Single event effects are stochastic events caused by a single particle striking a device. If enough energy is deposited in a sensitive device region, different kinds of malfunctions can take place, ranging from the corruption of the information stored in a memory bit to the catastrophic burnout of a power MOSFET.

In other cases, it is the progressive buildup of damage caused by several particles, during a prolonged exposure to ionizing radiation, that generates drifts in component parameters (for instance, a shift in the threshold voltage or an increase in power consumption). This second type of effects can involve damage to either dielectrics layers (total ionizing dose) or bulk semiconductor materials (displacement damage). Annealing effects can take place, resulting in improvement or, sometimes, worsening of the device degradation.

Usually, electronics response to radiation is tested through accelerated ground experiments, which allow engineers to simulate several years of exposure in radiation harsh environments in just a few hours, thanks to the use of radioactive sources or particle accelerators.

The knowledge of the mechanisms underlying the effects of radiation on electronic devices is of primary importance for developing suitable hardness assurance methodologies. Ground-based testing can only partially reproduce the spectrum and characteristics of radiation found in space, especially in terms of dose rates and particle energy. In addition, the ever growing cost associated with radiation testing, due to the increasing complexity and variability of advanced technologies, requires a deep understanding of basic mechanisms to make a good and effective use of beam time.

The aim of this book is to provide the reader with a wide perspective on the effects of ionizing radiation in modern semiconductor devices and on solutions to harden them. It includes contributions from top-level experts in the field of ionizing radiation effects, coming from industry, research laboratories, and academia. Thanks to the background material, case studies, and updated references, this book is suitable both for newcomers who want to become familiar with radiation effects and also for radiation experts who are looking for more advanced material.

The book is structured in the following way: The first two chapters provide background information on radiation effects, the underlying basic mechanisms, and the use of Monte Carlo techniques to simulate radiation transport and the effects of radiation on electronics. Then a series of chapters on state-of-the-art digital commercial devices are presented, including volatile and nonvolatile memories (SRAMs, DRAMs, and Flash memories) and microprocessors. Then the next set of chapters deal with hardening-by-design solutions in digital circuits, FPGAs, and mixed-analog circuits, including a case study on the read-out electronics of pixel detectors for high-energy physics applications. Finally, the last three chapters deal with radiation effects on imager devices (CMOS sensors and CCDs) and fiber optics.

MATLAB® is a registered trademark of The MathWorks, Inc. For product information, please contact:

The MathWorks, Inc.
3 Apple Hill Drive
Natick, MA 01760-2098 USA
Tel: 508-647-7000
Fax: 508-647-7001
E-mail: info@mathworks.com
Web: www.mathworks.com

Editors

Marta Bagatin received the Laurea degree (cum laude) in electronic engineering in 2006 and the PhD degree in information science and technology in 2010 both from the University of Padova, Italy. She is currently a postdoc at the Department of Information Engineering, University of Padova. Her research interests concern radiation and reliability effects on electronic devices, especially on nonvolatile semiconductor memories. Marta has authored or coauthored about 40 papers in peer-reviewed international journals on radiation and reliability effects, about 50 contributions at international conferences, and 2 book chapters. She regularly serves in the committee of international conferences on radiation effects such as the Nuclear and Space Radiation Effects Conference and Radiation Effects on Components and System, and as a reviewer for numerous scientific journals.

Simone Gerardin received the Laurea degree (cum laude) in electronics engineering in 2003 and the PhD degree in electronics and telecommunications engineering in 2007 both from the University of Padova, Italy. He is currently an assistant professor at the same university. His research is focused on soft and hard errors induced by ionizing radiation in advanced CMOS technologies, and on their interplay with device aging and ESD. Simone has authored or coauthored more than 60 papers published in international journals, more than 60 conference presentations, 3 book chapters, and 2 tutorials at international conferences on radiation effects. He is currently an associate editor for the *IEEE Transactions on Nuclear Science* and a reviewer for several scientific journals and member-at-large of the radiation effects steering group.

Krzysztof (Kris) Iniewski manages the R&D at Redlen Technologies Inc., a startup company in Vancouver, Canada. Redlen's revolutionary production process for advanced semiconductor materials enables a new generation of more accurate, all-digital, radiation-based imaging solutions. Kris is also the president of CMOS Emerging Technologies Research Inc. (http://www.cmosetr.com), an organization of high-tech events covering communications, microsystems, optoelectronics, and sensors. In his career, Dr. Iniewski held numerous faculty and management positions at the University of Toronto, University of Alberta, SFU, and PMC-Sierra Inc. He has published more than 100 research papers in international journals and conferences. He holds 18 international patents granted in the United States, Canada, France, Germany, and Japan. He is a frequently invited speaker and has consulted for multiple organizations internationally. He has written and edited several books for CRC Press, Cambridge University Press, IEEE Press, John Wiley & Sons, McGraw-Hill, Artech House, and Springer Science+Business Media. His personal goal is to contribute to healthy living and sustainability through innovative engineering solutions. In his leisure time, Kris can be found hiking, sailing, skiing, or biking in beautiful British Columbia. He can be reached at kris.iniewski@gmail.com.

Contributors

Jean-Luc Autran
Aix-Marseille University & CNRS
IM2NP (UMR 7334)
Faculté des Sciences–Service 142
Marseille, France

Marta Bagatin
Department of Information Engineering
University of Padova
Padova, Italy

Stefano Bettarini
Università degli Studi di Pisa
and
INFN
Pisa, Italy

Luciano Bosisio
Università degli Studi di Trieste
and
INFN
Trieste, Italy

Aziz Boukenter
Laboratoire Hubert Curien
Université de Saint-Etienne
Saint-Etienne, France

Michael P. Caffrey
Los Alamos National Laboratories
Los Alamos, New Mexico

Lawrence T. Clark
Arizona State University
Tempe, Arizona

Francesco Forti
Università degli Studi di Pisa
and
INFN
Pisa, Italy

Luigi Gaioni
Università degli Studi di Bergamo
and
INFN
Pavia, Italy

Gilles Gasiot
STMicroelectronics
Crolles, France

Simone Gerardin
Department of Information Engineering
University of Padova
Padova, Italy

Luigi Giacomazzi
CNR-IOM, Democritos National
    Simulation Center
Trieste, Italy

Sylvain Girard
Laboratoire Hubert Curien
Université de Saint-Etienne
Saint-Etienne, France

Vincent Goiffon
Institut Supérieur de l'Aéronautique
    et de l'Espace (ISAE-SUPAERO)
Université de Toulouse
Toulouse, France

Paul S. Graham
Los Alamos National Laboratories
Los Alamos, New Mexico

Steven M. Guertin
Jet Propulsion Laboratory/California
    Institute of Technology
Pasadena, California

Martin Herrmann
IDA
TU Braunschweig
Braunschweig, Germany

William Timothy Holman
Department of Electrical Engineering
   and Computer Science
Vanderbilt University
Nashville, Tennessee

James B. Krone
Los Alamos National Laboratories
Los Alamos, New Mexico

David Lee
Sandia National Laboratories
Albuquerque, New Mexico

Thomas Daniel Loveless
Electrical Engineering Department
College of Engineering and Computer
   Science
University of Tennessee
Chattanooga, Tennessee

Kevin Lundgreen
Raytheon Applied Signal Technology
Waltham, Massachusetts

Massimo Manghisoni
Università degli Studi di Bergamo
and
INFN
Pavia, Italy

Claude Marcandella
CEA, DAM, DIF
Arpajon, France

Layla Martin-Samos
Materials Research Laboratory
University of Nova Gorica
Nova Gorica, Slovenia

Soilihi Moindjie
Aix-Marseille University & CNRS
IM2NP (UMR 7334)
Faculté des Sciences–Service 142
Marseille, France

Keith S. Morgan
Los Alamos National Laboratories
Los Alamos, New Mexico

Fabio Morsani
Università degli Studi di Pisa
and
INFN
Pisa, Italy

Daniela Munteanu
Aix-Marseille University & CNRS
IM2NP (UMR 7334)
Faculté des Sciences–Service 142
Marseille, France

Youcef Ouerdane
Laboratoire Hubert Curien
Université de Saint-Etienne
Saint-Etienne, France

Philippe Paillet
CEA, DAM, DIF
Arpajon, France

Brian Pratt
L-3 Communications
Salt Lake City, Utah

Heather M. Quinn
Los Alamos National Laboratories
Los Alamos, New Mexico

Mélanie Raine
CEA, DAM, DIF
Arpajon, France

Irina Rashevskaya
Università degli Studi di Trieste
and
INFN
Trieste, Italy

Lodovico Ratti
Università degli Studi di Pavia
and
INFN
Pavia, Italy

Valerio Re
Università degli Studi di Bergamo
and
INFN
Pavia, Italy

Nicolas Richard
CEA, DAM, DIF
Arpajon, France

Giuliana Rizzo
Università degli Studi di Pisa
and
INFN
Pisa, Italy

Philippe Roche
STMicroelectronics
Crolles, France

Tarek Saad Saoud
Aix-Marseille University & CNRS
IM2NP (UMR 7334)
Faculté des Sciences–Service 142
Marseille, France

Gary M. Swift
Swift Engineering Research
San Jose, California

Gianluca Traversi
Università degli Studi di Bergamo
and
INFN
Pavia, Italy

Michael J. Wirthlin
Brigham Young University
Provo, Utah

1 Introduction to the Effects of Radiation on Electronic Devices

Simone Gerardin and Marta Bagatin

CONTENTS

1.1 INTRODUCTION

The presence of ionizing radiation may be a significant threat to the correct operation of electronic devices, both in the terrestrial environment, due to atmospheric neutrons and radioactive contaminants inside chip materials, or, to a much larger extent, in space, because of trapped particles, particles emitted by the Sun, and galactic cosmic rays. Artificial man-made radiation generated in biomedical devices, nuclear power plants, and high-energy physics experiments is another reason to carefully study radiation effects in electronic components.

The fundamental fact about ionizing radiation is that it deposits energy in the target material. As a result, radiation can cause a variety of effects: corruption in memory bits, glitches in digital and analog circuits, increase in power consumption, and speed reduction, in addition to complete loss of functionality in the most severe cases.

Analysis of radiation effects is necessary when designing electronic systems that must operate onboard satellites and spacecrafts, but it is also mandatory when

1

developing high-reliability systems to be used on the ground, such as bank servers, biomedical devices, avionics, or automotive components.

In this chapter we will describe the most relevant radiation environments, and then analyze the three main categories of radiation effects: total ionizing dose (TID), displacement damage (DD), and single event effects (SEEs). The first two are progressive drifts in electronic device parameters due to degradation of insulators and semiconductor materials, continuously hit by several ionizing particles, and occur mainly in space or due to artificial sources of radiations. In contrast, SEEs are due to the stochastic interaction of a single particle having high ionization power with sensitive regions of an electronic device, and occur both in space and in the terrestrial environment.

1.2 RADIATION ENVIRONMENTS

Electronic devices must often operate in environments with a significant presence of ionizing radiation. To ensure correct operation, one has to precisely know the features of the particular environment in which the component is expected to work. We will start this section illustrating the space environment, one of the harshest from a radiation standpoint. After that, we will consider the terrestrial environment characterized by neutrons and alpha particles. Finally, we will discuss man-made environments, such as nuclear power plants and high-energy physics experiments.

1.2.1 SPACE

As illustrated in Figure 1.1, there are three main sources of ionizing radiation in the space environment [1]:

1. Galactic cosmic rays
2. Particles generated during solar particle events
3. Particles trapped inside planets' magnetospheres

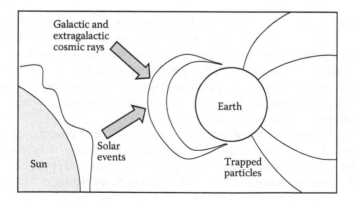

FIGURE 1.1 Schematic illustration of the three main sources of radiation in space: cosmic rays, particles generated during solar events, and particles trapped in the Earth's magnetosphere.

Galactic cosmic rays are known to originate from outside our solar system, but their source and acceleration mechanisms are not yet completely clear. For the most part they are made of protons, but they include all elements and can reach very high energies, up to 10^{11} GeV, which make them very penetrating and virtually impossible to shield with reasonable amounts of material. Fluxes of galactic cosmic rays are in the order of a few particles per square centimeter per second.

The second category of ionizing particles in space comes from the Sun. These particles include all naturally occurring elements, from protons to uranium, and their flux is dependent on the solar cycle and can reach values larger than 10^5 particles/cm^2/s with energy >10 MeV/nucleon. Solar activity is cyclic, alternating 7 years of high activity followed by 4 years of low activity. The changing number of sunspots is one of the most important manifestations of this cycle. During the declining phase of the solar maximum, solar particle events occur more frequently. These include coronal mass ejections and solar flares. The first are eruptions of plasma originating from a shock wave followed by an emission of particles. In contrast, solar flares take place when an increase in a coronal magnetic field causes a sudden burst of energy. In addition to solar particle events, a continuous, progressive loss of mass from the Sun occurs, consisting of protons and electrons that acquire enough energy to escape gravity. These particles feature an intrinsic magnetic field, which can interact with planets' magnetic fields. Interestingly, the solar cycle also modulates the galactic cosmic ray flux: the higher the activity of the Sun, the lower the flux of cosmic rays, thanks to the shielding effects of solar particles. In addition, the Sun interacts with planets' magnetospheres, in particular with the Earth's. Let us now focus on the Earth.

The magnetic field associated to the Earth (which has two components: an intrinsic one and an external one deriving from the solar wind) is able to capture charged particles. These particles, once confined to the Earth's magnetic field, move in a spiral, following field lines and bouncing from one pole to the other. Furthermore, they move longitudinally at a slower velocity in a direction dependent on the sign of their charge. Two distinct belts are formed by the particles trapped in the Earth's magnetic field: the outer belt, made for the most part of electrons, and the inner belt, consisting of both electrons and protons. Fluxes of electrons with energy above 1 MeV can reach 10^6 particles/cm^2/s; those of trapped protons can be as high as 10^5 particles/cm^2/s.

A peculiar feature of the Earth's radiation belts is the South Atlantic Anomaly (SAA), where the radiation belts come closest to Earth. The SAA is caused by the fact that the magnetic field axis forms an $11°$ angle with respect to the North-South axis, and its center is not located at the Earth's center, but is about 500 km away from it, causing a dip in the magnetic field over the South Atlantic area. The SAA is the area where most errors and malfunctions occur in satellites placed in low orbits.

Due to the complexity of the environment, the amount of ionizing particles hitting a system in space is difficult to assess and is highly dependent on the solar cycle and the orbit. In addition, the radiation dose received by a given electronic device also depends on its location inside the spacecraft or satellite because of the shielding effect of the surrounding material. In space, it is very important not to overdesign electronic systems due to the high cost of launching additional kilograms and the scarcity of power on board. Complex simulation tools and models have been made available to help designers to predict the dose and design systems with appropriate margins.

1.2.2 TERRESTRIAL ENVIRONMENT

Atmospheric neutrons and alpha particles from radioactive contaminants in chip materials are the two most important sources of soft errors in electronic devices at ground level [2,3].

Even though neutrons are not charged, they can indirectly ionize the target material because they are able to trigger nuclear reactions, giving rise to charged secondary byproducts. These, in turn, may deposit charge in sensitive volumes of electronic devices. If the deposited charge is collected by sensitive nodes, disturbances in the operation of devices can take place. Atmospheric neutrons originate from the interaction of cosmic rays with the outer layers of the atmosphere and are among the most abundant (indirectly) ionizing particles at sea level (Figure 1.2; Ref. [4]). Cosmic rays can be divided into primary cosmic rays (mainly protons and helium nuclei), coming from the space outside our solar system, and secondary cosmic rays, created from primary cosmic rays interacting with the atmosphere. As cosmic rays go through the layers of the atmosphere, they interact with nitrogen and oxygen atoms and generate a cascade of secondary particles. In this process, many different particles (protons, pions, muons, neutrons) and an electromagnetic component are produced. These particles can, in turn, have enough energy to create further particle cascades. As cosmic rays penetrate into the Earth's atmosphere, the number of particles first increases and then decreases when the shielding effect of the atmosphere dominates over the multiplication. The atmospheric neutron flux increases with altitude, as shown in Figure 1.2, with a peak around 15 km, which is why avionics is one of the

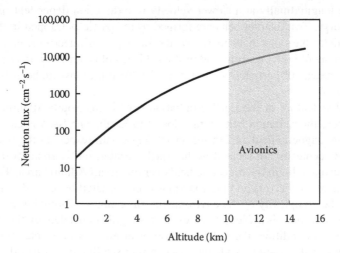

FIGURE 1.2 Flux of neutrons in the terrestrial environment as a function of the altitude. (Data from http://seutest.com/cgi-bin/FluxCalculator.cgi.)

applications where electronics is more threatened by neutrons. The dependence of the neutron flux on energy (E) displays a 1/E dependence. From the standpoint of radiation effects, there are mainly two types of neutron energy that are of interest:

- Thermal neutrons (energy about 25 meV), which feature a large interaction cross section with the boron isotope ^{10}B, often found in intermetal layers of integrated circuits or as dopant
- Neutrons with energy above 10 MeV, which can produce nuclear reactions with chip materials, such as silicon and oxygen, giving rise to charged byproducts

Besides depending on altitude, the neutron flux is also determined by other factors, such as solar activity, latitude, and atmospheric pressure. The reference neutron flux is that of New York City, which is about 14 n/cm^2/hour for neutrons with energy above 10 MeV. Tables are available to calculate neutron flux in other locations.

Alpha particles coming from the decay of radioactive contaminants in integrated circuit (IC) materials are the second source of radiation-induced effects in electronics operating at terrestrial level. A large part of the soft errors occurring at sea level are caused by the decay of elements, such as ^{238}U, ^{234}U, ^{232}Th, ^{190}Pt, ^{144}Nd, ^{152}Gd, ^{148}Sm, ^{187}Re, ^{186}Os, and ^{174}Hf, which are all alpha emitters. These elements can be either intentionally used in IC fabrication or unwanted impurities. Even though the generated alphas have a small ionizing power, soft errors induced by alphas are becoming increasingly more important than those induced by atmospheric neutrons because of the reduction in critical charge with each new generation as the circuit feature size scales down. A typical value for the alpha emission level in an integrated circuit is on the order of 10^{-3} alphas/cm^2/hour.

It is worth mentioning that the continuous shrinking of dimensions is opening the doors to new threats, such as muons. In fact, muons were considered innocuous for electronics until a few generations ago, but today the complementary metal-oxide semiconductor (CMOS) feature size has shrunk so much that even muons may produce soft errors in particular conditions [5].

1.2.3 MAN-MADE RADIATION

Some man-made radiation environments are very harsh in terms of ionizing radiation [6]. For instance, doses in excess of 100 Mrad(Si) are expected in the planned upgrade of the current Large Hadron Collider (LHC) at Conseil Européen pour la Recherche Nucléaire (CERN), Switzerland, one of the largest high-energy physics experiments (for comparison most National Aeronautics and Space Administration [NASA] missions in space are below 100 krad(Si)). As a consequence, these environments require custom-made electronics capable of withstanding high levels of radiation. This is usually achieved through dedicated libraries of rad-hard by design components, where the layout is carefully studied to avoid the problems of standard designs.

Ionizing radiation is also an issue in nuclear fission power plants and future fusion plants under development. For instance, in a fusion reactor such as the ITER, electronic systems for plasma control and diagnostics are placed near the vessel and bio-shield, where they are expected to be hit by large fluxes of neutrons (Deuterium-Tritium reactions produce 14-MeV neutrons), x rays and gamma rays, etc. To give an idea of the numbers, in the ITER environment doses in the range of 50 rad(Si) in one operating hour are expected [7].

1.3 TOTAL IONIZING DOSE EFFECTS

Total ionizing dose (TID) is the amount of energy deposited by ionization processes in the target material. TID is measured in rad, 1 rad corresponding to 100 ergs of energy deposited in one gram of material by the impinging radiation. As absorption depends on the target material, the radiation dose is usually indicated with the target material. The most commonly used units are rad and gray (Gy). One Gy corresponds to a deposition of 1 joule per kg of target material. One rad corresponds to 100 Gy. There are two main effects of TID on electronic devices [8,9]:

- Generation of defects in insulating layers
- Buildup of (positive) trapped charge in insulating layers

Because of TID, the metal-oxide-semiconductor field-effect transistor (MOSFET) experiences shifts in the threshold voltage, decreases in transconductance, and leakage currents. Technology scaling causes the gate oxide to become thinner and thinner, leading to a reduction in the amount of radiation-induced charge trapping and interface states. As a result, after the introduction of ultrathin gate oxides, total dose issues in low-voltage MOSFETs are mainly related to the thick lateral isolations and oxide spacers. TID effects in MOSFETs are time-dependent but not dose-rate-dependent.

In bipolar devices, charge trapping and defect formation can produce decreases in gain and leakage currents. A peculiar phenomenon occurring in bipolar devices is the enhanced low dose rate sensitivity (ELDRS): as the name suggests, degradation is larger at low dose rates than at high ones.

The basic mechanisms behind charge trapping and interface state generation in oxide layers are depicted in Figure 1.3, which shows the energy band diagram of a MOS capacitor on a p-substrate, biased at positive voltage. Radiation generates defects in insulating layers through indirect processes; that is, it does not directly break bonds, but releases positive particles (holes and hydrogen ions), which are responsible for the radiation response of the exposed devices.

When radiation impinges on a dielectric layer, it causes the generation of energetic electron-hole pairs. After a few picoseconds, the generated carriers thermalize, losing much of their energy. The electrons, thanks to their high mobility, are quickly swept toward the anode by the applied or built-in potential, whereas the heavier and slower holes move inside the oxide in the opposite direction. But before they do that, a large part of the e-h pairs recombine. The amount of recombination is given by the charge yield, which depends on the electric field, and the type and energy of the incident radiation.

The surviving holes may be trapped in preexisting deep traps while they migrate toward the cathode under the influence of the applied field. Hole transport occurs by

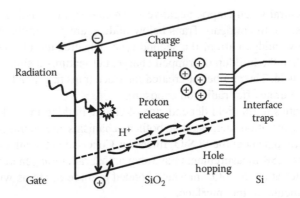

FIGURE 1.3 Energy band diagram of a MOS system on a p-substrate, biased at a positive voltage. (Adapted from O. Flament, J. Baggio, S. Bazzoli, S. Girard, J. Raimbourg, J. E. Sauvestre, and J. L. Leray, *Advancements in Nuclear Instrumentation Measurement Methods and their Applications [ANIMMA]*, 2009, p. 1.)

hopping through localized states. Since they are positively charged, holes are pushed towards the Silicon/Silicon Oxide (Si/SiO$_2$) or the gate/SiO$_2$ interface depending on the sign of the electric field. The positively charged holes introduce a local distortion in the electric field, which slows down their transport and makes it dispersive (i.e., occurring over many decades in time). This type of process is called polaron hopping and is highly dependent on temperature and external field. A polaron is the combination of the hole and the accompanying deformation of the electric field. If holes are transported to the Si/SiO$_2$ interface they can be trapped in defect sites whose density is typically higher close to the interface. The microscopic nature of these defects has been studied in detail. Electron spin resonance (ESR) has shown the presence of E' centers in silicon dioxide, a trivalent silicon associated to an oxygen vacancy, which is considered responsible for hole trapping in SiO$_2$. Vacancies are related to the out-diffusion of oxygen in the oxide and lattice mismatch at the surface. The amount of trapped charge depends on the number of holes that survive recombination, on the number of O vacancies, and on the field-dependent capture cross section of the traps. It is therefore very dependent on the quality of the oxide, with hardened ones showing orders of magnitude less radiation-induced charge trapping than soft oxides. Oxide hardness is strongly influenced by processing conditions: high temperature anneals, for instance, can increase the number of oxygen vacancies. Increasing the amount of hydrogen during processing also decreases oxide hardness, as discussed below.

During the polaron hopping process or when holes are trapped near the Si/SiO$_2$ interface, hydrogen ions (protons) are likely released (hopping is very slow and intrinsically localized, so the probability of such chemical effects is enhanced). Hydrogen ions arriving at the interface can generate interface traps. Protons, in fact, may react with hydrogen-passivated dangling bonds at the interface, causing the dangling bond to act as an interface trap. Interface traps may readily exchange carriers with the channel and are full or empty depending on the position of the Fermi level. Interface states are amphoteric: they are donor (positive when empty, neutral when charged) or acceptor (neutral when empty, negative when charged) depending on their position with respect to the midgap. Traps below midgap are predominantly of the first type; traps above midgap are of the second type. The creation of interface traps is much slower than the buildup of trapped charge, but features a similar dependence on the electric field. The number of created interface traps may need even thousands of seconds to saturate after radiation exposure.

The similarities in the field dependence of charge buildup in oxide traps and of interface trap generation suggest that both mechanisms are connected with hole transport and trapping near the Si/SiO_2 interface. Concerning the microscopic nature of these defects, ESR measurements have shown that radiation generates P_b centers, a trivalent center at the Si/SiO_2 interface bonded to three Si atoms with a dangling orbital perpendicular to the interface.

Annealing of charge in oxide traps starts immediately and is due to either tunneling or thermal processes. Indeed, the trapped charge can be neutralized by electrons thermally excited from the valence band, in which case the probability of these events is linked to the temperature and to the energy depth of the traps, with higher temperatures and shallower traps meaning faster annealing. On the other side, trapped charge can be neutralized by electrons tunneling through the oxide barrier; in this second case the process depends on the tunneling distance and trap energy spatial position, with thinner potential barriers (i.e., traps close to the interfaces), leading to faster annealing. Of course, the applied bias plays a fundamental role in determining the shape of the barrier and the direction of carriers.

In contrast, interface traps do not anneal at room temperature. Higher temperatures are required to reestablish the broken bonds. As a result, interface traps may play a predominant role in low dose-rate environments (such as space).

Trapped charge buildup, interface state generation, and anneal mechanisms are not instantaneous and have a strong time dependence. This time dependence can lead to apparent dose-rate effects (i.e., radiation can have a different impact depending on the rate at which radiation is delivered). At high dose rates and short times, annealing of charge in oxide traps is minimal and at the same time the number of interface traps has not reached saturation: as a result, the trapped charge contribution dominates over interface states. Instead, at low dose rate and long times, interfaces traps are prevalent. However, the key point is that if we give the same time to the trapped charge to anneal and to the interface traps to build up, the effects are independent of the dose rate. Thanks to this fact, radiation experiments on MOSFETs can be carried out at high dose rate, thus reducing the time needed for testing. If irradiation is then followed by moderate temperature annealing, one can bound the device response in a low dose-rate environment. Similar accelerated testing methods are also under development for bipolar devices, where, as we will see, the situation is made much more complex by true dose-rate effects.

1.3.1 MOSFETs

Positive charge trapping and generation of interface states can severely affect the behavior of a MOSFET [10,11]. This is shown in Figure 1.4, where the I_d-V_{gs} characteristic for an n-channel MOS transistor with thick gate oxide is depicted before and

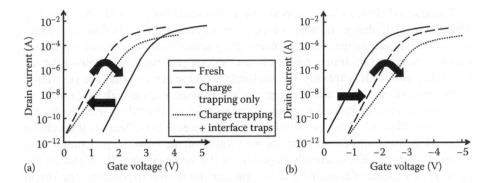

FIGURE 1.4 Effect of charge trapping and interface state formation on the I_d-V_{gs} characteristics of n-channel (a) and p-channel (b) MOSFETs.

after total ionizing dose exposure. As seen in the graphs, the effect of positive charge trapping in the gate oxide is to decrease the threshold voltage (i.e., to rigidly shift the I_d-V_{gs} curve toward lower values of V_{gs}). On the other hand, the formation of interface states decreases both the threshold voltage (by changing the subthreshold swing) and the carrier mobility (by adding Coulomb scattering centers). The behavior shown in Figure 1.4 is displayed by devices with thick gate oxide (>10 nm), such as power MOSFETs, where just a few krad(Si) can cause significant alterations. Ultrathin gate oxide transistors are affected by TID in a different way, as we will show later in detail.

Positive trapped charge and interface states cause additive effects in p-channel MOSFETs (Figure 1.4) because they both tend to shift the I_d-V_{gs} characteristic toward higher V_{gs}. On the other hand, the effects tend to cancel in n-channel MOSFET. This is the reason for the rebound effect: due to the different time constants of charge buildup and interface state formation, the threshold voltage first increases and then decreases during exposure to ionizing radiation. Detrapping and neutralization of trapped positive charge cause interface traps to be more important in low dose-rate environments (e.g., space) with respect to high dose ones.

Flicker noise, also known as low-frequency noise (LFN), is also affected by TID exposure. LFN is caused by trapping and detrapping of charge in defect centers located in the gate oxide, whose number can increase because of exposure to radiation, leading to fluctuations in carrier density and mobility.

Technology evolution, in particular thinning of the gate oxide, has been beneficial for total dose issues in low-voltage CMOS electronics. The effects and the amount of charge trapping are less and less severe as the oxide thickness is scaled down. The following formula, valid at low dose and for relatively thick oxides, expresses the dependence of the threshold voltage shift on the oxide thickness:

$$\Delta V_T = -\frac{Q_{OX}}{C_{OX}} \propto t_{ox}^2$$

The trapped charge Q_{OX} is proportional to the square of the oxide thickness t_{OX}. This decrease of charge trapping is even more accentuated for ultrathin gate oxides ($t_{OX} < 10$ nm) because electrons can more easily tunnel from the channel or the gate into the oxide and neutralize trapped holes. Silicon gate oxides for state-of-the-art low-voltage MOSFETs are only 1–2 nm thick: for these devices, both the generation of oxide trap charge and interface traps are not an issue, even at high total doses.

Unfortunately, the reduction of the gate oxide thickness has also some drawbacks. Radiation-induced leakage current (RILC) is an increase in the leakage through the gate oxide after exposure to particles with low ionizing power (e.g., gamma rays, electrons, x-rays). RILC linearly depends on the received total dose and on the applied bias (hence electric field in the gate oxide) during irradiation. The origin of this phenomenon has been found in inelastic trap assisted tunneling through the thin gate dielectrics. RILC may not be an issue in logic circuits, where it generally leads to a small increase in power consumption, but it can be a critical issue for flash memories, which are based on charge storage on an electrically insulated electrode (floating gate). For flash devices, the loss of charge due to RILC may degrade memory cell retention.

In modern low-voltage MOSFETs, TID issues come from the lateral isolation local oxidation isolation structure, LOCOS, in older devices, and Shallow Trench Isolation, STI, in state-of-the-art ones. Not only are these insulating layers still quite thick (100–1000 nm), hence prone to charge trapping, but they are also generally deposited (not thermally grown, as with gate oxides). In other words, the quality of these oxides is lower than the gate oxides, meaning that they are more susceptible to TID effects.

The effect of positive charge trapping in the STI following TID exposure in a modern low-voltage transistor is illustrated in Figure 1.5. The drain current of a MOSFET can be considered as the superposition of the current of the drawn MOSFET and of two parasitic MOSFETs, which have the same gate and channel of the drawn transistor and whose gate oxide is the lateral isolation (Figure 1.5a). These parasitic transistors are off at normal operating voltage, yet due to positive charge trapped in the lateral isolation, they may experience a decrease in their threshold voltage and start to conduct current in parallel to the drawn transistor (Figure 1.5b). These effects are visible only in n-channel MOSFETs (due to the sign of the radiation-induced threshold shifts) and are most evident when high electric fields are applied to the STI.

In addition to intradevice leakage, interdevice leakage can also occur. A conducting path can be formed between adjacent transistors when charge trapping in the isolation causes the inversion of the region underneath. This can lead to a dramatic increase in the static power consumption of a circuit.

During the last decade, some innovative solutions have been introduced to face roadblocks in Moore's law. Nitridation has been applied to the gate oxide to avoid the penetration of boron from the polysilicon gate in p-channel MOSFETs. Interestingly, nitrided oxides have shown a higher radiation hardness with respect to conventional oxides because of the barrier effect of the nitrogen layer against hydrogen penetration, which causes a beneficial reduction in the interface traps generation.

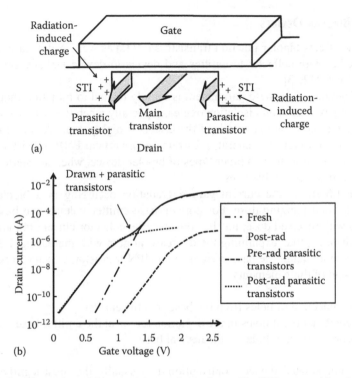

FIGURE 1.5 Illustration of the parasitic transistors formed by STI spacers (a) and effects of their onset induced by TID exposure (b).

The impossibility of thinning conventional SiO$_2$ gate oxides much below 2 nm due to excessive leakage has been overcome with the use of high-k materials, for instance hafnium oxide, which has been commercially introduced from the 45-nm technology node. This has greatly alleviated the issues of undesired leakage current from the gate electrode thanks to the use of thicker layers that do not negatively affect the channel control because of their higher dielectric constant. As we mentioned, the thicker the oxide, the more severe are the TID effects. Considerable radiation-induced charge trapping has been observed in hafnium oxide (HfO$_2$) capacitors with thick oxides irradiated with x-rays. Yet, charge trapping in thinner and more mature oxides, suitable for integration in advanced technologies, appears much less critical.

Silicon on insulator (SOI) technology has been recently used for mainstream products, while only a few years ago it was only used in niche applications, such as in the rad-hard market. The beneficial aspects in terms of single-event effect radiation susceptibility of this technology are offset by some negative aspects concerning TID sensitivity. In fact, positive charge trapping and interface state generation in the thick buried oxide (BOX) leads to leakage currents in partially depleted devices and to variations in the front gate characteristics in fully depleted MOSFETs due to the coupling between the front and the back channel.

1.3.2 BIPOLAR DEVICES

Total dose affects bipolar junction transistors (BJTs) as well. A decrease in current gain together with collector-to-emitter and device-to-device leakage are the most critical effects [12,13].

The degradation of these parameters is mainly related to radiation-induced degradation of passivation and isolation oxides, especially when these are close to critical device regions. The magnitude of the effects is highly dependent on the type of bipolar transistor (vertical, lateral, substrate, etc.): vertical PNP transistors are less sensitive to TID effects than other types of bipolar device, whereas lateral BJTs are among the most susceptible ones.

Figure 1.6 shows the current gain degradation occurring in a bipolar device exposed to total ionizing dose. For a given base-emitter voltage, the base current increases with received dose; for the converse, the collector current remains practically unchanged, thus explaining the decrease in gain with increasing TID. Let us analyze in more detail the base current in an NPN transistor, which can be thought of as the sum of three elements:

1. Back injection of holes from the base into the emitter
2. Recombination of holes in the depletion region at the emitter-base junction
3. Recombination of holes in the neutral base

In an unirradiated device, contribution 1 is usually the most significant. TID exposure causes the second term to grow and eventually to dominate due to an increase in the velocity of surface recombination and in the emitter-base depletion region surface width.

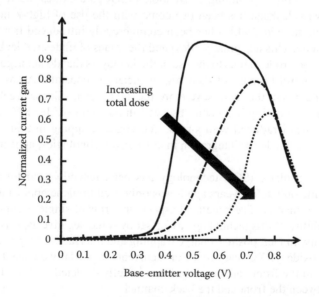

FIGURE 1.6 Degradation of the current gain in a BJT exposed to total ionizing dose.

The reason for the increased recombination is the formation of interface traps at the surface of the depletion region of the base-emitter junction. Traps located in the middle of the gap are very efficient recombination centers able to exchange carriers between the conduction and valence bands. In addition, in NPN bipolar devices, positive net charge trapped in the oxides after total dose irradiation tends to increase the base depletion region, further increasing recombination. In contrast, in lateral PNPs, positive charge trapped in the passivation oxide decreases base recombination (while interface trap formation increases the surface recombination velocity). Vertical PNPs are harder from a radiation standpoint, as positive trapped charge tends to drive the n-doped base to accumulation and the highly doped p emitter to slight depletion, causing a reduction in the size of the emitter-base depletion region, thus decreasing recombination.

ELDRS is a phenomenon occurring in many bipolar devices, which exhibit a larger degradation at a low dose rate compared to a high dose rate. This means that, contrary to MOS components, the effects of total dose in BJTs are dose-rate-dependent. ELDRS highly complicates the interpretation of radiation tests results and their extrapolation to operating conditions. Space is a low dose-rate environment (typical dose rates are in the mrad(SiO_2)/s range), while laboratory testing is done at high dose rates (to save time, normally higher than 10 rad(SiO_2)/s). As a result, there is a high risk of underestimating the degradation in space. This is illustrated in Figure 1.7, where the normalized current gain degradation of a bipolar device is plotted as a function of the dose rate. As seen, the degradation at dose rates peculiar of space is twice the degradation observed during accelerated laboratory testing.

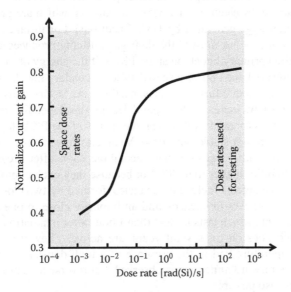

FIGURE 1.7 Dependence on dose rate of the current gain degradation in a BJT exposed to TID due to ELDRS effects. (Adapted from H. L. Hughes and J. M. Benedetto, *IEEE Trans. Nucl. Sci.*, Vol. 50, No. 3, p. 500, June 2003.)

Different models have been proposed in the literature concerning the physical origin of ELDRS. According to the space charge model, the reduced degradation at a high dose rate is due to the large amount of generated positive charge, which acts as a barrier for holes and hydrogen migrating towards the interface. Another model explains ELDRS with the competition between trapping and recombination of radiation-induced carriers due to electron traps: at a low dose rate, there are few free carriers in the conduction and valence bands, hence trapping dominates; at a high dose rate, there is a higher density of free carriers, hence recombination is more relevant.

1.4 DISPLACEMENT DAMAGE

Displacement damage (DD) is related to the displacement of atoms from the lattice of the target material by impinging particles due to Coulomb interactions and nuclear reactions with the target nuclei [14]. DD can be produced by energetic neutrons, protons, heavy ions, electrons, and (indirectly) photons.

The displacement creates a vacancy in the lattice and an interstitial defect in a nonlattice position. The combination of a vacancy and an interstitial defect is called a Frenkel pair. Annealing after the creation of the Frenkel pair leads to recombination (i.e., the pair disappears), or to a lesser extent, to the creation of more stable defects. The ability of an energetic particle to create DD is determined by its nonionizing energy loss (NIEL) coefficient. NIEL measures the amount of energy lost per unit of length by the impinging particle through nonionizing processes.

Different arrangements of defects can originate from DD depending on the features of the impinging radiation (energy, etc.): either point defects that are isolated defects (caused, for instance, by electrons at 1 MeV), or clusters, which are groups of defects close to each other (e.g., generated by 1-MeV neutrons). For instance, with particles such as neutrons or protons, most of the damage is usually produced by the first displaced atom (called primary knock-on atom, PKA). If the energy of the PKA is higher than a certain threshold, the PKA is able to displace secondary knock-on atoms (SKA), which, in turn, can generate further defects, resulting in clusters. This is illustrated in Figure 1.8, which shows increasingly larger defect groups depending on the energy of the impinging protons (bottom axis) and on the energy of the PKA (top axis).

Lattice defects are not stable with time. Vacancies move across the lattice until they become stable either because they recombine soon after creation (this happens with a probability higher than 90%) or because they evolve into other kinds of defects. More stable defects include divacancies (formed by two close vacancies) or defect-impurity complexes (a vacancy and an impurity close to one another). Both short-term annealing, which lasts for less than 1 hour after irradiation, and long-term annealing, which goes on for several years, are active. They are usually accelerated at high temperature and in the presence of a high density of free carriers. In most of the cases forward annealing is observed, but reverse annealing (degradation enhancement) is also possible.

Nominally identical devices irradiated with different particles may show different features: for instance, lack of dependence on impurity types and oxygen concentration in samples irradiated with fission neutrons and strong impurity dependence in

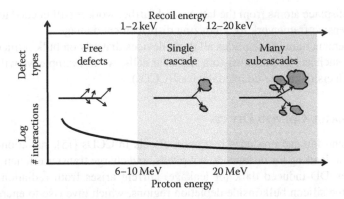

FIGURE 1.8 Types of defects and number of interactions due to nonionizing processes as a function of the energy of the impinging proton, or equivalently, of the recoil energy of the secondary knock-on atom. (Adapted from R. L. Pease, *IEEE Trans. Nucl. Sci.,* Vol. 50, No. 3, p. 539, June 2003.)

samples irradiated with electrons. The differences between these two sets of irradiated samples have been attributed to the formation of clusters of defects as opposed to isolated point defects. Clusters are believed to be more effective than point defects in reducing recombination lifetime for a given total number of defects. Indeed, clusters enhance recombination by creating a potential well in which minority carriers recombine. In addition, the formation of divacancies is much more probable with clusters due to the close proximity of defects, and dominates over impurity-based defects. With point defects, there is no enhancement on the recombination and divacancies and impurity-related defects are both important.

Although successful, cluster models are not entirely consistent with the results obtained using the NIEL concept. NIEL is used to correlate the damage produced by different particles. It is the sum of elastic (Coulomb and nuclear) and inelastic nuclear interactions that produce the initial Frenkel pairs and phonons. It can be calculated analytically from first principles using cross sections and kinematics. Over the years the calculations have been improved, and despite some shortcomings, this approach is very useful because it allows one to reduce the amount of testing by extrapolating the results obtained with a single particle at a single energy to many other conditions. The basic idea is that the number of electrically active stable defects, which give rise to parameter degradation, scales with the amount of energy deposited through nonionizing energy loss. Several experimental data findings support this conclusion. This result bears many consequences. Since NIEL and damage are proportional, we can conclude that the amount of generated defects that survive recombination is independent of the PKA energy. Further, one must assume that radiation-induced defects impact on the device characteristics in the same manner and have the same characteristics (in terms of energy levels) regardless of the initial PKA energy and of the spatial distribution (isolated defects versus clusters). Even though these considerations suggest that cluster models are not necessary to explain the experimental results, there are cases in which there is no proportionality between NIEL and damage, for instance at low particle energies close to the minimum energy needed to displace atoms from the lattice, and further work is still needed to develop a comprehensive framework for analyzing displacement damage.

Displacement damage degrades all those devices that rely on bulk semiconductor properties, such as bipolar transistors and solar cells. As an example we will investigate DD effects in charge-coupled devices (CCDs).

1.4.1 CHARGE-COUPLED DEVICES

Displacement damage produces two main effects in CCDs [15]: radiation-induced dark current, leading for instance to hot pixels, and charge transfer efficiency (CTE) degradation. DD-induced dark (or leakage) current arises from radiation-induced defects in the silicon bulk inside depletion regions, which give rise to energy levels near midgap and thermal generation of carriers.

The generation of dark current in a CCD develops in the following way. The first energetic particle to hit the array induces displacement damage and dark current in one pixel. As more particles impinge, additional pixels are damaged. When the particle fluence reaches a high-enough level, each pixel has been hit by more than one particle. In this way, the CCD exhibits radiation-induced dark current increase in all pixels. There is then a distribution of dark-current magnitudes over those pixels, with a tail that includes multiple events or events that produce dark currents much higher than the mean. The pixels in the tail are usually referred to as hot pixels or dark current spikes.

A second major effect of DD on CCDs is the charge transfer efficiency (CTE) degradation. This results in a loss of signal charge during transfer operations. To express the decrease in the CTE, the concept of charge transfer inefficiency (CTI) (CTI = 1 − CTE) is commonly used in the literature. Radiation-induced CTI in an irradiated CCD increases linearly with incident particle fluence and is proportional to the deposited displacement damage dose. The mechanism underlying this phenomenon is the introduction of temporary trapping centers in the forbidden energy gap by the impinging radiation. Those centers are able to trap charge located in the buried channel, causing a reduction in the signal-to-noise ratio. CTE degradation in an irradiated CCD device is influenced by many parameters, such as clock rate, background charge level, signal charge level, irradiation temperature, and measurement temperature.

1.5 SINGLE-EVENT EFFECTS

A single-event effect (SEE) is caused by the passage of a single, highly ionizing particle (heavy ion) through sensitive regions of a microelectronic device. Depending on the consequences a SEE has on the device, it may be classified as soft (no permanent damage, only loss of information, e.g., a soft error in a memory latch) or hard (irreversible physical damage, e.g., rupture of the gate dielectric). Some other SEEs, such as single-event latch-up, may or may not be destructive depending on how quickly the power supply is cut after the occurrence of the event.

In contrast to TID and DD effects we discussed in the previous sections, which are cumulative and build up over time, a SEE can occur stochastically at any time in a microelectronic device. SEEs are related to the short time response to radiation (< nanoseconds) and only a tiny part of a device (~ tens of nanometers) is affected, corresponding to the position of the particle strike.

What follows is a list and brief description of the main SEEs [16,17].

- Soft effects (nondestructive)
  Single-event upset (SEU): a corruption of a single bit in a memory due to a single ionizing particle. It is also known as soft error. The correct logic value can be usually restored by simply rewriting the bit
  Multiple-bit upset (MBU): a corruption of two or more adjacent bits due to the passage of a single particle
  Single-event transient (SET): a voltage/current transient induced by an ionizing particle in a combinatorial or analog part of a circuit. The radiation-induced transient can propagate and be latched by a memory element, resulting in a soft error
  Single-event functional interrupt (SEFI): a corruption in the controlling state machine of a chip that leads to functional interruptions. Depending on the type of interruption, SEFIs can be recovered repeating the operation, resetting, or power cycling the device
- Hard effects (destructive)
  Single-event gate rupture (SEGR): an irreversible rupture of the gate oxide of a transistor occurring especially in power MOSFETs
  Single-event burnout (SEB): a burnout of a power device due to the activation of parasitic bipolar structures occurring, for instance in IGBTs or power MOSFETs
- Effects that may or may not be destructive
  Single-event latch-up (SEL): the radiation-induced activation of parasitic bipolar structures inherently present in CMOS structures that lead to a sudden increase in the supply current
  Single-event snapback (SES): a regenerative feedback mechanism sustained by impact ionization occurring in SOI devices

The most important figure for SEEs is the rate of occurrence (i.e., how many events take place per hour/day/year) in a particular environment. An environment-independent way to characterize SEEs is the cross section, σ, defined as the number of observed events divided by the particle fluence received by the device. The cross section is a function of the linear energy transfer (LET) of the impinging particle, which measures the energy loss per unit of length (i.e., the ability of a particle to ionize the material it traverses). The LET is usually normalized by the density of the target material and is measured in $MeV \cdot mg^{-1} \cdot cm^2$. σ increases for increasing LET and typically follows a Weibull cumulative probability distribution. A σ versus LET curve is characterized by two main parameters: the threshold LET (i.e., the minimum LET that is able to generate an SEE) and the saturation LET (LET at which the cross section saturates). The threshold LET is usually associated with the concept of critical charge; that is, the minimum amount of charge that must be collected at a given node of a circuit to generate an event.

SEEs can be generated not only by directly ionizing particles (e.g., heavy ions), but also by indirect ionization. Neutrons and protons, for instance, can generate secondary particles through nuclear interactions, and these particles, in turn, can trigger the event. In recent technologies, particles with lower and lower LET are sufficient to generate SEEs, and SEU from direct proton ionization have been recently reported.

Static random access memory (SRAM) cells and latches in digital circuits are the memory elements most sensitive to SEUs and MBUs. Dynamic random access memory (DRAM) cells are quite robust thanks to the beneficial effect of scaling, which has reduced the cell area without decreasing the cell capacitance at a corresponding rate. Floating gate cells were once considered immune to SEEs, but now have become sensitive as well, as a result of the aggressive scaling. SETs are an issue in circuits working at gigahertz frequency, where the fast clock greatly enhances the probability of latching radiation-induced transients. SEFIs occur in all (e.g., field programmable gate array [FPGA], microcontroller, flash) but the simplest circuits. In the following we will present SEUs in SRAM cells as a case study.

1.5.1 SEUs in SRAMs

In this section we will examine one of the most common SEEs, the SEU in an SRAM cell [16–19]. SRAMs closely follow Moore's law and have become the preferred benchmark to study the soft error sensitivity of a technology.

In general, the charge generated by an ionizing particle must be collected at a sensitive region of a circuit to generate a disturbance. Reverse-biased pn junctions are among the most efficient regions in collecting charge, thanks to the large depletion region and high electric field. Figure 1.9 schematically shows the upset of an SRAM cell. An ionizing particle strikes one of the reverse-biased drain junctions, for instance the drain of the OFF NMOSFET in the cross-coupled inverter pair in the cell (see Figure 1.9). As a consequence, electron-hole pairs are generated and collected by the depletion region of the drain junction. This causes a transient current, which flows through the struck junction, while the restoring transistor (the ON PMOS

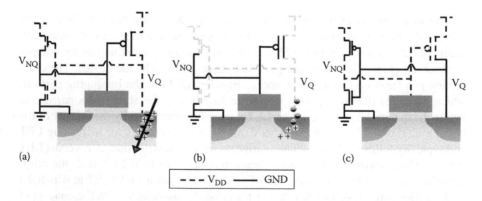

FIGURE 1.9 SEU in a SRAM cell: (a) a heavy ion strikes the drain of the OFF n-MOSFET, (b) charge is collected and voltage drops at V_Q, and (c) feedback is triggered and SEU occurs.

in the same inverter) sources current in an attempt to balance the particle-induced current. However, since the restoring PMOS has a finite amount of current drive and a finite channel conductance, the voltage drops at the struck node. If the voltage goes below the switching threshold and the drop lasts for a long enough time, the feedback causes the cell to change its initial logic state, creating a SEU (bit flip).

Many factors determine the occurrence of a SEU: radiation transport through the back-end layers before the reverse-biased junction, charge deposition, and charge collection. In addition, the circuit response is of primary importance.

Charge deposition is mainly determined by the LET of the ionizing particle (obviously, the higher the LET, the larger the deposited charge). The amount of collected charge is also impacted by the ion incidence angle: the larger the angle with respect to normal incidence, the larger the collected charge. The cosine law states that the effective LET of the impinging particle is inversely proportional to the cosine of the incidence angle. However, this is valid only with thin enough sensitive volumes.

Let us now discuss charge collection. Figure 1.10 is a schematic illustration of the temporal evolution of generation and collection phenomena in a reversed-biased pn junction. Immediately after the particle strike, a track of electron-hole pairs is created through the depletion region (Figure 1.10a). The electric field separates the e-h pairs and gives rise to a drift current. Since the track is highly conductive, it creates a distortion in the junction potential, extending the field lines deep into the substrate (Figure 1.10b). This is called a funnel effect, due to the shape of the field lines, and causes an increase of the region where charge can be collected by drift, increasing SEU sensitivity. However, the funnel effect plays a significant role in charge collection only in junctions where the bias is kept fixed, whereas it has a smaller impact in circuits where the junction bias is allowed to change (e.g., SRAM cells). Finally, when the first phase dominated by drift of carriers in the depletion region is over, carrier diffusion from around the junction still sustains a current through the struck node, although of smaller magnitude. In fact, charge closer than a diffusion length from the drift region can be collected (Figure 1.10c) by the junction. In short, after charge is generated by the impinging particle, the drift and funnel determine the shape of the transient current at earlier times and the slower diffusion process dominates the response at later times.

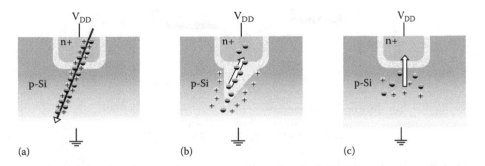

FIGURE 1.10   Schematic illustration of charge generation (a) and collection (drift [b] and diffusion [c]) process steps following an ion strike on a reverse-biased pn junction.

Charge collection mechanisms can be even more complicated for deeply scaled circuits. The alpha-particle source-drain penetration effect (ALPEN) can originate from a grazing alpha-particle strike through the drain and source of a transistor. This can create a disturbance in the potential of the channel, possibly turning on an off device.

Parasitic bipolar effects may increase charge collection as transistors are scaled down. This occurs when electron-hole pairs are generated inside a well by the ionizing particle and the potential of the well itself is altered. For example, if an NMOSFET is located in a p-well inside an n-substrate, the generated carriers can be collected either at the drain/well junction or at the well/substrate one. The source/well junction becomes forward-biased thanks to diffusing holes that raise the potential of the p-well. Bipolar amplification occurs in the parasitic bipolar structure (the source is the emitter, the well is the base, and the drain is the collector), increasing the transient current at the drain node and the possibility of giving rise to an SEU.

The final element is the response of the circuit. The faster the cell feedback time, the shorter the duration of a spurious voltage pulse that is able to flip the cell. The weaker the restoring PMOSFET, the larger the voltage amplitude of the particle-generated pulse. In other words, a slower cell and a restoring PMOS with high conductance decrease the cell susceptibility to SEUs.

MBUs, that is the corruption of multiple memory bits due to a single particle, are a serious concern when designing error correcting code schemes and technology scaling is making things worse. As the physical dimensions of transistors are reduced to just a few nanometers, the size of the electron-hole pairs cloud created by the impinging ions becomes comparable or even larger than the size of a device [19]. As a result, multiple nodes are involved at the same time in charge collection processes and charge sharing occurs between adjacent nodes. Scaling has brought about a great enhancement of multiple bit upsets. Figure 1.11 shows the soft error rate (SER) per bit and MBU probability as a function of feature size for SRAMs irradiated with atmospheric neutrons. As seen in the graph, for this particular manufacturer, SER

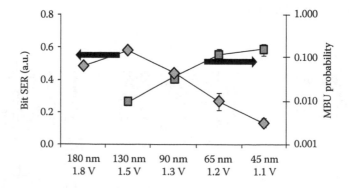

FIGURE 1.11 Neutron bit SER and multiple cell upset probability in SRAMs due to atmospheric neutron strikes. (Adapted from N. Seifert, B. Gill, K. Foley, and P. Relangi, *Proceedings of the IEEE International Reliability Physics Symposium [IRPS]*, 2008, p. 181.)

decreases for decreasing feature size (even though the system SER stays more or less constant due to the increasing number of memory element per chip), while the multiple bit upset probability monotonically increases (and this is a general conclusion valid for all vendors).

A typical neutron cross section for SRAM cells is $\sim 10^{-14}$ cm^2, which at New York City (NYC) corresponds to a bit error rate of $\sim 10^{-13}$ error/bit/hour and $>3 \cdot 10^{-11}$ error/bit/hour at the altitude in which commercial aircrafts fly. The error rate in space varies greatly, depending on the memory, the orbit, and solar cycle.

1.6 CONCLUSIONS

Atmospheric neutrons and alpha particles coming from radioactive contaminants threaten the operation of chips at sea level. Electronics in satellites and spacecrafts must deal with a large amount of radiation, originating from radiation belts, solar activity, and galactic cosmic rays.

Radiation effects in electronic components range from soft errors, in which loss of information and no permanent damage is produced, to parametric shifts and destructive events. They can be categorized in total ionizing dose, displacement damage, and single event effects. The first two classes are cumulative and occur primarily in harsh natural environments such as space or due to man-made radiation sources; SEEs also occur at sea level. Design of critical applications must carefully consider radiation effects to ensure the required reliability levels.

REFERENCES

1. J. L. Barth, C. S. Dyer, and E. G. Stassinopoulos, Space, Atmospheric, and Terrestrial Radiation Environments, *IEEE Trans. Nucl. Sci.*, Vol. 50, No. 3, p. 466, June 2003.
2. R. C. Baumann, Radiation-Induced Soft Errors in Advanced Semiconductor Technologies, *IEEE Trans. Dev. Mat. Rel.*, Vol. 5, No. 3, p. 305, Sept. 2005
3. JEDEC standard JESD-89A, available online at http://www.jedec.org/download/search/JESD89A.pdf.
4. http://seutest.com/cgi-bin/FluxCalculator.cgi.
5. B. D. Sierawski, B. Bhuva, R. Reed, K. Ishida, N. Tam, A. Hillier, B. Narasimham, M. Trinczek, E. Blackmore, W. Shi-Jie, and R. Wong, Bias Dependence of Muon-Induced Single Event Upsets in 28 nm Static Random Access Memories, *IEEE International Reliability Physics Symposium 2014*, pp. 2B.2.1–2B.2.5.
6. O. Flament, J. Baggio, S. Bazzoli, S. Girard, J. Raimbourg, J. E. Sauvestre, and J. L. Leray, Challenges for Embedded Electronics in Systems Used in Future Facilities Dedicated to International Physics Programs, *Advancements in Nuclear Instrumentation Measurement Methods and their Applications (ANIMMA)*, 2009, p. 1.
7. M. Bagatin, A. Coniglio, M. D'Arienzo, A. De Lorenzi, S. Gerardin, A. Paccagnella, R. Pasqualotto, S. Peruzzo, and S. Sandri, Radiation Environment in the ITER Neutral Beam Injector Prototype, *IEEE Trans. Nucl. Sci.*, Vol. 59, No. 4, p. 1099, June 2012.
8. T. R. Oldham and F. B. McLean, Total Ionizing Dose Effects in MOS Oxides and Devices, *IEEE Trans. Nucl. Sci.*, Vol. 50, No. 3, p. 483, June 2003.
9. J. R. Schwank, M. R. Shaneyfelt, D. M. Fleetwood, J. A. Felix, P. E. Dodd, P. Paillet, and V. Ferlet-Cavrois, Radiation Effects in MOS Oxides, *IEEE Trans. Nucl. Sci.*, Vol. 55, No. 4, p. 1833, Aug. 2008.

10. H. L. Hughes and J. M. Benedetto, Radiation Effects and Hardening of MOS Technology: Devices and Circuits, *IEEE Trans. Nucl. Sci.,* Vol. 50, No. 3, p. 500, June 2003.

11. P. E. Dodd, M. R. Shaneyfelt, J. R. Schwank, and J. A. Felix, Current and Future Challenges in Radiation Effects on CMOS Electronics, *IEEE Trans. Nucl. Sci.,* Vol. 57, No. 4, p. 1747, Aug. 2010.

12. R. L. Pease, Total Ionizing Dose Effects in Bipolar Devices and Circuits, *IEEE Trans. Nucl. Sci.,* Vol. 50, No. 3, p. 539, June 2003.

13. R. D. Schrimpf, Gain Degradation and Enhanced Low-Dose-Rate Sensitivity in Bipolar Junction Transistors, *Int. J. High Speed Electron. Syst.,* Vol. 14, p. 503, 2004.

14. J. R. Srour, C. J. Marshall, and P. W. Marshall, Review of Displacement Damage Effects in Silicon Devices, *IEEE Trans. Nucl. Sci.,* Vol. 50, No. 3, p. 653 June 2003.

15. J. C. Pickel, A. H. Kalma, G. R. Hopkinson, and C. J. Marshall, Radiation Effects on Photonic Imagers—A Historical Perspective, *IEEE Trans. Nucl. Sci.,* Vol. 50, No. 3, p. 671, June 2003.

16. P. E. Dodd and L. W. Massengill, Basic Mechanisms and Modeling of Single-Event Upset in Digital Microelectronics, *IEEE Trans. Nucl. Sci.,* Vol. 50, pp. 583–602, 2003.

17. D. Munteanu and J.-L. Autran, Modeling and Simulation of Single-Event Effects in Digital Devices and ICs, *IEEE Trans. Nucl. Sci.,* Vol. 55, No. 4, Aug. 2008.

18. K. P. Rodbell, D. F. Heidel, H. H. K. Tang, M. S. Gordon, P. Oldiges, and C. E. Murray, Low-Energy Proton-Induced Single-Event-Upsets in 65 nm Node, Silicon-on-Insulator, Latches and Memory Cells, *IEEE Trans. Nucl. Sci.,* Vol. 54, No. 6, p. 2474, 2007.

19. N. Seifert, B. Gill, K. Foley and P. Relangi, Multi-Cell Upset Probabilities of 45nm High-k + Metal Gate SRAM Devices in Terrestrial and Space Environments, *Proceedings of the IEEE International Reliability Physics Symposium (IRPS)*, 2008, p. 181.

2 Monte Carlo Simulation of Radiation Effects

Mélanie Raine

CONTENTS

2.1 INTRODUCTION

Development of the Monte Carlo method as we know it today originally started immediately after the Second World War, in the frame of the Manhattan Project at Los Alamos, for the simulation of radiation transport [1]. Its inception is also closely linked with the introduction of the first digital computers, without which the method would probably have remained theoretical without any application. Monte Carlo simulations have since been used to study the effects of radiation in semiconductor devices as soon as a relevant problem appears, with the discovery of single-event effects in the second half of the 1970s [2] and for which this method is ideally suited, as will be explained in this chapter.

Rather than explaining the theory behind these computational techniques, this chapter aims at defining the perimeter of use of Monte Carlo simulations for the specific problem of radiation effects in semiconductor devices. Through analyzing how this method works, we will identify its areas of applications, what it can really bring to the study of radiation effects in electronics, and determine which effects it is best suited for.

A brief historical section will first describe the origins of the Monte Carlo method. We will define this method before determining what can be achieved regarding the analysis of radiation effects in electronics using these simulations and what the limitations to consider are. Finally, some examples of Monte Carlo radiation transport simulation codes will be presented, giving examples of their application for electronic devices.

2.2 A BRIEF HISTORY OF THE MONTE CARLO METHOD

The term Monte Carlo method applies to any method aiming at calculating a numerical value using probabilistic techniques. While the Monte Carlo method was formally defined and given its name in the 1940s, this kind of calculation technique can be traced back to 1777, when the Comte de Buffon posed Buffon's needle problem, describing a way of evaluating the number π based on the realization of repeated experiments [3]. The term "Monte Carlo" itself was invented by Nicholas Metropolis in 1947, alluding to games of chance played in the city of Monte Carlo, to apply to a statistical approach suggested by Stanislaw Ulam to solve the problem of neutron diffusion in fissionable materials [1] in the context of the Manhattan Project. A description of the method was first published in 1949 in an article authored by Nicholas Metropolis and Stanislaw Ulam [4].

This early development of the method was also closely linked with the development of computers. Indeed, statistical sampling techniques, while potentially powerful to solve a number of problems, were not actually used because of the duration and tediousness of the calculations. The combined existence of particular people (e.g., Enrico Fermi, Stanislaw Ulam, John von Neumann, Nicholas Metropolis, and Edward Teller), a given problem to solve (the simulation of neutron histories), and the technology to solve it at Los Alamos in the 1950s would change that. The first computer used for Monte Carlo simulation of neutron histories was actually an analog one: the Fermiac (Figure 2.1). It allowed an efficient graphical simulation of neutron histories inputting the cross sections manually, using a piece of paper with the geometry drawn on it.

Simulations at Los Alamos were later performed with the Electronic Numerical Integrator and Computer (ENIAC), the first general-purpose electronic computer (Figure 2.2). It was 80 feet long, weighed 30 tons, and had more than 17,000 tubes. Its clock was 5 kHz and it ended up with a 100-word core memory. It was later replaced by the Mathematical Analyzer, Numerical Integrator, and Computer (MANIAC), which started being built in 1952 under the direction of Metropolis, based on the von Neumann architecture.

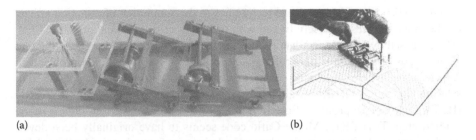

(a)   (b)

FIGURE 2.1 (a) Enrico Fermi's Fermiac in the Bradbury Museum, Los Alamos, New Mexico (photo by Mark Pelligrini [CC-BY-SA-1.0]) and (b) in action (photo by N. Metropolis, The beginning of the Monte Carlo method, *Los Alamos Science*, Vol. 15, pp. 125–130, 1987, public domain).

FIGURE 2.2 The ENIAC in the University of Pennsylvania, Philadelphia. (U.S. Army photo, public domain.)

An interesting illustration of the concomitant evolution of Monte Carlo simulation and available computing power is the first article reporting a numerical Monte Carlo simulation of photon transport in 1954 [5]: 67 photon histories were generated with a desk calculator. Such a simulation was possible at the time because photon transport involves only a small number of events in each trajectory. A photon is indeed absorbed after a single photoelectric or pair-production interaction or after only a few Compton interactions (around 10). With today's available computational power, detailed simulation of photon transport is a simple routine task than can be achieved on a personal computer.

This concomitant evolution of computational power and Monte Carlo simulation never stopped: problems solved by Monte Carlo techniques are getting more and more complicated with the possibility to use more and more powerful computers with different architectures. Monte Carlo codes were first adapted to vector computers in the 1980s, then to clusters and parallel computers in the 1990s, and to teraflop systems in the 2000s. Recent advances include hierarchical parallelism, combining threaded calculations on multicore processors with message-passing among different nodes [6]. After only half a century of evolution, even a simple laptop computer can now store a million times more information than ENIAC did and is hundreds of thousand times faster.

The application of Monte Carlo techniques to the simulation of radiation effects in electronic devices is closely linked with the study of single event effects (SEEs), effects caused by a single incident particle. Indeed, we will see in the following that Monte Carlo simulation is particularly well suited to solve this kind of problem. In its pioneering article evidencing cosmic rays as the root cause for upsets observed in satellite electronics [2], Binder already used an electron transport Monte Carlo code (BETA II) to obtain profiles of electron energy depositions at various depths in the studied chip. This BETA Monte Carlo code seems to have originally been developed as early as 1969 for ionizing dose calculations performed at the Jet Propulsion Laboratory (JPL) for the Voyager project [7], although there is not much information on this code in the literature. The development of Monte Carlo methods specifically dedicated to the analysis of radiation effects in electronics really started in the 1980s [8–10], shortly after the first reports of anomalies in electronics directly related to single particles from natural radiation environments: cosmic rays in space [2,11], radioactive decay of impurities in the packaging materials of commercial electronics [12,13], and atmospheric neutrons [14]. The development of radiation transport tools for the simulation of SEEs still goes on today; an anthology of this development can be found in [7], presenting chronologically various Monte Carlo tools developed in the radiation effects community.

2.3   DEFINITION OF THE MONTE CARLO METHOD

The Monte Carlo method aims at giving a numerical solution to a problem, modeling the interaction of objects with other objects or with their environment based on known relationships. The solution is then obtained by a computational algorithm, relying on repeated random sampling of these relationships until convergence of the results is obtained. A survey of the technical details of different Monte Carlo methods can be found in [15].

While first developed to solve particle physics problems, the Monte Carlo method described above can be applied to a wide range of applications, such as computational biology, finance and business, the entertainment industry, social science, traffic flow, population growth, and fluid mechanics.

An essential characteristic of Monte Carlo simulations is the use of random numbers and random variables. Random-sampling algorithms are generally based on the use of random numbers uniformly distributed in the interval [0,1]. Such numbers are not so easy to generate with a computer [16]. Indeed, they are usually generated using a deterministic algorithm, often with a periodic sequence, and are thus rather called pseudorandom numbers. With today's available computational facilities, the period must be large enough to prevent using all available numbers in a single simulation run. A critical review of random-number generators has been published in [17], recommending the use of sophisticated algorithms.

Once the random number generator is chosen, the algorithms in the Monte Carlo method must include

- A description of the system to be simulated
- Tables of interactions and relationships

- Methods for sampling random quantities from probability distributions
- Scoring, accumulation of results

In principle, such algorithms are as accurate as the uncertainties in the tables of interactions and relationships as long as the sampling and accumulation are correct. In practice, all results are affected by statistical uncertainties because of the random nature of these simulations. These uncertainties can be reduced at the expense of increasing the sampled population, and as a consequence, the computation time.

2.4   INTEREST OF THE MONTE CARLO METHOD FOR THE SIMULATION OF RADIATION EFFECTS IN SEMICONDUCTOR DEVICES

Radiation effects in semiconductor devices may be divided into three main categories:

1. SEEs, or effects directly or indirectly induced by a single-incident particle
2. Total ionizing dose (TID) effects, which are cumulative effects induced by the interaction of many ionizing particles depositing energy in the device
3. Displacement damage dose (DDD) effects, which are cumulative effects induced by the accumulation of atomic displacements induced by many incident particles in a single device

For all these effects, the simulation can actually be divided into two separate problems: (1) the radiation transport and interaction into matter and (2) the electrical effect of these radiation interactions in the semiconductor device. Indeed, radiation transport and interaction phenomena generally happen in less than 100 fs. The result from this simulation step will be the trajectory of the incident particle in matter along with the trajectories of all generated secondaries and the energy deposited by all these particles following interactions. Because of the short duration of these transport phenomena, the final state is generally considered as the initial condition of the next simulation step aiming at deducing the radiation effects in semiconductor devices (i.e., the electrical effect resulting from radiation transport and interaction). The timescale of these events is of the order of pico- to nanoseconds.

As may be deduced from its history, the Monte Carlo method is ideally suited for radiation transport and interaction simulation. The translation of the effect at the material level into an electrical effect is generally much less obvious. In the following, we will analyze each category of effects to identify the contribution that Monte Carlo methods can bring to each of them.

2.4.1   SINGLE-EVENT EFFECTS

The term "single-event effects" (SEEs) regroups all the effects on electronic systems that are due to single energetic particles. There are many such effects, which can be divided into soft errors (such as single-event upsets or single-event transients) and hard errors (such as single-event burnout or single-event gate rupture). For all these different denominations and results, the electrical effect is actually due to the collection of energy deposited in the sensitive area of a device by a single-incident particle. The simulation of such effects can then be described by the following steps:

1. Description of the device geometry and physical environment, including the sensitive area and all surrounding structures and materials that can be in the trajectory of incoming particles (this can include the spacecraft shielding, electronic packaging, metallization overlayers, etc.)
2. Description of the radiation environment
3. Transport of the incident radiation through the device geometry and surrounding structures
4. Energy deposition, with a particular focus on the sensitive area
5. Conversion of energy into electrical charges
6. Charge transport, recombination, and collection in the semiconductor device
7. Electrical response of the device or circuit of interest

The use of a Monte Carlo simulation of radiation transport is suitable to solve steps 1 to 4. Although the simpler approximation of the rectangular parallelepiped (RPP) [18] and integral rectangular parallelepiped (IRPP) models of energy deposition and charge collection have been used for many years [19], their simplicity itself introduces limitations that prevent them for being effective in a number of applications. In short, the RPP model assumes that the sensitive volume can be approximated by a rectangular parallelepiped (Figure 2.3a), that the deposited energy can be calculated from the linear energy transfer (LET) of the incident particle, which is constant in the volume, that the particle trajectory in the volume is a straight line, and that the electrical effect only depends on the energy deposited in the volume and is triggered above a given value expressed as a critical charge or threshold LET. From these hypotheses, the single-event rate can simply be calculated through geometrical considerations on the chord-length distribution of possible trajectories of the incident particle in the RPP volume. In the IRPP model, the critical charge characterizing the electrical effect is no longer fixed to a single value; the observed integral upset cross

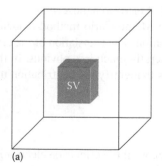
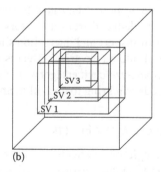

(a)   (b)

FIGURE 2.3 Example of (a) RPP geometry with a single rectangular parallelepiped sensitive volume (SV) and (b) nested sensitive volumes, with three different SVs arranged concentrically.

section is instead used as a measure of the critical charge's variability. In [20], Weller et al. proposed a very good comparison of the different methods to evaluate SEE rates, giving indications on when the Monte Carlo approach should be used rather than simple analytical models.

Step 5 is a crucial step, a transition in which the deposited energy distribution calculated at step 4 needs to be converted into charges that will be transported, recombined, and collected in the semiconductor device at step 6, leading to a particular electrical response (step 7). This energy-to-charge conversion is generally performed analytically, using simple relations giving the mean energy needed to produce an electron-hole pair in a given material [21]. This step is often considered as a formality and is rarely discussed in this type of simulation [22,23]. With the constant decrease in critical charge; that is, the number of electrons needed to trigger an electrical effect in more and more integrated devices—recent papers considered critical charges equivalent to only hundreds of electrons for advanced technologies [24]—the average nature of this conversion factor may need to be questioned for accurate estimation of generated charge similarly to what is done in the field of radiation detectors [25,26]. The Monte Carlo treatment of the problem generally stops before this conversion step, although some example of Monte Carlo simulation of charge transport can be found in the literature.

The modeling of charge transport in semiconductor materials at step 6 consists of solving the Boltzmann transport equation. This can be done either through stochastic methods, using Monte Carlo simulation, or through deterministic approaches. The Monte Carlo method assumes that the motion of charge carriers under the influence of an electric field consists in free flights interrupted by scattering mechanisms. This method requires a strong knowledge of the band structure [27]. The drift-diffusion model and the hydrodynamic model are two examples of deterministic approaches. These latter are the most commonly used and are both based on simplified approximations valid for long-channel devices. These simplified transport models are used in particular in technology computer-aided design (TCAD) simulators [28]. Their popularity compared to Monte Carlo full-band calculations can be explained by the fact that in most electronic devices, a large number of charges moving in an electric field is involved, thus reducing the stochastic nature of the involved processes. This allows modeling charge transport through deterministic equations without appealing to the more computationally intensive Monte Carlo methods. Monte Carlo may, however, be required for specific applications involving the transport of a reduced number of charges in small dimensions with ballistic or quasi-ballistic regimes of transport [29], the extreme case being the single-electron transistor [30].

Moreover, when using deterministic equations, a unique deposited energy distribution averaged over several incident particles is generally used. This also leads to losing part of the stochastic nature of the energy loss process, such as the fluctuations in deposited energy that may become nonnegligible for thin active layer devices [31,32] and the variations in the shape of the distribution itself from particle to particle [24,33].

Much simpler methods that do not require a detailed simulation of charge transport are also often used to assess radiation effects in semiconductor devices. Based on the knowledge of the technology and on the results of previous detailed TCAD simulations for example, different sensitive volumes may be defined in the device geometry (Figure 2.3b) with different charge collection efficiencies. Depending on the deposited energy, these different sensitive volumes will lead to different values of collected charge. A critical charge value can then be determined among the values, leading to an electrical effect. For a review of this composite-sensitive volume approximation, see [20] and [7] for examples of application.

2.4.2 TOTAL IONIZING DOSE EFFECTS

Total ionizing dose (TID) effects are induced by the accumulation of energy deposited by multiple ionizing particles into the oxide layers of a device. Again, this deposited energy needs to be converted into generated charge that will recombine or get trapped at preexisting defects in the material or generate new defects. An important step of the calculation is the evaluation of the fractional charge yield remaining after recombination as a function of the electric field in the insulating material and on the incoming particle [34–36]. The electrical effect in this case is due to the density of trapped charges in the insulating layers. The Monte Carlo part of TID effects simulations, if any, is generally limited to a dose calculation. The first part of the simulation is then very similar to the Monte Carlo simulation performed for SEE studies (steps 1 to 4), except that the deposited energy is here recorded as a sum for a large population of incident particles, while SEE will focus on the effect of each particle individually. This type of calculation assumes that energy deposits are strictly independent. In this case, the computation of the cumulated deposited energy, or dose, by Monte Carlo techniques is mainly useful for complex geometries or geometries involving small dimensions compared to the extent of the incident particles cascade (i.e., cases when the stochastic nature of particle transport is not completely erased by the large number of particles involved). Monte Carlo simulation has also been used for the evaluation of the charge yield [37].

2.4.3 DISPLACEMENT DAMAGE DOSE EFFECTS

Finally, displacement damage dose (DDD) effects are induced by atomic displacements in the active layer of a device, creating defects in the crystal lattice (vacancies and interstitials) of the semiconductor material. While these effects are generally considered as dose effects resulting from the cumulative damage induced by multiple interactions, single-particle displacement damage effects may also be considered independently [38–40].

The simulation of displacement damage effects can be described by the following steps:

1–4. Similar steps as for SEE simulation, corresponding to the system description and particle transport
5. Conversion of energy deposition into a change in the atomic structure
6. Consequence of this change in the atomic structure on the electronic structure of the material
7. Further evolution of the damage structure

Similar to the distribution of ionizing energy simulated for SEE and TID, the result of Monte Carlo simulation of particle transport for DDD calculation (steps 1 to 4) gives the distribution of nonionizing energy loss [41] in the limits of what can be achieved with the binary collision approximation (BCA). This approximation, used in the majority of Monte Carlo radiation transport codes, assumes that the interaction between the incident particle and the crystal lattice can be reduced to the interaction of the incident particle and the knock-on atom, ignoring modifications in the surrounding lattice.

In step 5, the damage is then quantified in terms of vacancies using simple proportionality with deposited energy such as a modified Kinchin-Pease approach [42]. However, such models do not support complex damage structures; in particular they do not treat collective motions of particles in the lattice during relaxation of the system. Molecular dynamics methods are then needed to track the evolution of changes, although this tracking is limited to very short periods of time (up to the microsecond time range).

An additional step (step 6) is then necessary to translate the change in atomic structure into an electrical structure of the damage and get the device electrical response. Previous atomic structures of defects have to be related to energy levels in the bandgap of the semiconductor, leading to electrical effects. While experimental measurements or detailed *ab initio* methods (such as the density functional theory [DFT] or the GW approximation) may be used for isolated defects, the determination of the electronic structure of defect clusters is much more complex and often remains only qualitative [43].

Finally, the damage evolution over long periods of time (step 7) has to be considered. Kinetic Monte Carlo models may be involved in this step, but their range of application remains mostly limited to isolated defect structures. Figure 2.4, extracted from [44], illustrates these different aspects of displacement damage simulation, evidencing again the restricted contribution of Monte Carlo simulation to particle transport.

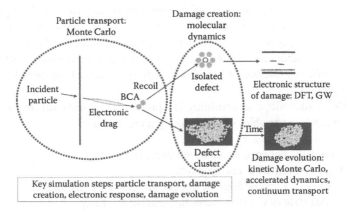

FIGURE 2.4 Overview of several important aspects of displacement damage simulation. (From J. R. Srour and J. W. Palko, in *Proc. Short Course of the Nuclear and Space Radiation Effects Conference*, San Francisco, 2013.)

2.5 MONTE CARLO SIMULATION OF RADIATION TRANSPORT

As seen in the previous section, in practice, Monte Carlo algorithms are generally limited to the first part of the problem, simulating only the radiation transport in matter. The second part of the problem, assessing the electrical effect induced by radiation, is usually solved by other different methods. In the following, we will thus mainly focus on the main features and limitations of Monte Carlo simulation of radiation transport and interactions.

2.5.1 DEFINITION OF THE MONTE CARLO METHOD
FOR RADIATION TRANSPORT AND INTERACTION

The trajectory of a particle in matter may be seen as a random sequence of free flights ending with an interaction event where the particle may change its direction, lose energy, and/or produce secondary particles. Monte Carlo simulation of radiation transport consists in the numerical generation of these random histories.

As stated in the general definition, the first ingredient of a Monte Carlo algorithm is a description of the system to be simulated. In the case of radiation transport, this corresponds to a description of the geometry and materials in which the particles will propagate and a description of the incident particle beam.

In general, materials taken into account in Monte Carlo simulations of radiation transport are considered as homogeneous random-scattering media with a uniform density of randomly distributed molecules. While this is a satisfying representation of gases, liquids, and amorphous solids, this approximation must be kept in mind when simulating crystalline media: one must be aware that the atomic arrangement of the material is completely ignored in this kind of simulation.

The second main ingredient is the tables of interactions and relationships that govern the studied system. In the case of radiation transport, these are a set of differential cross sections (DCSs), one for each interaction mechanism taken into account in the simulation, which are used as input data in Monte Carlo codes. These DCSs may have different origins: they can be directly determined through analytical formulations computed in the code or they can be made available in the form of tabulated data, coming themselves from theoretical calculations, from experimental data, or from a combination of both. A single code can combine these different possibilities for different interaction processes or different models proposed for a single interaction process.

These DCSs then allow determining the probability density functions (PDFs) of the different random variables characterizing a particle track:

- The free paths between successive interaction events
- The type of interaction taking place
- The energy loss and/or angular deflection following the interaction
- The initial state of emitted secondary particles, if any

Once the PDFs are known, the random trajectories are generated by using appropriate sampling methods. In a complete Monte Carlo simulation, these trajectories and discrete interactions should be simulated for all incident and secondary particles produced during the simulation. For more information regarding the basic concepts of Monte Carlo simulation of radiation transport, see the first sections of [45].

2.5.2 PARTICLES AND INTERACTIONS TO CONSIDER

The particles worth considering for electronic applications are photons (optical photons, gamma rays, and x-rays), beta particles (electrons and positrons), neutrons, protons, and heavy ions (alphas and heavier particles). Other types of particles, such as pions, mesons, neutrinos, or quarks are generally ignored in the study of radiation effects in semiconductor devices, although recent work showed possible effects induced by previously ignored particles, such as muons [46].

When entering matter, particles will interact in different ways depending on their nature and energy. The different possible interactions can be divided into four categories: elastic interactions with the target atom nucleus, inelastic interactions with the target atom nucleus, elastic interactions with orbital electrons, and inelastic interactions with orbital electrons. In terms of electrical effects, these categories are generally redistributed into only two denominations: ionizing interactions, corresponding to the last category, and atomic displacements, which regroup the first three. A review of the different interaction processes and their implementation as models in radiation transport codes is available in [47].

Because the trajectory of each particle is followed step by step in the simulation, Monte Carlo codes have inherent advantages over deterministic codes for radiation transport. In particular, any information about the details of the transport can be extracted from the simulation. In contrast, deterministic codes only provide average general information. For example, a deterministic code can only calculate the average energy deposited in a detector by an incident beam, while a Monte Carlo code can provide information on the energy deposited by each particle, the distribution of this energy, the physics process involved in each energy deposit, and so forth. In Monte Carlo calculations, the position, energy, and angle of each particle also vary continuously, while deterministic codes solving a transport equation by finite difference techniques will end up with discrete values in their prediction, thus inducing uncertainties in the calculation results.

On the other hand, a detailed description of all interaction events experienced by a particle with the explicit generation and transport of all secondaries is not always possible because of simulation time or memory constraints. In fact, it is not even always needed. Depending on the purpose and configuration of the simulation, choices may need to be made by the user to speed up the calculation and avoid recording unnecessary information. The focus may be on certain types of particles or particular interactions, ignoring other particles or favoring a particular interaction process. The optimization may also simply consist in simplifying the simulated geometry. In the following, we will explore some examples of common simplifications or methods often used in simulations to optimize the accuracy versus simulation time ratio.

2.5.3 Electron Transport: Condensed History and Cutoff Energy

A good example of simplification is the case of electron transport. We said before that the simulation of photon transport remains relatively easy, with a small number of events in each trajectory. In contrast, electron and positron transport can be quite complicated. Indeed, the main processes involved in electron transport are

- Ionization, which is the loss of energy through interaction with orbital electrons of target atoms and ejection of secondary electrons that will also need to be followed in a complete simulation
- Elastic interaction, which is a change of direction without any energy loss

With these processes, the energy lost in a single interaction is very small (of the order of a few tens of eV). Moreover, as the electron energy decreases, elastic interactions will become the dominant process, meaning that the electron will experience a lot of interactions before losing any energy. Simulating the trajectory of electrons until they completely stop is thus very inefficient in terms of computing time. This is the reason for the early introduction of the condensed history Monte Carlo technique by M. J. Berger in 1963 [48]. Because most interactions lead to very small changes in energy and/or direction, the idea is to combine the effect of many small-change collisions into a single, large-effect, virtual interaction. The PDF for these interactions are provided by multiple-scattering theories.

An additional feature to facilitate electron transport consists in introducing a cutoff energy below which secondary particles are not generated and the energy loss is continuous along the particle path, and above which the energy loss is discrete, with the explicit production of secondary particles. In standard models, this threshold energy is generally set at 1 keV. For the simulation of radiation effects in semiconductor devices, this limit should be carefully chosen depending on the size of the studied electronic device. For large devices, the simulation of secondary electrons may not even be needed; in this case the corresponding energy is simply deposited along the path of the incident particle and corresponding only to the continuous part of the energy loss process. In practice, the energy is actually distributed around the path of the incident particle by secondary electrons traveling away from their parent particle, introducing a lateral dimension to the distribution of deposited energy. Figure 2.5 shows examples of this deposited energy distribution around the path of different heavy ions with different energies in silicon (for details on the simulation of these ion tracks, see [49]). It should be noted that this kind of representation is averaged over several incident particles. With the continuous scaling of semiconductor devices, the lateral extension D of the distribution (from hundreds of nanometers to tens of micrometers, depending on the energy of the incident particle, as shown in Figure 2.5) becomes larger than the dimensions L of a single device (Figure 2.6).

The shape of the distribution starts having an impact on simulation results roughly for devices with characteristic dimensions smaller than 0.25 μm [50,51]. More and more detailed simulation of secondary electrons is then needed [24,49]. That is why some effort has been directed toward the development of new models for generation and transport of low-energy electrons in materials relevant for electronics

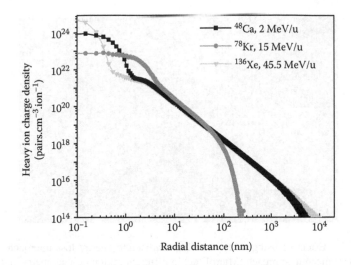

FIGURE 2.5 Examples of averaged deposited energy distributions around of the path of different heavy ions with different energies in silicon. For details on the simulation of these ion tracks, see [49].

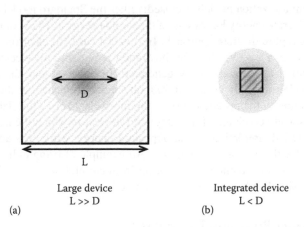

FIGURE 2.6 Illustration of the (a) lateral extension D of the distribution of deposited energy compared to the (b) dimension L of the sensitive volume for a large device and for an integrated device.

application [52–54], using a pure discrete approach with the explicit production of all secondary electrons.

This distinction between a mixed continuous/discrete energy loss approach and pure discrete models is illustrated in Figure 2.7. In this example, simulated with the Monte Carlo code Geant4 [55], a cube of silicon is irradiated with a single-incident 1 MeV/nucleon ^{10}N ion. The ion trajectory is represented as a straight central horizontal line from left to right in Figure 2.7, while secondary electrons appear as small tracks emitted from the ion trajectory. In this simulation, different regions appear, in

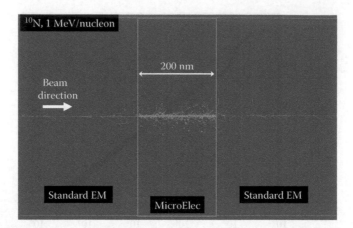

FIGURE 2.7   Effect of using mixed continuous/discrete energy loss approach (Standard EM) and pure discrete approach (MicroElec) in different region of a geometry and obtained with the microelectronics advanced example provided in Geant4 [55]. The two-dimensional projected track of a 1 MeV/nucleon ^{10}N ion is shown. MicroElec processes are activated only in the central 200-nm slab.

which different ionization models are used: either the Standard models, mixing continuous and discrete energy losses with a threshold energy at 1 keV, or the MicroElec models, using a pure discrete approach. In this last case, all electrons are explicitly generated down to a few eV, while for the Standard models, only electrons with energies larger than 1 keV are explicitly generated. In this example, the more detailed MicroElec models are only used in a 200 nm thick slab located at the center of the geometry, leading to the production of many more electrons. This slab could correspond to the sensitive volume of an integrated device, in which the user application requires a detailed simulation of the deposited energy, while less accurate results are sufficient for the rest of the geometry. An example of using either a mixed or a discrete approach for the energy deposition in integrated devices and the resulting impact on the simulated device response can be found in [56].

2.5.4   VARIANCE REDUCTION TECHNIQUES

Another way to optimize the efficiency of a simulation (i.e., to reduce the statistical uncertainty without increasing the computer simulation time) is to use variance-reduction techniques. Detailed simulations that do not use these techniques are called "analog." The choice of a variance reduction technique can be very dependent on the problem to solve. These methods can be classified as follows: population control methods, modified sampling methods, and partially deterministic methods. Some examples are given next among the simplest techniques. For more details, a review of variance reduction methods in radiation transport can be found in [57].

Population control methods consist of artificially increasing (or decreasing) the number of particles in spatial or energy regions that are important (or unimportant) for the problem to solve. An example of this kind of technique is the geometry splitting and Russian roulette. Geometry splitting consists of assigning values of importance to different regions in the geometry. Assuming the user is particularly interested in what happens in region A, other regions will have greater importance values the closer they get to A. The idea is to favor the flux of particles toward the region of interest and ignore the particles that tend to leave this region. When a particle leaves a region of importance, I_1, and enters a region of importance, I_2, the particle is split or rouletted following the ratio I_2/I_1. In other words, the weight of particles is modified. At the beginning, each incident particle has a weight equal to one. In a complete analog simulation, each secondary particle will have the same weight as the primary one. When biasing the simulation, particles that enter regions closer to the region A of interest ($I_2/I_1 > 1$) are transformed by splitting them: a particle with weight w will be transformed into N identical particles with weights w/N. If I_2/I_1 is a whole number, $N = I_2/I_1$. If $I_2/I_1 = 2.75$ for example, N will be equal to 3 with 75% probability and equal to 2 with 25% probability. On the contrary, when a particle moves away from the region of interest ($I_2/I_1 < 1$), Russian roulette is applied and the particle is killed with a probability ($1-I_2/I_1$). If it survives, its weight is increased by a factor I_1/I_2. This allows keeping the total weights of all involved particles constant.

Modified sampling methods artificially increase the likelihood of events that enhance the probability for a particle to reach the scoring region. A solution can be to use forced collisions. This is particularly useful when the interest is directed toward rare interactions, such as neutron interactions, or if the interest is in interaction A when interactions B and C are far more probable. The variance-reduction method then consists of artificially increasing the interaction probability of A to force the collision.

Finally, partially deterministic methods consist of replacing the random-walk process by a deterministic process to move particles from one region to another.

2.5.5 CONCLUSION ON THE USE OF MONTE CARLO SIMULATION FOR RADIATION TRANSPORT

A general recommendation when using Monte Carlo simulation for radiation transport is to carefully analyze the problem before starting any coding. The first question to ask would be: Are Monte Carlo methods really needed to solve the problem, or can the answer be obtained through simpler analytical techniques? In practice, the efficiency of Monte Carlo simulation versus analytical calculation depends to a great extent on the details of the given problem. Monte Carlo methods are particularly indicated for high-dimensionality problems, for example when complicated three-dimensional (3-D) geometries need to be handled or when mixed radiation fields and interaction processes are involved. The popularity of the Monte Carlo method then comes from its intuitiveness: for the user, it is conceptually easy to understand the process of following the trajectories of each individual particle.

When choosing to use Monte Carlo, carefully setting the problem may allow optimizing the accuracy versus simulation time ratio. Simplifying the geometry is often a good way to speed up the simulation. Analyzing the geometry related to the problem should allow identifying regions that need very detailed description and regions that can be simplified or the level of details needed to define the materials: Is it needed to include elements that are present in very small proportions in the simulated material? Will they change the result of the simulation?

Similarly, do all interactions processes or all secondary particles need to be included? Some examples of this issue have been discussed in Section 2.5.3. An additional method that can be applied is the so-called range-rejection method. It simply consists of stopping a particle when it cannot leave (or reach) the region of interest.

To sum up, knowing and understanding the problem at hand before starting to code can save much development and computation time. One does not always need a very detailed simulation when a simple calculation makes the job. As well, the user also needs to know when his or her code is no longer sufficient or not detailed enough to solve the problem. A simulation will always give an answer; it is up to the user to be able to evaluate its pertinence for a specific application.

2.6 EXAMPLES OF MONTE CARLO TOOLS

Below are some examples of general-purpose Monte Carlo codes dedicated to radiation transport. The codes presented here have a wide user community and continue to be actively developed. All have been used at some point for the study of radiation effects on electronics.

Specific development of tools dedicated to the simulation of radiation effects in semiconductor devices have also been conducted over the years; an anthology can be found in [7], for application to the study of SEEs. They will not be detailed here.

2.6.1 MCNP

The Monte-Carlo N-Particle Transport Code (MCNP) was developed by Los Alamos National Laboratory and is a direct descendent of pioneering work by von Neumann, Ulam, Fermi, Richtmyer, and Metropolis. It is a general purpose 3-D transport code written in Fortran90. Given its original goal of simulating neutron-induced fission, it was originally restricted to the transport of neutrons, photons, and electrons, with a limited energy range (up to 20 MeV for neutrons). An extended version (MCNP Extended (MCNPX)) was also developed in parallel, taking into account additional particle types (protons, alphas, and some heavy ions) and a larger energy range. The last MCNP version, MCNP6, released in August 2013, is a fusion of the previous MCNP5 version and MCNPX [58]. This code is distributed by the Radiation Safety Information Computational Center (RSICC) [59] and the Organisation for Economic Co-operation and Development Nuclear Energy Agency (OECD/NEA) databank [60], under export control regulations, giving access to precompiled executables and source code files.

Example of applications of MCNP to the study of SEEs include [61–63]. Not surprisingly, given its strong neutron physics heritage, all these studies are dedicated to SEE induced by atmospheric or terrestrial neutrons. The new capabilities of the last version, with more diverse particle species taken into account, may change that in future work.

2.6.2 GEANT4

Geometry and Tracking 4 (Geant4) is an open-source toolkit for the simulation of the passage of particles through matter [64,65]. This code was originally developed for the simulation of high-energy physics experiments at CERN, but it quickly widened to a very broad range of science fields (nuclear, accelerator, space, and medical physics in particular). It is actually a redesign of the Fortran-based Geant3 simulation program for a modern object-oriented environment based on C++. It is now developed by an international collaboration of physics programmers and software engineers from a number of institutes and universities in Europe, Japan, Canada, and the United States. Since version 10.0 was released in December 2013, Geant4 includes support for multithreaded applications, with parallelism achieved through dispatching events to different threads. Geant4 can be downloaded from its website [66] as precompiled binaries or source code.

Since the first version was released in 1998, Geant4 has been used extensively in the analysis of radiation effects on electronics, and many dedicated tools have been built based on its libraries of C++ classes. Examples include MULASSIS [67], MRED [20], and GRAS [68]. This popularity is due, among other reasons, to the flexibility of the code—allowing since the first version to simulate a very large range of particles and energies—its object-oriented design and the availability of the source code.

2.6.3 FLUKA

Fluktuierende Kaskade (FLUKA) is a fully integrated particle physics Monte Carlo simulation package written in Fortran77 [69,70]. First generations of the code were originally devoted to high-energy accelerator shielding calculations, but it finally evolved into a multipurpose multiparticle code applied in a very wide range of fields and energies. The software is sponsored and copyrighted by INFN and CERN and is not under a General Public License regime. It can be used freely for scientific and academic purposes, or after agreement for commercial purposes, provided explicit signature of the license. The code is distributed in precompiled binary form; the complete source code is available on request although the user is not allowed to modify the code.

Examples of applications of FLUKA to the study of radiation effects on electronic devices include [71,72]. FLUKA has also been extensively used to characterize the radiation field in the mixed environment of the Large Hadron Collider (LHC) and assess its effect on electronics [73].

2.6.4 PHITS

Particle and Heavy-Ion Transport code System (PHITS) is a general-purpose Monte Carlo particle transport simulation code written in Fortran77 [74]. It aims at studying high-energy heavy ion transport for a wide variety of applications, from radioactive beam facilities for nuclear physics to the effect of cosmic ray radiation on the human body or electronics. It was developed under collaboration between several institutes in Japan and Europe. Like MCNP, PHITS is subject to export control law for nuclear-related technology; it is distributed by the Research Organization for Information Science and Technology (RIST) in Japan [75], RSICC in the United States and Canada [59], and the OECD/NEA databank [60], giving access to pre-compiled executables along with source code files.

Examples of applications of PHITS to the study of SEEs include [76,77].

2.7 SUMMARY

A historical perspective on the Monte Carlo simulation of radiation effects has been presented with a focus on the application of this method for the study of electronics under irradiation. After giving a general definition of the method, its utility for the study of radiation effects in semiconductor devices was analyzed. This type of study is actually a problem in two parts: radiation transport and effect on electronics. While Monte Carlo methods are ideally suited for the simulation of radiation transport and interaction, the translation of the effect at the material level into an electrical effect is much less straightforward and is generally solved by other means.

The main features and limitations of Monte Carlo simulations were discussed, along with common solutions to speed up the calculations. Finally, some examples of Monte Carlo radiation transport simulation codes were presented along with examples of their application for electronics devices.

REFERENCES

1. N. Metropolis, The beginning of the Monte Carlo method, *Los Alamos Science*, Vol. 15, pp. 125–130, 1987.
2. D. Binder, E. C. Smith and A. B. Holman, Satellite anomalies from galactic cosmic rays, *IEEE Transactions on Nuclear Science*, Vol. 22, pp. 2675–2680, 1975.
3. G. Comte de Buffon, Essai d'arithmétique morale, *Histoire naturelle, générale et particulière: Supplément*, Vol. 4, pp. 46–109, 1777.
4. N. Metropolis and S. Ulam, The Monte Carlo method, *Journal of the American Statistical Association*, Vol. 44, pp. 335–341, 1949.
5. E. Hayward and J. H. Hubbell, The albedo of various materials for 1-Mev photons, *Physical Review*, Vol. 93, pp. 955–956, 1954.
6. F. B. Brown, Recent advances and future prospects for Monte Carlo, *Progress in Nuclear Science and Technology*, Vol. 2, pp. 1–4, 2011.
7. R. A. Reed et al., Anthology of the development of radiation transport tools as applied to single event effects, *IEEE Transactions on Nuclear Science*, Vol. 60, pp. 1876–1911, 2013.
8. G. A. Sai-Halasz and M. R. Wordeman, Monte Carlo modeling of the transport of ionizing radiation created carriers in integrated circuits, *IEEE Electron Device Letters*, Vol. 1, pp. 210–212, 1980.
9. P. J. McNulty, G. E. Farrell and W. P. Tucker, Proton-induced nuclear reactions in silicon, *IEEE Transactions on Nuclear Science*, Vol. 28, pp. 4007–4012, 1981.
10. G. R. Srinivasan, Modeling the cosmic-ray-induced soft-error rate in integrated circuits: An overview, *IBM Journal of Research and Development*, Vol. 40, pp. 77–89, 1996.

11. J. C. Pickel and J. T. Blandford Jr., Cosmic ray induced errors in MOS memory cells, *IEEE Transactions on Nuclear Science*, Vol. 25, pp. 1166–1171, 1978.
12. T. C. May and M. H. Woods, A new physical mechanism for soft errors in dynamic memories, in Proc. IEEE Reliability Physics Symposium, 1978.
13. T. C. May and M. H. Woods, Alpha-particle-induced soft errors in dynamic memories, *IEEE Transactions on Electron Devices*, Vol. 26, pp. 2–9, 1979.
14. J. F. Ziegler and W. A. Lanford, The effect of cosmic rays on computer memories, *Science*, Vol. 206, pp. 776–788, 1979.
15. M. H. Kalos and P. A. Whitlock, *Monte Carlo Methods*, Vol. 1. New York: Wiley, 1986.
16. P. Hellekalek, Good random number generators are (not so) easy to find, *Mathematics and Computers in Simulation*, Vol. 46, pp. 485–505, 1998.
17. F. James, A review of pseudorandom number generators, *Computer Physics Communications*, Vol. 60, pp. 329–344, 1990.
18. E. L. Petersen, J. C. Pickel, J. H. Adams Jr. and E. C. Smith, Rate prediction for single event effects—A critique, *IEEE Transactions on Nuclear Science*, Vol. 39, pp. 1577–1599, 1992.
19. E. L. Petersen, Soft errors results analysis and error rate prediction, in Proc. NSREC Short Course, Tucson, AZ, 2008.
20. R. A. Weller, M. H. Mendenhall, R. A. Reed, R. D. Schrimpf, K. M. Warren, B. D. Sierawski and L. W. Massengill, Monte Carlo simulation of single event effects, *IEEE Transactions on Nuclear Science*, Vol. 57, pp. 1726–1746, 2010.
21. C. A. Klein, Bandgap dependence and related features of radiation ionization energies in semiconductors, *Journal of Applied Physics*, Vol. 39, pp. 2029–2038, 1968.
22. M. Murat, A. Akkerman and J. Barak, Spatial distribution of electron-hole pairs induced by electrons and protons in SiO2, *IEEE Transactions on Nuclear Science*, Vol. 51, pp. 3211–3218, 2004.
23. M. Murat, A. Akkerman and J. Barak, Electron and ion tracks in silicon: Spatial and temporal evolution, *IEEE Transactions on Nuclear Science*, Vol. 55, pp. 3046–3054, 2008.
24. M. P. King, R. A. Reed, R. A. Weller, M. H. Mendenhall, R. D. Schrimpf, M. L. Alles, E. C. Auden, S. E. Armstrong and M. Asai, The impact of delta-rays on single-event upsets in highly scaled SOI SRAMs, *IEEE Transactions on Nuclear Science*, Vol. 57, pp. 3169–3175, 2010.
25. F. Gao, L. W. Campbell, Y. Xie, R. Devanathan, A. J. Peurrung and W. J. Weber, Electron-hole pairs created by photons and intrinsic properties in detector materials, *IEEE Transactions on Nuclear Science*, Vol. 55, pp. 1079–1085, 2008.
26. R. D. Narayan, R. Miranda and P. Rez, Monte Carlo simulation for the electron cascade due to gamma rays in semiconductor radiation detectors, *Journal of Applied Physics*, Vol. 111, pp. 064910, 2012.
27. K. Hess, ed., *Monte Carlo Device Simulation: Full Band and Beyond*: Kluwer Academic Publishers, 1991.
28. Synopsys [Online]. Available: http://www.synopsys.com/tools/tcad/Pages/default.aspx.
29. J. Saint Martin, A. Bournel and P. Dollfus, On the ballistic transport in nanometer-scaled DG MOSFETs, *IEEE Transactions on Electron Devices*, Vol. 51, pp. 1148–1155, 2004.
30. C. Wasshuber, H. Kosina and S. Selberherr, SIMON—A simulator for single-electron tunnel devices and circuits, *IEEE Transactions on Computer-Aided Design of Integrated Circuits and Systems*, Vol. 16, pp. 937–944, 1997.
31. M. A. Xapsos, Applicability of LET to single events in microelectronic structures, *IEEE Transactions on Nuclear Science*, Vol. 39, pp. 1613–1621, 1992.

32. M. Raine, M. Gaillardin, P. Paillet, O. Duhamel, S. Girard and A. Bournel, Experimental evidence of large dispersion of deposited energy in thin active layer devices, *IEEE Transactions on Nuclear Science*, Vol. 58, pp. 2664–2672, 2011.

33. M. Raine, G. Hubert, P. Paillet, M. Gaillardin and A. Bournel, Implementing realistic heavy ion tracks in a SEE prediction tool: comparison between different approaches, *IEEE Transactions on Nuclear Science*, Vol. 59, pp. 950–957, 2012.

34. F. B. McLean and T. R. Oldham, Basic mechanisms of radiation effects in electronic materials and devices, HDL-TR-2129, Harry Diamond Labs, Adelphi, MD, 1987.

35. A. Javanainen, J. R. Schwank, M. R. Shaneyfelt, R. Harboe-Sorensen, A. Virtanen, H. Kettunen, S. M. Dalton, P. E. Dodd and A. B. Jaksic, Heavy-ion induced charge yield in MOSFETs, *IEEE Transactions on Nuclear Science*, Vol. 56, pp. 3367–3371, 2009.

36. M. R. Shaneyfelt, D. M. Fleetwood, J. R. Schwank and K. L. Hughes, Charge yield for 10-keV X-ray and cobalt-60 irradiation of MOS devices, *IEEE Transactions on Nuclear Science*, Vol. 38, pp. 1187–1194, 1991.

37. M. Murat, A. Akkerman and J. Barak, Charge yield and related phenomena induced by ionizing radiation in SiO2 layers, *IEEE Transactions on Nuclear Science*, Vol. 53, pp. 1973–1980, 2006.

38. J. R. Srour and R. A. Hartmann, Effects of single neutron interactions in silicon integrated circuits, *IEEE Transactions on Nuclear Science*, Vol. 32, pp. 4195–4200, 1985.

39. P. W. Marshall, C. J. Dale, E. A. Burke, G. P. Summers and G. E. Bender, Displacement damage extremes in silicon depletion regions, *IEEE Transactions on Nuclear Science*, Vol. 36, pp. 1831–1839, 1989.

40. E. C. Auden, R. A. Weller, M. H. Mendenhall, R. A. Reed, R. D. Schrimpf, N. C. Hooten and M. P. King, Single particle displacement damage in silicon, *IEEE Transactions on Nuclear Science*, Vol. 59, pp. 3054–3061, 2012.

41. R. A. Weller, M. H. Mendenhall and D. M. Fleetwood, A screened coulomb scattering module for displacement damage computations in Geant4, *IEEE Transactions on Nuclear Science*, Vol. 51, pp. 3669–3678, 2004.

42. M. J. Norgett, M. T. Robinson and I. M. Torrens, A proposed method of calculating displacement dose rates, *Nuclear Engineering Design*, Vol. 33, pp. 50–54, 1975.

43. J. R. Srour and J. W. Palko, Displacement damage effects in irradiated semiconductor devices, *IEEE Transactions on Nuclear Science*, Vol. 60, pp. 1740–1766, 2013.

44. J. R. Srour and J. W. Palko, Displacement damage effects in devices, in Proc. Short Course of the Nuclear and Space Radiation Effects Conference, San Francisco, 2013.

45. F. Salvat, J. M. Fernandez-Varea and J. Sempau, PENELOPE-2008: A code system for Monte Carlo simulation of electron and photon transport, in Proc. OECD-NEA Workshop, Barcelona, Spain, 2008.

46. B. D. Sierawski et al., Muon-induced single-event upsets in deep-submicron technology, *IEEE Transactions on Nuclear Science*, Vol. 57, pp. 3273–3278, 2010.

47. P. Truscott, Radiation transport models and software, in Proc. Short Course of the 10th Radiation Effects on Components and Systems (RADECS) Conference, Bruges, Belgium, 2009.

48. M. J. Berger, Monte Carlo calculation of the penetration and diffusion of fast charged particles, in *Methods in Computational Physics*, Vol. 1. New York: Academic Press, 1963, pp. 135.

49. M. Raine, M. Gaillardin, J.-E. Sauvestre, O. Flament, A. Bournel and V. Aubry-Fortuna, Effect of the ion mass and energy on the response of 70-nm SOI transistors to the ion deposited charge by direct ionization, *IEEE Transactions on Nuclear Science*, Vol. 57, pp. 1892–1899, 2010.

50. M. Raine, G. Hubert, M. Gaillardin, L. Artola, P. Paillet, S. Girard, J.-E. Sauvestre and A. Bournel, Impact of the radial ionization profile on SEE prediction for SOI transistors and SRAMs beyond the 32 nm technological node, *IEEE Transactions on Nuclear Science*, Vol. 58, pp. 840–847, 2011.

51. V. Ferlet-Cavrois et al., Analysis of the transient response of high performance 50-nm partially depleted SOI transistors using a laser probing technique, *IEEE Transactions on Nuclear Science*, Vol. 53, pp. 1825–1833, 2006.

52. A. Akkerman, M. Murat and J. Barak, Monte Carlo calculations of electron transport in silicon and related effects for energies of 0.02–200 keV, *Journal of Applied Physics*, Vol. 106, pp. 113703, 2009.

53. A. Valentin, M. Raine, J.-E. Sauvestre, M. Gaillardin and P. Paillet, Geant4 physics processes for microdosimetry simulation: Very low energy electromagnetic models for electrons in silicon, *Nuclear Instruments and Methods in Physics Research B*, Vol. 288, pp. 66–73, 2012.

54. A. Valentin, M. Raine, M. Gaillardin and P. Paillet, Geant4 physics processes for microdosimetry simulation: Very low energy electromagnetic models for protons and heavy ions in silicon, *Nuclear Instruments and Methods in Physics Research B*, Vol. 287, pp. 124–129, 2012.

55. M. Raine, M. Gaillardin and P. Paillet, Geant4 physics processes for silicon microdosimetry simulation: Improvements and extension of the energy-range validity up to 10 GeV/nucleon, *Nuclear Instruments and Methods in Physics Research B*, Vol. 325, pp. 97–100, 2014.

56. M. Raine, A. Valentin, M. Gaillardin and P. Paillet, Improved simulation of ion track structures using new Geant4 models—Impact on the modeling of advanced technologies response, *IEEE Transactions on Nuclear Science*, Vol. 59, pp. 2697–2703, 2012.

57. A. F. Bielajew and D. W. O. Rogers, Variance-reduction techniques, in *Monte Carlo Transport of Electrons and Photons*, T. M. Jenkins, W. R. Nelson, and A. Rindi, eds. New York: Plenum, 1988, pp. 407–419.

58. T. Goorley et al., Initial MCNP6 release overview, *Nuclear Technology*, Vol. 180, pp. 298–315, 2012.

59. RSICC [Online]. Available: https://rsicc.ornl.gov/.

60. OECD/NEA databank Online. Available: https://oecd-nea.org/dbprog/.

61. C. H. Tsao, R. Silberberg and J. R. Letaw, A comparison of neutron-induced SEU rates in Si ans GaAs devices, *IEEE Transactions on Nuclear Science*, Vol. 35, pp. 1634–1637, 1988.

62. G. Gasiot, V. Ferlet-Cavrois, J. Baggio, P. Roche, P. Flatresse, A. Guyot, P. Morel, O. Bersillon and J. Du Port de Pontcharra, SEU sensitivity of bulk and SOI technologies to 14 MeV neutrons, *IEEE Transactions on Nuclear Science*, Vol. 49, pp. 3032–3037, 2002.

63. O. Flament, J. Baggio, C. D'Hose, G. Gasiot and J. L. Leray, 14 MeV neutron-induced SEU in SRAM devices, *IEEE Transactions on Nuclear Science*, Vol. 51, pp. 2908–2911, 2004.

64. S. Agostinelli et al., GEANT4—A simulation toolkit, *Nuclear Instruments and Methods in Physics Research A*, Vol. 506, pp. 250–303, 2003.

65. J. Allison et al., Geant4 developments and applications, *IEEE Transactions on Nuclear Science*, Vol. 53, pp. 270–278, 2006.

66. Geant4 [Online]. Available: http://geant4.web.cern.ch/geant4.

67. F. Lei, P. Truscott, C. S. Dyer, B. Quaghebeur, D. Heynderickx, P. Nieminen, H. Evans and E. Daly, MULASSIS: A Geant4-based multilayered shielding simulation tool, *IEEE Transactions on Nuclear Science*, Vol. 49, pp. 2788–2793, 2002.

68. G. Santin, V. Ivanchenko, H. Evans, P. Nieminen and E. Daly, GRAS: A general-purpose 3-D modular simulation tool for space environment effects analysis, *IEEE Transactions on Nuclear Science*, Vol. 52, pp. 2294–2299, 2005.
69. A. Ferrari, P. R. Sala, A. Fasso and J. Ranft, FLUKA: A multi-particle transport code, *CERN-2005-10*, 2005.
70. G. Battistoni, S. Muraro, P. R. Sala, F. Cerutti, A. Ferrari, S. Roesler, A. Fasso and J. Ranft, The FLUKA code: Description and benchmarking, in Proc. Hadronic Shower Simulation Workshop, Fermilab, 2007.
71. M. Huhtinen and F. Faccio, Computational method to estimate single event upset rates in an accelerator environment, *Nuclear Instruments and Methods in Physics Research A*, Vol. 450, pp. 155–172, 2000.
72. S. Koontz, B. Reddell and P. Boeder, Calculating spacecraft single event environments with FLUKA: Investigating the effects of spacecraft material atomic number on secondary particle showers, nuclear reactions, and linear energy transfer (LET) spectra, internal to spacecraft avionics materials, at high shielding mass, in Proc. IEEE Radiation Effects Data Workshop, 2011.
73. M. Brugger, Radiation effects, calculation methods and radiation test challenges in accelerator mixed beam environments, in Proc. Short Course of the Nuclear and Space Radiation Effects Conference, Paris, 2014.
74. T. Sato et al., Particle and Heavy Ion Transport Code System PHITS, version 2.52, *Journal of Nuclear Science and Technology*, Vol. 50, pp. 913–923, 2013.
75. RIST [Online]. Available: http://www.rist.or.jp/nucis/.
76. H. Kobayashi, N. Kawamoto, J. Kase and K. Shiraish, Alpha particle and neutron-induced soft error rates and scaling trends in SRAM, in *Proc. IEEE Reliability Physics Symposium*, Montreal, Canada, 2009.
77. S. Abe, Y. Watanabe, N. Shibano, N. Sano, H. Furuta, M. Tsutsui, T. Uemura and T. Arakawa, Multi-scale Monte Carlo simulation of soft-errors using PHITS-HyENEXSS code system, *IEEE Transactions on Nuclear Science*, Vol. 59, pp. 965–970, 2012.

3 A Complete Guide to Multiple Upsets in SRAMs Processed in Decananometric CMOS Technologies

Gilles Gasiot and Philippe Roche

CONTENTS

3.1 INTRODUCTION

Susceptibility to radiation environment of advanced electronic devices is often responsible for the highest failure rate of all reliability concerns (electromigration, gate rupture, negative bias temperature instability [NBTI], etc.). In modern static random access memories (SRAMs), the two predominant single-event effects (SEEs) are the single-event upset (SEU) and multiple upsets (MUs). MUs are topological errors in neighboring cells. If the cells belong to the same logical word they are called multiple-bit upsets (MBUs), otherwise they are labeled as multiple-cell upsets (MCUs). MUs have received increased scrutiny in recent years [1–8] because they are uncorrectable by simple error-correcting code (ECC) schemes and therefore threaten the efficiency of error detection and correction (EDAC).

As technologies scale down, the amount of transistors per mm^2 doubles at each generation while the radioactive feature size (ion track diameter) is constant. This is illustrated in Figure 3.1 with three-dimensional (3-D) technology computer-aided design (TCAD) simulation showing an ion impacting a single cell in 130 nm while several are impacted in 45 nm. Moreover, SRAM's ability to store electrical data (critical charge) is reduced as technology feature size and power supply are jointly decreased. The probability that a particle upsets more than a single cell is therefore increased [9–11].

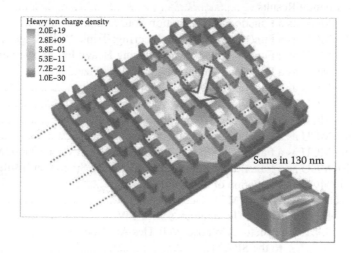

FIGURE 3.1 3-D TCAD simulation of ion impact (single LET) in a single SRAM bitcell in 130 nm and 12-SRAM bitcells in 45 nm.

The mechanism for MCU occurrence in SRAM arrays is more than "enough energy was deposited to upset two cells" and depends on the radiation used. Directly ionizing radiation from single particles (alpha particles, ions, etc.) deposits charges diffusing in wells that can be collected by several bitcells. This phenomenon is enhanced by using tilted particles either naturally (alpha particles whose emission angle is random from the radioactive atom) or artificially (heavy ions can be chosen during experimental tests from 0° to 60°). Nonionizing radiation such as neutrons and protons can have different MCU occurrence mechanisms (Figure 3.2). A nonionizing particle can produce one or more secondary products. Several cases have to be considered: two secondary ions from two nucleons upset two or more bitcells, two secondary ions from a single nucleon upset two or more bitcells, and a single secondary ion from a single nucleon upsets two or more bitcells (in this case the phenomenon is close to the previously described direct ionizing mechanism). It has been shown that the type 1 mechanism was negligible but that the type 2 and 3 mechanisms coexist [12]. However, the proportion of MCUs due to these two mechanisms has never been precisely assessed.

One of the first experimental evidence of MBUs was reported in 1984 in a 16 × 16 bit bipolar RAM under heavy ion irradiation [13]. It is noteworthy that as many as 16 bit errors in columns from a single ion strike were detected. This means that 6% of the entire memory array was in error from a single particle strike. Since this first experimental evidence, multiple bit errors were detected in several device types, such as DRAM [14], polysilicon load SRAM [15], and antifuse-based FPGA [16], and under various radiation types, such as protons [17], neutrons [18], and laser [19].

The goal of this work is first to experimentally quantify MCU occurrence as a function of several parameters such as radiation type, test conditions (temperature, voltage, etc.), SRAM architecture, and so forth. These results will be used to sort by order of importance the parameters driving MCU susceptibility. Second, three-dimensional TCAD simulations will be used to investigate the mechanisms leading to MCU occurrence and to determine the most sensitive location to trigger a 2-bit MCU as well as the cartography of MCU sensitive areas.

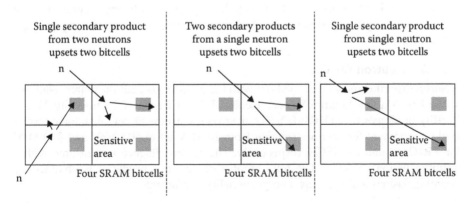

FIGURE 3.2 Scheme of neutron interaction that can cause MCU in SRAM array. (Derived from F. Wrobel et al., *IEEE Transactions on Nuclear Science,* Vol. 48, No. 6, pp. 1946–1952, Dec. 2001.)

3.2 EXPERIMENTAL SETUP

The design of the experiment included different test patterns and supply voltages. The test procedure was compliant with the JEDEC SER test standard JESD89 [20] for alpha and neutrons and ESA test standard number 22900 for heavy ions and protons [21].

3.2.1 IMPORTANCE OF A TEST ALGORITHM FOR COUNTING MULTIPLE UPSETS

When experimentally measuring MCUs, it is mandatory to distinguish (1) multiple independent failures from a cluster of nearest neighbor upset from a single multicell upset caused by a single energetic particle, and (2) signature of errors due to a hit in redundancy latch or sense amplifier that may upset an entire row or column from a MCU signature. A test algorithm allows separating independent events due to multiple particle hits from single events that upset multiple cells. Dynamic testing of memory usually involves writing once and then reading continuously at a specified operating frequency at which events are recorded one at a time. This gives a real insight on MCU shapes and occurrence. However, with static testing of memory, the test pattern is written once and stored for an extended period before reading the pattern back out. The result is a failure bit mapping in which events due to multiple particle hits and single events that upset multiple cells cannot be distinguished. However, statistical tools can be applied to quantify rate of neighboring upsets due to several ions [22,23]. One of these tools is described in detail in Annex 1.

3.2.2 TEST FACILITY

3.2.2.1 Alpha Source

The tests were performed with an alpha source, which is a thin foil of Americium-241 that has an active diameter of 1.1 cm. The source activity was 3.7 MBq as measured on February 1, 2002. The alpha particle flux was precisely measured in March 2003 with a Si detector that was placed at 1 mm from the source surface. Since the atomic half-life of Am241 is 432 years, the activity and flux figures are still very accurate. During soft error rate (SER) experiments, the Americium source lies above the chip package in the open air.

3.2.2.2 Neutron Facilities

Neutron experiments were carried out with the continuous neutron source available at the Los Alamos Neutron Science Center (LANSCE) and Tri University Meson Facility in Vancouver (TRIUMF). The neutron spectrums closely match the terrestrial environment for energies ranging from 10 MeV up to 500 MeV and 800 MeV for TRIUMF and LANSCE, respectively. The neutron fluence is measured with a uranium fission chamber. The total number of neutrons produced is obtained by counting fissions and applying a proportionality coefficient.

3.2.2.3 Heavy-Ion Facilities

The heavy-ion tests were conducted using the Radiation Effect Facility (RADEF) [24] cyclotrons. The RADEF facility is located in the Accelerator Laboratory at the

University of Jyväskylä, Finland (JYFL). The facility includes beam lines dedicated to proton and heavy-ion irradiation studies of semiconductor materials and devices. The heavy ion line consists of a vacuum chamber with component movement apparatus inside and ion diagnostic equipment for real-time analysis of beam quality and intensity. The cyclotron used at JYFL is a versatile, sector-focused accelerator for producing beams from hydrogen to xenon. The accelerator is equipped with three external ion sources. There are two electron cyclotron resonance (ECR) ion sources designed for high-charge-state heavy ions. Heavy ions used at the RADEF facility have stopping ranges in silicon much larger than the whole stack of back-end metallization and passivation layers (~10 μm).

3.2.2.4 Proton Facility

Proton irradiations were performed at the Proton Irradiation Facility (PIF) at PSI (Paul Scherrer Institute, Switzerland). This institute was constructed for the testing of spacecraft components. The main advantages of PIF are that irradiation takes place in the air, the flux/dosimetry is about 5% absolute accuracy, and beam uniformity is higher than 90%. The experiments have used the low-energy PIF line whose energy range is 6 to 71 MeV and maximum proton flux is 5E8 p/cm^2/sec.

3.2.3 TESTED DEVICES

Most of the data presented in this work was obtained using a single testchip (Figure 3.3). This testchip embeds three different bitcell architecture, two single-port (SP) and one dual-port (DP). It was manufactured in a 65-nm commercial complementary metal-oxide semiconductor (CMOS) technology with a low-power (LP) process option. The main features of tested devices are summarized in Figure 3.4. Each bitcell was processed with and without the triple-well process option.

A triple-well layer consists of either a N+ or P+ buried layer in, respectively, a p- or n-doped substrate. As most devices are processed in a P-substrate, triple wells

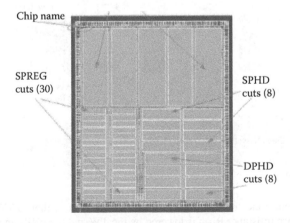

FIGURE 3.3 Floorplan of the test vehicle designed and manufactured in a 65-nm CMOS technology.

Bitcell	Bitcell area	Capacity	DNW
Single port SRAM high density	0.52 μm²	2 Mb	No
Single port SRAM high density	0.52 μm²	2 Mb	Yes
Single port SRAM standard density	0.62 μm²	2 Mb	No
Single port SRAM standard density	0.62 μm²	2 Mb	Yes
Dual port SRAM high density	0.98 μm²	1 Mb	No
Dual port SRAM high density	0.98 μm²	1 Mb	Yes

FIGURE 3.4 Content of the test vehicle. Three different bitcell architectures were embedded. Every bitcell is processed with and without a TW layer.

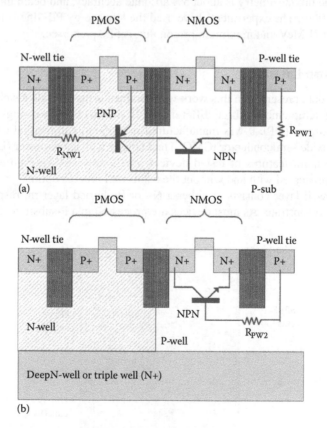

(a)

(b)

FIGURE 3.5 Schematic cross section of a CMOS inverter (a) without a TW and (b) with a TW. The PNP base resistance R_{NW1} is lowered by the TW and the PNP cannot be triggered. Conversely, the TW layer pinches the P-well and increases the NPN base resistance R_{PW2}, and the NPN triggering is facilitated.

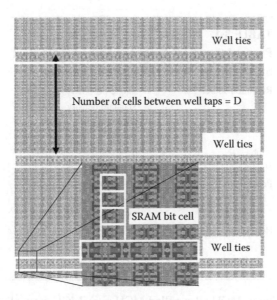

FIGURE 3.6 Layout of an SRAM cell array showing the periodical distribution of the well tie rows every 32 cells.

are often referred to as a deep N-well or n+ buried layer (Figure 3.5). For years triple-well (TW) layers have been used to electrically isolate the P-well and to reduce electronic noise from the substrate. The TW is biased through the N-well contacts/ties connected to VDD while the P-wells are grounded. The well ties are regularly distributed along the SRAM cell array as depicted in Figure 3.6. The triple-well process option has two main effects on the radiation susceptibility. First, it allows for decreasing the SEL sensitivity since the PNP base resistance is strongly reduced (Figure 3.1). TW accordingly makes the latch-up thyristor more difficult to trigger on. In the literature, full latch-up immunity is reported even under extreme conditions (high voltage, high temperature, and high LET) [25,26]. Second, this buried layer allows for concurrently decreasing the SEU/SER sensitivity since the electrons generated deep inside the substrate are collected by the TW layer and then evacuated through the N-well ties. The improvement of the SER using TW is reported in several papers [27–29]. However, other research teams have published an increased SER sensitivity due to the TW in a commercial CMOS 0.15-μm technology [30,31].

3.3 EXPERIMENTAL RESULTS

MCUs were recorded during the SER experiments on the 65-nm SRAM, but no MBU was ever detected as the tested memory uses bit interleaving or scrambling. All the MCU percentages reported in this work were computed by dividing the number of upsets from MCU by the total number of upsets (single-bit upsets [SBUs] plus MCUs). Note that in the literature, events are sometimes used instead of upsets [31], so the MCU percentages are in this case significantly underestimated. Unless otherwise specified, tests were performed at room temperature in dynamic mode with checkerboard and uniform test patterns. In addition to the usual MCU percentages, we report in this work

TABLE 3.1

Percentage of MCU for the Same Single-Port SRAM under Several Radiation Sources

Radiation source	Single-port SRAM Standard density CKB pattern No TW
Alpha	0.5%
Neutron	21% at LANSCE
Proton	4% at 10 MeV
	20% at 40 MeV
	25% at 60 MeV
Heavy ion	0% at 5.85 MeV/cm².mg
	87% at 19.9 MeV/cm².mg
	99.8% at 48 MeV/cm².mg

the failure rates due to MCU (also called MCU rate). MCU rates allow quantitatively comparing MCU occurrence between different technologies and test conditions.

3.3.1 MCU AS A FUNCTION OF A RADIATION SOURCE

The four radiation sources have different interaction modes that are either directly ionizing (alpha and heavy ions) or nonionizing (neutron and protons). However, it is of interest to compare the MCU percentage from these radiations on the same test vehicle. The test vehicle chosen is a single-port SRAM of standard density processed without a TW. MCU percentages are synthesized as seen in Table 3.1, which shows that alpha particles lead to the lower MCU occurrence. Moreover, heavy ions lead to the higher MCU percentages while neutrons and protons are similar. Heavy ions have the harshest radiation MCU-wise.

3.3.2 MCU AS A FUNCTION OF WELL ENGINEERING: TRIPLE-WELL USAGE

Table 3.2 synthesizes and compares MCU rates and percentage for the standard-density single-port SRAMs processed with and without a triple well. Table 3.3

TABLE 3.2

MCU Rates and Percentages of a Single-Port SRAM Processed with and without TW[a]

	MCU Rate	% MCU
SP SRAM standard density (no TW)	100 (norm)	21
SP SRAM standard density (TW)	1000	76

[a] MCU rate is normalized to its value without TW.

TABLE 3.3

MCU Percentages and Rates after Neutron Irradiation at Nominal Voltage and Room Temperature for Two Different Test Patterns

Technology	Bitcell Area	CKB Pattern	
		MCU %	MCU Rate (au)
Bulk	2.5 μm²	16.90	100
SOI	2.5 μm²	2.10	10

indicates first that the usage of TW increases the MCU rate by a decade and the MCU percentage by a factor ×3.6. Usage of MCU rate is mandatory since MCU percentages can lead to incomplete information. As presented in Figure 3.7, devices without TW have a lower number of bits involved per MCU event (≤8) compared to those with TW. This figure also indicates that for SRAMs with a triple-well, 3-bit and 4-bit MCU events are more likely than 2-bit events.

The effect of a triple-well layer on MCU percentages under heavy ions is reported in Figure 3.8. The SRAM under test is a high-density SP SRAM. For the smallest LET, MCUs represent 90% of the events with a TW but less than 1% without a TW. For LET_{eff} higher than 5.85 MeV/cm².mg there is no SBU in the SRAM with a TW. For LET upper than 14.1, all the MCU events induce more than five errors with a TW. With a TW, the significant increase in the MCU amount and order causes an increase in the error cross section.

Whatever the radiation source, the usage of a TW strongly increases the occurrence of MCU. This increase is so high that it can be seen in the total bit error rate for neutrons and error cross section for heavy ions.

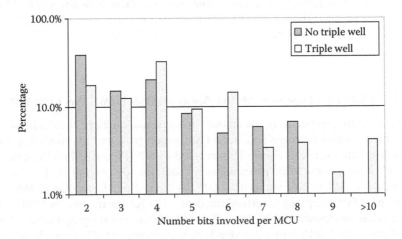

FIGURE 3.7 Number of bits involved in MCU events for high-density SP SRAM under neutron irradiation.

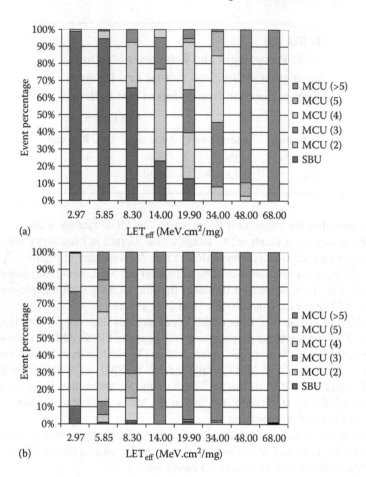

(a)

(b)

FIGURE 3.8 Proportion for single and multiple events for (a) high-density SP SRAM without a TW option and (b) high-density SP SRAM with a TW option. (From D. Giot, P. Roche, G. Gasiot, J.-L. Autran and R. Harboe-Sørensen, *IEEE Transactions on Nuclear Science*, Vol. 55, No. 4, 2007.)

3.3.3 MCU AS A FUNCTION OF TILT ANGLE DURING HEAVY ION EXPERIMENTS

Figure 3.7 shows, respectively, the amount of single and multiple bit fails induced by a given ion species (nitrogen (N), neon (Ne), argon (Ar), and krypton (Kr)) whose tilt angle is either vertical (Figure 3.9a) or tilted by 60° (Figure 3.9b). Tilt angle from 0° to 60° increases the MBU percentages for each ion species. For N, the MBU% is increased from 0% to 30% with a tilt = 60°. For Ne and Ar, the amount of MBU fails is doubled at 60° compared to vertical incidence. For Kr, the increase of MBU% with the tilt is less pronounced (+10% from 0° to 60°) because of the progressive substitution of low order MBUs (order 2, order 3) by higher order MBUs (order 5, order >5).

On average the amount of bit fails due to MBU is doubled for 60° tilt compared to normal incidence [32].

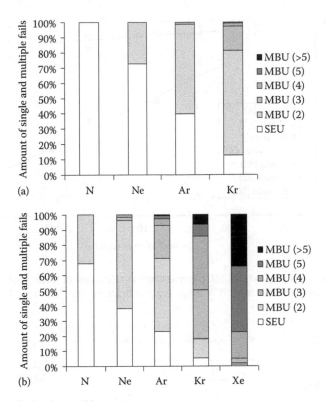

FIGURE 3.9 Amount of bitfails due to single and multiple events in 90-nm SP SRAM: (a) with heavy ion beam not tilted and (b) with heavy ion beam tilted at 60°.

3.3.4 MCU AS A FUNCTION OF TECHNOLOGY FEATURE SIZE

Figure 3.10 shows the experimental neutron MCU percentages as a function of technology feature size and compares data from this work with data from the literature. This data shows that technologies with TW have MCU percentages higher than 50% while technologies without have MCU percentage lower than 20%. Data from the literature fits either our set of data with TW or without TW. Consequently, Figure 3.8 suggests that MCU percentages can be sorted with a criterion of TW usage. Moreover, the MCU percentages both increase with and without TW when the technologies scale down, this slope being higher without TW since MCU percentages were very low (~1% in 150 nm) for old technologies.

3.3.5 MCU AS A FUNCTION OF DESIGN: WELL TIE DENSITY

TCAD simulations on 3-D structures built from the layout of the tested SRAMs have been performed as shown in Section 3.4. Simulation results for the ratio between drain-collected charge with and without TW is plotted in Figure 3.11. This figure indicates first that the collected charge with TW is higher than without regardless of the well tie frequency. Second, the charge collection increase ranges from ×2.5 to ×7

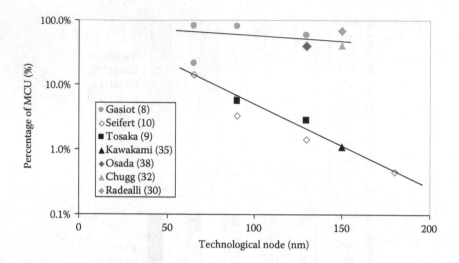

FIGURE 3.10 Neutron-induced MCU percentages as a function of technological node from this work and from the literature. TW usage is not indicated in the data from the literature.

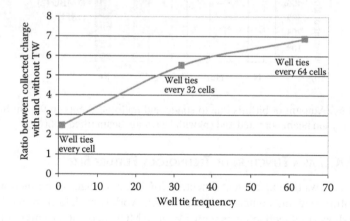

FIGURE 3.11 Simulation results for the ratio between collected charge by the N-OFF drain with and without a TW. This ratio is plotted as a function of well ties frequency.

for the highest and the lowest well tie frequency, respectively. This demonstrates that when TW is used, increasing the well tie frequency mitigates the bipolar effect and therefore the MCU rate and SER.

3.3.6 MCU as a Function of Supply Voltage

The effect of supply voltage on the radiation susceptibility is well known: the higher the voltage, the lower the susceptibility since the charge storing the information is increased proportionally to the supply voltage. However, the effect of the supply voltage on the MCU rate is not documented. Experimental measurements were performed at LANSCE on an HD SRAM processed with and without

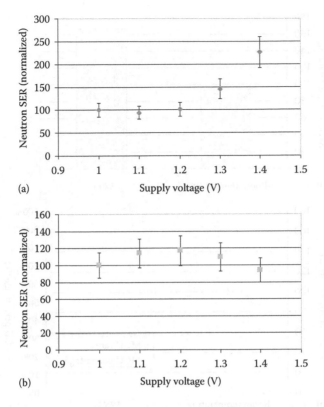

FIGURE 3.12   MCU rate as a function of supply voltage for the HD SRAM processed (a) without a TW and (b) with a TW process option. MCU rates are normalized to their value at 1 V.

a TW option at different supply voltage ranging from 1 V to 1.4 V. Results are synthesized in Figure 3.12. It shows that when the supply voltage is increased the device with TW MCU rate remains constant within the experimental uncertainty. However, a different trend is observed for the device without a TW layer. When the supply voltage is increased the MCU rate is constant from 1.0 V to 1.2 V and then increases from 1.3 V to 1.4 V. The MCU rate increase is 220% for V_{DD} equal to 1.4 V.

3.3.7   MCU as a Function of Temperature

High temperature constraint is associated with high-reliability applications such as automotive. Some papers have quantified the temperature effect on SER or heavy ion susceptibility [33,34]. At the time of this writing no reference can be found in the literature experimentally measuring temperature effect on the MCU rate. Experimental measurements were performed at LANSCE on an HD SRAM processed with and without a TW option at room temperature and 125°C. Results are synthesized in Figure 3.13. It demonstrates that the MCU rate increases by

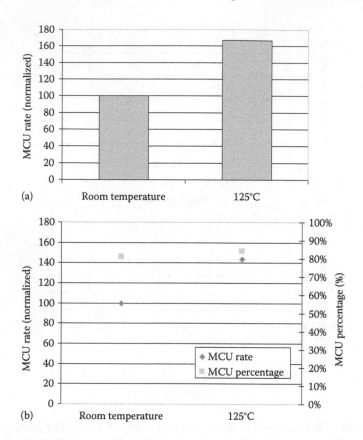

(a)

(b)

FIGURE 3.13 MCU rate as a function of temperature for the HD SRAM processed (a) without a TW and (b) with a TW process option. MCU rates are normalized to their value at room temperature. Figure 3.13b also displays the MCU percentages.

65% for the device without a TW and by 45% for the device with a TW. Note that the usage of MCU percentage would have been misleading since the MCU percentage is constant between room temperature and 125°C for the device with a TW.

3.3.8 MCU AS A FUNCTION OF BITCELL ARCHITECTURE

Figure 3.14 synthesizes MCU rates for high-density and standard-density single-port SRAMs as well as a dual-port SRAM (eight transistors). These SRAMs were processed without a TW. Figure 3.14 indicates that the higher the density, the higher the MCU rate. A decrease in the bitcell area by a factor ×2 (HD SP SRAM compared to DP SRAM) induces a decrease in the MCU rate by a factor ×3.

Effect of bitcell architecture on MCU percentages under heavy ions is reported in Figure 3.15. The devices under test are high-density SP SRAMs (Figure 3.15a) and standard-density SP SRAMs (Figure 3.15b). Figure 3.15a and b show the respective amount of SBU and MCU events for experimental ion LET ranging from 2.97 to

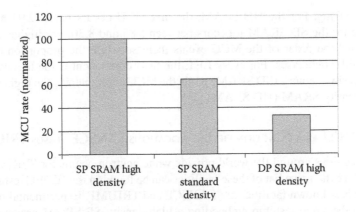

FIGURE 3.14 MCU rate comparison for several bitcell architectures. The devices under test were processed without a TW. SP = single port, DP = dual port (8-transistor SRAM).

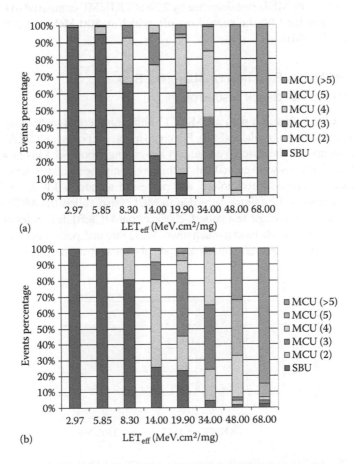

FIGURE 3.15 Amount of bitfails due to single and multiple upsets for (a) high-density SP SRAM and (b) standard density SP SRAM.

68 MeV/cm².mg. For the HD SRAM, the first MCU occurs below 2.97 MeV/cm². mg while for the SD SRAM it occurs between 5.85 and 8.30. For higher LET, the amount and the order of the MCU events increase while the proportion of single events (SBU) decreases. For every LET, the SBU component is the highest for the lowest density memory (SD SRAM) while the MCU component is the highest for the highest density SRAM (HD SRAM) [35].

3.3.9 MCU AS A FUNCTION OF TEST LOCATIONS LANSCE VERSUS TRIUMF

Several facilities around the world provide white neutron beams for SER character-ization. An exhaustive list of these facilities can be found in the JEDEC test standard [20]. The best-known facilities are LANSCE and TRIUMF. Experimental measure-ments on the same testchip embedding a high-density SP SRAM processed with a TW option were performed at these two facilities. The MCU percentages were perfectly equal to 76% for both facilities. The MCU rates are reported in Figure 3.16, which shows that the MCU rate decrease by 22% at TRIUMF compared to LANSCE. This can be explained by the energy cut-off, which is 800 MeV at LANSCE and 500 MeV at TRIUMF.

3.3.10 MCU AS A FUNCTION OF SUBSTRATE: BULK VERSUS SILICON ON INSULATOR

SRAMs were manufactured with a CMOS 130-nm commercial technology that was either bulk or silicon on insulator (SOI). For comparison purposes both SRAM designs are strictly identical. The testchip contains 4 Mb of single-port SRAMs in which two different bitcell designs were embedded. In this work only the standard-density SRAM will be reported. The bulk technology was processed without a TW layer. Table 3.3 therefore synthesizes the failure rates due to MCU (also called the MCU rate) and MCU percentage for a single test pattern checker board (CKB). It is noteworthy from Table 3.2 that SOI SRAMs have a much lower MCU rate and percentage compared to bulk. More parameters (pattern, bitcell area, supply voltage) were studied in [36].

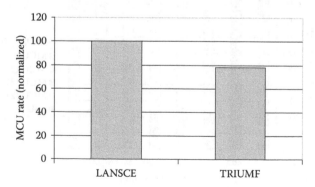

FIGURE 3.16 MCU rate comparison between LANSCE and TRIUMF white neutron beam sources. The device under test is a high-density SRAM processed with a TW.

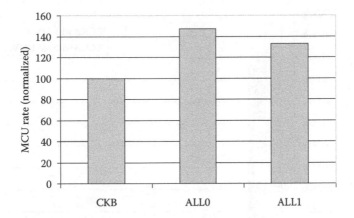

FIGURE 3.17 MCU rate comparison for several test patterns. Note that test patterns are physical. The device under test is a high-density SRAM processed without a TW. CKB = checkerboard, ALL0 and ALL1 = uniform of 0 and 1, respectively.

3.3.11 MCU AS A FUNCTION OF TEST PATTERN

A high-density SRAM was measured at LANSCE with several test patterns using a dynamic test algorithm. Results are synthesized in Figure 3.17, which shows that uniform patterns have a higher MCU rate than the CKB. To understand the reason of this discrepancy, it is necessary to plot the topological shape of experimental 2-bit MCU events as a function of pattern filling the memory during the testing (Figure 3.18a and b). The prevailing shape for 2-bit MCU and a checkerboard pattern is diagonal-adjacent while it is column-adjacent with a uniform pattern (as observed in [37]). Three-dimensional TCAD simulations have shown that 2-bit MCU threshold LET is the lowest for 2 bitcells in column (see [32] and Section 3.4.2). It is therefore consistent that uniform patterns have a higher MCU rate since their error clusters are the easiest to trigger.

It is also noteworthy that in Figure 3.18a and b TW usage did not modify the prevailing shape of MCU either for a checkerboard or for a uniform pattern.

3.4 3-D TCAD MODELING OF MCU OCCURRENCE

Section 3.3 clearly highlighted the importance of a TW in the MCU response. In this section, 3-D TCAD simulations are set up to analyze the increased MCU occurrence when a TW is used. All 3-D SRAM structures in this part were built using a methodology described in [38] and the tool suite v10.0 from the Sentaurus Synopsys package [39]. Cell boundaries are defined from the CAD layout and technological process steps. One-dimensional doping profiles are precisely modeled from secondary ion mass spectrometry (SIMS) profiles. Cell boundaries are defined from the CAD layout and technological process steps. One-dimensional doping profiles are included to define N-well, P-well (with a 4-μm epi layer thickness) and active regions of transistors (Figure 3.19). Mesh refinements are included

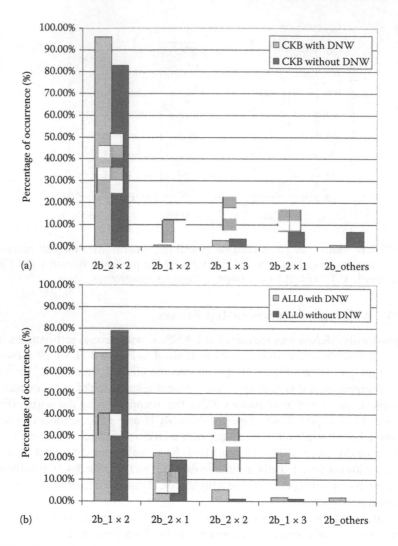

FIGURE 3.18 2-bit MCU cluster shape on high-density SP SRAM processed with or without a TW after neutron irradiation when the test pattern is (a) a checkerboard or (b) a uniform pattern.

in regions of interest: channels, lightly doped drain (LDD), junction boundaries (to tackle short channel effects) and around ion track (to allow accurate generation of carriers in silicon). Wire connections between the different electrodes of the cell are modeled in the SPICE domain (mixed-mode TCAD simulations) to reduce the CPU burden. The parasitic circuit capacitances due to metallization layers are also taken into account.

Device simulations with ion impacts are performed using the Sentaurus Device simulator. For this purpose, several physical models are activated: drift diffusion for carriers' transport, Shockley-Read-Hall and Auger for recombination, electric field and doping-dependant models for mobility, and the heavy ion module for carrier deposition along particles' tracks. The heavy ion generation model uses a Gaussian radial distribution of charges with a fixed characteristic radius of 0.1 μm and a Gaussian time distribution centered at 1 ps. An additional assumption consists of taking a constant LET along the track because of the low diffusion depth of transistor active areas (~0.2 μm). Properties of boundaries are defined by the Neumann reflective conditions [38,39].

3.4.1 Bipolar Effect in Technologies with TW

For an in-depth analysis of the MCU phenomenon, 3-D device simulations were performed on full SRAM bit cells. Ion strikes were located in the most sensitive MCU location (source of the SRAM) for different distances from the well taps with and without TW. It is noteworthy that Osada et al. [40] already tried to model the effect of the parasitic bipolar amplification on the MCU. A more simple mix of device (two-dimensional [2-D] uniformly extended) and circuit simulations was used but not for the worst-sensitive location for MCU occurrence [32].

3.4.1.1 Structures Whose Well Ties Are Located Close to the SRAM

Figure 3.19 presents the 3-D SRAM bit cell made up of six transistors (6T), two P-wells, one N-well, and three well ties. The well ties are as close as possible to transistors. The simulation results of these structures are presented in Figure 3.20, which compares source and drain currents after an ion impact in the source

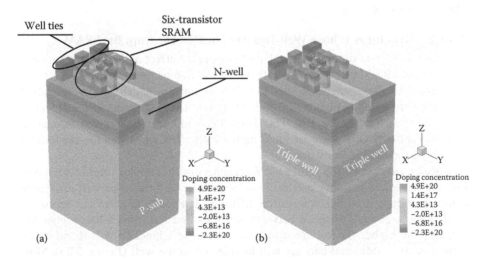

FIGURE 3.19 Full 3-D structures of the 65-nm 6T SRAM located as close as possible to the well ties (a) without a TW and (b) with a TW. Two NMOSs are embedded per P-well (one is a part of the inverter, the other is an access transistor).

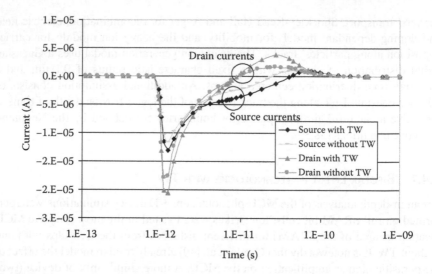

FIGURE 3.20 Full 3-D TCAD simulations results on the structure presented in Figure 3.9 (6T SRAM very close to the well taps) show a limited bipolar effect due to the presence of the TW layer. Heavy ion LET is 5.5 fC/μm.

at 1 ps. The charge collected at the N-OFF drain is slightly higher with the TW when well ties are located close to the SRAM transistors. With the TW a limited bipolar effect (see Section 3.4.1.2 for details on bipolar triggering) is observed for structures close to the ties. These simulation results are consistent with the experimental results presented in [22,30] that have shown that MCU occurrence is less likely close to well ties.

3.4.1.2 Structures Whose Well Ties Are Located Far from the SRAM

A second set of 3-D structures were built to model the effect of the spacing between well ties and SRAM cells with and without the TW doping profiles. Figure 3.21a and b illustrates four structures dedicated to well tie frequency modeling. The simulation results are presented in Figure 3.22 for ion features (LET and strike location) identical to those used in Figure 3.20. The charge injected by the source and the charge collected at the N-OFF drain are much higher with a TW when well ties are located away from the SRAM transistors.

Injected carriers by the source are forerunners of the bipolar transistor triggering. Ion-deposited majority carriers flow toward the well ties. The well resistance causes a voltage drop beneath source diffusion. If enough carriers are deposited or if there is enough distance between well ties and ion impact (the higher the distance the higher the voltage drop), the source-well junction will therefore be turned on and additional carriers will be injected in the well (Figure 3.23). Most of these additional carriers will be collected at the drain junction and thus increasing the collected charge at the drain. The additional charge collection due to the source injection and to the parasitic bipolar action is responsible for the bitcell

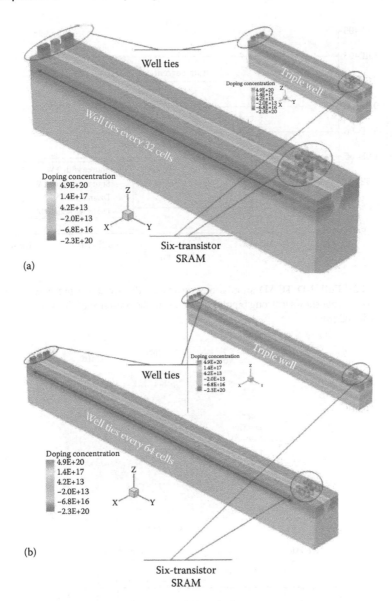

FIGURE 3.21 Full 3-D structures of the 65-nm 6T SRAM whose well ties are located (a) 32 cells and (b) 64 cells away from the well taps without a TW. The same structures with TW are shown in the upper-right inserts.

upset. Moreover, voltage drop in the well can turn on several sources along the well, which will upset several bitcells and be responsible for the MCU pattern experimentally reported in Section 3.3.11.

The simulations have shown that with a TW a strong bipolar effect (electron injection from the sources) is observed for structure away from the ties. These simulation

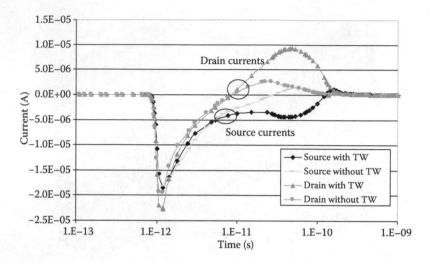

FIGURE 3.22 Full 3-D TCAD simulations results on the structure presented in Figure 3.11a. Source current shows a strong bipolar effect due to the presence of the TW layer. Heavy ion LET is 5.5 fC/μm.

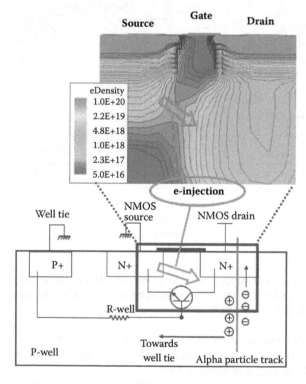

FIGURE 3.23 Illustration of the carrier injected by the source and triggering of the parasitic bipolar transistor after an alpha particle strike in the drain. Insert is from device simulation of the 65-nm 3-D structure.

results are consistent with the experimental results presented in [22,30] that have shown that MCU occurrence is more likely away from well ties.

3.4.2 REFINED SENSITIVE AREA FOR ADVANCED TECHNOLOGIES

We now aim to show by means of 3-D TCAD simulations that a bitcell SEE sensitive area is not restricted to the area of reverse-biased junctions. Figure 3.24 shows the 3-D TCAD final structures of two SP bitcells arranged in column (a) and row (b). These continuous TCAD domains include 710,000 and 580,000 elements, respectively. The double bitcell structures are dedicated to double MBU studies. CPU burden is respectively around 1 week to simulate a double SRAM structures with up-to-date high-performance workstations.

Figure 3.25 shows an area of four SP bitcells. Two bit cells of the same column share the sources of their MOS transistors whereas two bitcells of the same row do not share any P/N junction and are isolated with shallow trench isolation (STI). At first order, a MBU of two adjacent cells is horizontal, vertical, or diagonal (configuration 1, 2, and 3 in Figure 3.25). The third case of diagonal double MBU was not simulated. Indeed, a diagonal MBU would provide a higher MBU LETth than one computed for a row MBU because of the longer distance between the adjacent SEU-sensitive areas (both are separated with STI) (Table 3.4).

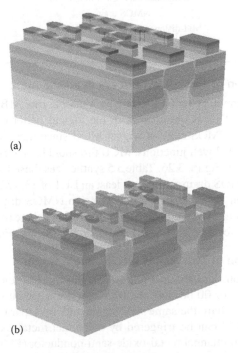

(a)

(b)

FIGURE 3.24 SRAM 3-D structures (STI not displayed for clarity): (a) double 6T bitcells in a column and (b) double 6T bitcells in a row.

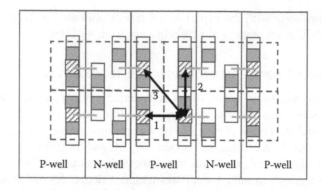

FIGURE 3.25 Four contiguous SRAM bitcells: dashed rectangles are bitcells. Connected striped and white squares are drains of NMOS and PMOS transistors, respectively. Single gray and white squares are gates and sources of NMOS and PMOS.

TABLE 3.4

Simulated MCU Threshold LET for Two Single-Port SRAMs Arranged in a Row and in a Column

TCAD Structure	Ion Location	LETth (MeV.cm²/mg)
Double-row MBU	NMOS drain	13.5 ± 0.5
	Mid-distance between NMOS drains	8.5 ± 0.5
Double-column MBU	NMOS drain	11.5 ± 0.5
	Mid-distance between NMOS drains	3.75 ± 0.25
	Mid-distance between PMOS drains	5.25 ± 0.25

3.4.2.1 Simulation of Two SRAM Bitcells in a Row

The most efficient memory pattern to trigger a double row MBU is to reverse-bias neighboring drains. This is obtained with the logical pattern "01" (Figure 3.26). In row configuration PMOS can not trigger MCU since they are separated by two reverse biased N-well/P-well junctions. MCU threshold LET were computed for two ion locations shown in Figure 3.26. Table 3.5 synthesizes these LETth and shows that an ion crossing a NMOS drain requires at least an LET of 13.5 MeV.cm²/mg to create a MCU while an ion at mid-distance between two NMOS drains requires a lower LET (8.5 MeV.cm²/mg). Grey area in Figure 3.26 shows the extrapolated spread out of the sensitive area for row MBU until a LET of 13.5 MeV.cm²/mg.

3.4.2.2 Simulation of Two SRAM Bitcells in a Column

For the configuration depicted in Figure 3.27, the most efficient memory pattern to induce MBU is 11 or 00 because the transistors of adjacent bitcells (particularly SEU-sensitive areas) share the same well region and are separated by the same distance. Note that MCU can be triggered by n-channel metal-oxide-semi-conductor (NMOS) as well as p-channel metal-oxide-semi-conductor (PMOS).

MCU threshold LET were computed for three ion locations schematized in Figure 3.27. MCU LETth values are synthesized in Table 3.5. As already observed for

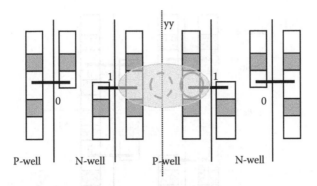

FIGURE 3.26 Scheme of the layout for two SRAM bitcells arranged in row. Solid circle is an ion impact in the NMOS drain (most sensitive SBU location) while dashed circles are the impact at mid-distance between two NMOS drains. Gray area is the spread of MCU-sensitive area at a LET of 13.5 MeV.cm²/mg.

TABLE 3.5
Relative Neutron MCU Rate Variation as a Function of Several Parameters

Parameter	Details in Section	Relative MCU Rate
SOI Substrate (1)[a]	3.3.10	10
Bitcell architecture	3.3.8	30
Reference 65-nm single-port SRAM without TW	–	100
Test location	3.3.9	125
Test pattern	3.3.11	145
Temperature	3.3.7	165
Supply voltage	3.3.6	230
TW usage	3.3.2	1000

[a] (1) experimental results in 130 nm technology.

row configuration, the lowest LETth is obtained for an ion impact at mid-distance between NMOS drains (3.75 MeV.cm²/mg). However, MCU LETth for an impact at mid-distance between PMOS drains is slightly higher (5.25 MeV.cm²/mg). Gray areas in Figure 3.10b show the extrapolated spread out of the sensitive area for column MCU until a LET of 11.5 MeV.cm²/mg.

3.4.2.3 Conclusions and SRAM-Sensitive Area Cartography

Despite a lower distance between two adjacent SEU-sensitive areas, the row MCU LETth is twice as high compared to column MBU LETth. This is explained by the incidence of the ion that crosses through 0.3 μm of STI in the first case (dashed circle on Figure 3.26, whereas it directly strikes the active area of the NMOS transistor in the second case (dashed circle in Figure 3.27). As a consequence, there is less silicon volume for a carrier's deposition in the case of row MBU. Row and column LETth

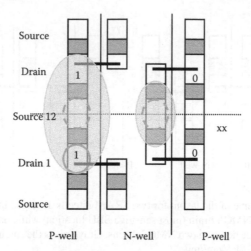

FIGURE 3.27 Scheme of the layout for two SRAM bitcells arranged in a column. Plain circle is an ion impact in the NMOS drain (most sensitive SBU location) while dashed circles are the impact at mid-distance between two NMOS or PMOS drains. Gray area is the spread of MCU-sensitive area at a LET of 11.5 MeV.cm²/mg.

show that the layout of the memory cells (STI regions, silicon regions) strongly impacts their sensitive area.

SEE-sensitive area cartography as a function of ion LET can be drawn from TCAD results shown in Sections 3.4.2.1 and 3.4.2.2. This cartography is shown in Figure 3.28. It is noteworthy that the double-MBU-sensitive area extends beyond a single bitcell area.

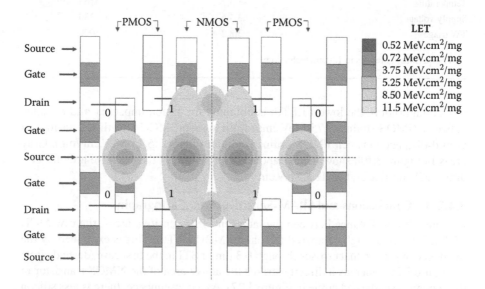

FIGURE 3.28 SEE-sensitive area cartography as a function of ion LET.

3.5 GENERAL CONCLUSION: SORTING OF PARAMETERS DRIVING MCU SENSITIVITY

SEE testings carried out with alpha, neutrons, heavy ions, and protons on several SRAMs are reported in this work. These SRAMs were processed by STMicroelectronics in a CMOS 65-nm technology and embedded in several test vehicles. MCU percentages and MCU rates were given as a function of a dozen of parameters. These parameters are either technological (feature size, process option, etc.) or design (bitcell architecture, well tie density, etc.) or related to experimental test conditions (supply voltage, temperature, test pattern, etc.). Table 3.5 synthesizes the relative neutron MCU rate variations as a function of these parameters. It is noteworthy that the use SOI substrate is the solution that decreases the MCU rate the most by taking advantage of its fully isolated transistors. The parameter that worsens the MCU rate more is the use of a TW layer process option. On the other hand, it must be be remembered that the use of a TW allows suppressing the SEL occurrence even in harsh environment (high temperature, high voltage, heavy ions).

3.5.1 SEE-SENSITIVE AREA CARTOGRAPHY

Full 3-D structures were built from layout of 65-nm SRAM bitcells. The use of TCAD structures whose SRAM bitcells are located away from the well ties was mandatory to confirm that the bipolar effect enhances the collected charge with a TW. The simulations have additionally confirmed that bipolar effect is reduced by increasing the well tie frequency and therefore efficiently mitigate MCU and SER.

Other 3-D structures embedding two SRAM bitcells were built. Bitcells were arranged either in a column or a row to reproduce the actual SRAM array. Simulation of these structures has allowed building a SEE-sensitive area cartography as a function of ion LET. This cartography shows that the sensitive area extends beyond a single bitcell area.

3.6 ANNEX 1

After radiation testings with static algorithm, bitmap error can have thousands of SEUs. With such a high density of SEUs the key question is therefore: how many upsets are "true" MCUs (i.e., several SEUs simultaneously created by a single ion), and how many are "false" MCUs (i.e., sequentially created in the same vicinity by different ion strikes)? [41]

MCU rates and shapes depend on the test pattern filling the memory. It was experimentally verified that checkerboard, All1, and All0 test patterns have similar MCU rates. The following analyses and MCU counting are given for a CKB pattern. A cell spacing criterion (k) is chosen when analyzing a postirradiation error bitmap for MCU detection. This criterion corresponds to the upset-to-upset spacing (maximum number of cells between two SEUs in the X and Y directions to count an MCU). The effect of this criterion on the number of counted MCUs is illustrated in Figure 3.29. This figure points out that the MCU number (0 or 1 bitflip) and type (two or three cells) is a function of the cell spacing (CS) criterion value: the larger this value (5, 6...), the

	Cell spacing criterion	MCU detected
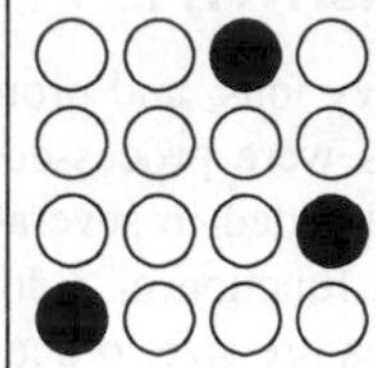 ○○●○ ○○○○ ○○○● ●○○○	k = 1	No MCU
	k = 2	One MCU of two cells
	k = 3	One MCU of three cells

FIGURE 3.29 Illustration of the impact of CS criterion on the MCU detection efficiency.

higher the MCU number. However, a large k value would lead to count as MCU two single SEU in neighboring cells created by two different events, i.e., not simultaneously generated. This would lead to a large overestimation of the MCU rates.

For this reason, Equation 3.3 is proposed for quantifying the rates of false MCU in order to correct raw experimental data to count only the true MCUs. The formula is further detailed in Appendix 3A. We believe that this result should be useful in hardness assurance processes for the total number of fails to target before stopping the irradiation and for the choice of the radiation source intensity (here a radioactive alpha source).

The probability to count a false MCU is given by

$$P = E_{SRP} \times \frac{AdjCell}{Nbit} \tag{3.1}$$

where E_{SRP} is the number of SEUs recorded after irradiation (from a single readout period), $AdjCell$ is the number of cells around each SEU that are inspected to detect a MCU (this number is function of the CS criterion [Table 3.6]), and $Nbit$ is the total number of bits in the memory array.

The probability that a MCU occurred is the complementary probability that no MCU occurred ($n = 0$) and is given using the cumulative Poisson probability by

$$MCU_{proba} = 1 - \sum_{i=0}^{n} \frac{e^{-P} \times P^i}{i!} = 1 - e^{-P} \text{ for } n = 0 \tag{3.2}$$

TABLE 3.6

Number of Adjacent Cells Inspected for MCU around Each SEU as a Function of the CS Criterion

CS Criterion	k = 1	k = 3	k = 5	k = 8
No. of adjacent cells = AdjCell	8	48	120	288

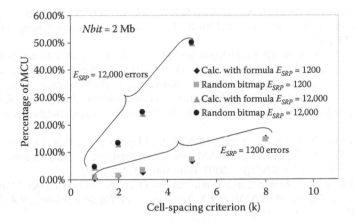

FIGURE 3.30 Comparison of MCU percentages obtained from either a randomly generated bitmap or from Equation 3.3 for a 2-Mb memory array (*Nbit* = 2 Mb).

Multiplying this probability by the total number of SEUs gives the number of MCUs. This number divided by the total number of SEUs is the percentage of MCUs. Using Equations 3.1 and 3.2, the percentage of false MCUs (SEUs from two different events are counted as one MCU) is

$$\text{false MCU}\% = 1 - e^{-E_{SRP} \times \frac{AdjCell}{Nbit}} \tag{3.3}$$

In order to double check the relevance of this model, MCU percentages obtained from Equation 3.3 are compared to MCU percentages counted from randomly generated error bipmaps (Figure 3.30). This figure shows that whatever the CS criterion, the MCU percentages match perfectly.

Equation 3.3 is very convenient as it is easy to use and can be used for different devices (SRAM, DRAM, etc.) and many radiation sources (alpha, neutron, heavy ions, etc.).

REFERENCES

1. X. Zhu, X. Deng, R. Baumann and S. Krishnan, A Quantitative Assessment of Charge Collection Efficiency of N+ and P+ Diffusion Areas in Terrestrial Neutron Environment, *IEEE Transactions on Nuclear Science*, Vol. 54, No. 6, pp. 2156–2161, Part 1, Dec. 2007.
2. A.D. Tipton et al., Device-Orientation Effects on Multiple-Bit Upset in 65 nm SRAMs, *IEEE Transactions on Nuclear Science*, Vol. 55, No. 6, Part 1, pp. 2880–2885, Dec. 2008.
3. V. Correas et al., Simulations of Heavy Ion Cross-Sections in a 130 nm CMOS SRAM, *IEEE Transactions on Nuclear Science*, Vol. 54, No. 6, Part 1, pp. 2413–2418, Dec. 2007.
4. D.G. Mavis et al., Multiple Bit Upsets and Error Mitigation in Ultra-Deep Submicron SRAMS, *IEEE Transactions on Nuclear Science*, Vol. 55, No. 6, Part 1, pp. 3288–3294, Dec. 2008.

5. F.X. Ruckerbauer and G. Georgakos, Soft Error Rates in 65 nm SRAMs—Analysis of New Phenomena, presented at 13th IEEE International On-Line Testing Symposium (IOLTS 2007).

6. D. Heidel et al., Single-Event-Upset and Multiple-Bit-Upset on a 45 nm SOI SRAM, presented at IEEE International Conference NSREC, Québec City, July 20–24, 2009.

7. S. Uznanski, G. Gasiot, P. Roche, J.-L. Autran and R. Harboe-Sørensen, Single Event Upset and Multiple Cell Upset Modeling in a Commercial CMOS 65 nm SRAMs, was presented at IEEE International RADECS Conference, Bruges, Belgium, Sept. 14–18, 2009.

8. G. Gasiot, D. Giot and P. Roche, Multiple Cell Upsets as the Key Contribution to the Total SER of 65 nm CMOS SRAMs and Its Dependence on Well Engineering, 44th Annual International NSREC 2007, Honolulu, HI, July 2007.

9. T. Merelle et al., Monte-Carlo Simulations to Quantify Neutron-Induced Multiple Bit Upsets in Advanced SRAMs, *IEEE Transactions on Nuclear Science,* Vol. 52, No. 5, pp. 1538–1544, Oct. 2005.

10. Y. Tosaka et al., Comprehensive Study of Soft Errors in Advanced CMOS Circuits with 90/130 nm Technology, IEEE International Electron Devices Meeting IEDM Conference, Technical Digest, 2004.

11. N. Seifert et al., Radiation-Induced Soft Error Rates of Advanced CMOS Bulk Devices, presented at IRPS Conference, San Jose, CA, 2005.

12. F. Wrobel et al., Simulation of Nucleon-Induced Nuclear Reactions in a Simplified SRAM Structure: Scaling Effects on SEU and MBU Cross Sections, *IEEE Transactions on Nuclear Science,* Vol. 48, No. 6, pp. 1946–1952, Dec. 2001.

13. T.L. Criswell, P.R. Measel and K.L. Wahlin, Single Event Upset Testing with Relativistic Heavy Ions, *IEEE Transactions on Nuclear Science,* Vol. NS-31, No. 6, Dec. 1984.

14. J.A. Zoutendyk, H.R. Schwartz and R.K. Watson, Single-Event Upset (SEU) in a DRAM with on-Chip Error Correction, *IEEE Transactions on Nuclear Science,* Vol. NS-34, No. 6, Dec. 1987.

15. Y. Song, K.N. Vu, J.S. Cable, A.A. Witteles, W.A. Kolasinski, R. Koga, J.H. Elder, J.V. Osborn, R.C. Martin and N.M. Ghoniem, Experimental and Analytical Investigation of Single Event Multiple Bit Upsets in Polysilicon Load 64k NMOS SRAMs, *IEEE Transactions on Nuclear Science,* Vol. 35, No. 6, p. 1673, 1988.

16. J.J. Wang et al., Single Event Upset and Hardening in 0.15 µm Antifuse-Based Field Programmable Gate Array, *IEEE Transactions on Nuclear Science,* Vol. 50, No. 6, Dec. 2003.

17. R.A. Reed et al., Heavy Ion and Proton-Induced Single Event Multiple Upset, *IEEE Transactions on Nuclear Science,* Vol. 44, No. 6, Part 1, pp. 2224–2229, Dec. 1997.

18. N. Seifert, B. Gill, K. Foley and P. Relangi, Multi-cell Upset Probabilities of 45 nm High-k + Metal Gate SRAM Devices in Terrestrial and Space Environments, *IEEE International Reliability Physics Symposium IRPS,* April 27–May 1, 2008, pp. 181–186.

19. O. Musseau et al., Analysis of Multiple Bit Upsets (MBUs) in CMOS SRAM, *IEEE Transactions on Nuclear Science,* Vol. 43, No. 6, Part 1, pp. 2879–2888, Dec. 1996.

20. JEDEC Standard No. JESD 89, Measurement and Reporting of Alpha Particles and Terrestrial Cosmic Ray-Induced Soft Errors in Semiconductor Devices, Aug. 2001.

21. Single Event Effects Test Method and Guidelines, European Space Agency, ESA/SCC Basic Specification No. 22900, 1995.

22. G. Gasiot, D. Giot and P. Roche, Alpha-Induced Multiple Cell Upsets in Standard and Radiation Hardened SRAMs Manufactured in 65 nm CMOS Technology, *IEEE Transactions on Nuclear Science,* Vol. 53, No. 6, pp. 3479–3486, Dec. 2006.

23. E.H. Cannon, M.S. Gordon, D.F. Heidel, A.J. Klein Osowski, P. Oldiges, K.P. Rodbell and H.H.K. Tang, Multi-Bit Upsets in 65 nm SOI SRAMs, presented at the IRPS conference, Phoenix, AZ, May 2008.

24. A. Virtanen, R. Harboe-Sorensen, H. Koivisto, S. Pirojenko and K. Rantilla, High Penetration Heavy Ions at the RADEF Test Site, presented at RADECS, 2003.
25. H. Puchner, R. Kapre, S. Sharifzadeh, J. Majjiga, R. Chao, D. Radaelli and S. Wong, Elimination of Single Event Latchup in 90 nm SRAM Technologies, *IEEE International Reliability Physics Symposium Proceedings,* pp. 721–722, March 2006.
26. P. Roche and R. Harboe-Sorensen, Radiation Evaluation of ST Test Structures in commercial 130 nm CMOS Bulk and SOI, in Commercial 90 nm CMOS Bulk and in Commercial 65 nm CMOS Bulk and SOI, European Space Agency QCA Workshop, January 2007.
27. T. Kishimoto et al., Suppression of Ion-Induced Charge Collection against Soft-Error, in *Procedings of the 11th International Conference on Ion Implantation Technology,* Austin, TX, Jun. 16–21, 1996, pp. 9–12.
28. D. Burnett et al., Soft-Error-Rate Improvement in Advanced BiCMOS SRAMs, in *Proc. 31st Annual International Reliability Physics Symposium,* Atlanta, GA, Mar. 23–25, 1993, pp. 156–160.
29. P. Roche and G. Gasiot, invited review paper, Impacts of Front-End and Middle-End Process Modifications on Terrestrial Soft Error Rate, *IEEE Transactions on Device and Materials Reliability,* Vol. 5, No. 3, pp. 382–396, Sept. 2005.
30. H. Puchner et al., Alpha-Particle SEU Performance of SRAM with Triple Well, *IEEE Transactions on Nuclear Science,* Vol. 51, No. 6, pp. 3525–3528, Dec. 2004.
31. D. Radaelli et al., Investigation of Multi-Bit Upsets in a 150 nm Technology SRAM Device, *IEEE Transactions on Nuclear Science,* Vol. 52, No. 6, pp. 2433–2437, Dec. 2005.
32. D. Giot, G. Gasiot and P. Roche, Multiple Bit Upset Analysis in 90 nm SRAMs: Heavy Ions Testing and 3D Simulations, presented at the RADECS Conference, Athens, Greece, September 2006.
33. D. Tryen, J. Boch, B. Sagnes, N. Renaud, E. Leduc, S. Arnal and F. Saigne, Temperature Effect on Heavy-ion Induced Parasitic Current on SRAM by Device Simulation: Effect on SEU Sensitivity, *IEEE Transactions on Nuclear Science,* Vol. 54, No. 4, pp. 1025–1029, 2007.
34. M. Bagatin, S. Gerardin, A. Pacagnella, C. Andreani, G. Gorini, A. Pietropaolo, S.P. Platt and C.D. Frost, Factors Impacting the Temperature Dependence of Soft Errors in Commercial SRAMs, *IEEE Transactions on Nuclear Science,* Vol. 47, No. 6, 2008.
35. D. Giot, P. Roche, G. Gasiot, J.-L. Autran and R. Harboe-Sørensen, Heavy Ion Testing and 3D Simulations of Multiple Cell Upset in 65nm Standard SRAMs, *IEEE Transactions on Nuclear Science,* Vol. 55, No. 4, 2007.
36. G. Gasiot, P. Roche and P. Flatresse, Comparison of Multiple Cell Upset Response of BULK and SOI 130 NM Technologies in the Terrestrial Environment, presented at the IRPS Conference, Phoenix, AZ, May 2008.
37. Y. Kawakami et al., Investigation of Soft Error Rate Including Multi-Bit Upsets in Advanced SRAM Using Neutron Irradiation Test and 3D Mixed-Mode Device Simulation, *IEDM Technical Digest,* 2004.
38. Ph. Roche et al., SEU Response of an Entire SRAM Cell Simulated as One Contiguous Three Dimensional Device Domain, *IEEE Transactions on Nuclear Science,* Vol. 45, No. 6, pp. 2534–2543, Dec. 1998.
39. Synopsys Sentaurus TCAD tools, http://www.synopsys.com/products/tcad/tcad.html.
40. K. Osada et al., Cosmic-Ray Multi-Error Immunity for SRAM, Based on Analysis of the Parasitic Bipolar Effect, *2003 Symposium on VLSI Circuit Digest of Technical Papers.*
41. A.M. Chugg, A Statistical Technique to Measure the Proportion of MBU's in SEE Testing, *IEEE Transactions on Nuclear Science,* Vol. 53, No. 6, pp. 3139–3144, Dec. 2006.

4 Radiation Effects in DRAMs

Martin Herrmann

CONTENTS

4.1 INTRODUCTION

Processors need memory that provides fast random access for both read and write operations (random access memory [RAM]). The two most common memory technologies for this purpose are static RAM (SRAM), which uses a flip-flop to store data, and dynamic RAM (DRAM), which uses a capacitor. Both are volatile memory; that is, the memory contents are only preserved as long as the memory device is powered (as opposed to nonvolatile memory like flash or electrically erasable programmable read-only memory [EEPROM]). Compared to SRAM, DRAM has the advantage of small cells (6 to 8 F^2, as opposed to >100 F^2 for SRAM [1]). On the other hand, the dynamic nature of DRAM requires periodic refresh operations.

    Apart from processor main memory, DRAM devices are also used for mass memory modules [2], where random access is not a requirement. In such an application,

DRAM competes with other memory technologies (mainly NAND flash) in terms of cost, power requirements, access speed, and radiation sensitivity.

Radiation sensitivity is particularly important in space applications. For example, SDRAM is less sensitive to total ionizing dose (TID) than NAND flash, making it a viable candidate for high-TID applications like the European Space Agency's (ESA's) upcoming JUICE mission to Jupiter [3]. But even for terrestrial applications, the sensitivity to radiation has to be considered. Radiation effects have been demonstrated for commercial flight altitudes [4] as well as on ground level [5]. Radiation-induced errors can have consequences ranging from none (if the data stored at the affected memory location is not used any more) to program crashes to silent corruption of data.

DRAMs, and particularly the newer-generation types with synchronous interface (SDRAMs), are very complex devices; Ladbury et al. [6] even likens a SDRAM to "a microcontroller with a large memory array." Apart from the actual array, DRAM devices contain various pieces of circuitry that can be the source of radiation-induced errors:

- Sense amplifiers
- Row and column address decoders
- Control logic and state machines

Therefore, a particle strike can have many different consequences depending on the location of the strike.

The complexity of the internal logic states and state transitions during operation also causes highly dynamic radiation effects: in addition to the strike *location*, the strike *time* is also relevant for the error patterns, with a resolution down to a single clock cycle. For example, an upset in the refresh logic might have no effect if the refresh logic is idle at the time of the strike or can cause data corruption if the strike occurs during a refresh operation.

The complexity of DRAM devices makes the development of rad-hard parts expensive. Both terrestrial and space applications therefore have to resort to commercial off-the-shelf (COTS) parts. Unfortunately, the internal design of DRAM devices is not typically disclosed by the manufacturers [7]. Therefore, and due to the complexity, exhaustive radiation testing of modern DRAM devices is not possible. Instead, DRAMs are usually tested by operating them in a manner similar to the target application and observing the errors caused by irradiation. Once error patterns have been identified, they can be examined in more detail. Typically, experiments are designed to test for multiple effects at once to maximize the results determined from the available test time. Test procedures must be designed in a way to minimize errors masking other kinds or errors.

Often, the root cause for an error pattern cannot be determined due to lack of knowledge about the internal structure of the device. It is also possible that a specific error pattern is hard to reproduce because it only occurs under very rare conditions. These are unfortunate effects of the complexity of modern DRAM devices.

4.2 DRAM BASICS

4.2.1 PRINCIPLE OF OPERATION

The term *dynamic RAM* denotes memory devices that store data as charge in a capacitor. The basic cell consists of one transistor and one capacitor (this is also called the *1T1C* cell), as shown in Figure 4.1. The capacitor stores the data as $U_{cell} = +V_{DD}/2$ for one of the two possible logic values and $U_{cell} = -V_{DD}/2$ for the other. The bottom node of the capacitor is connected to $V_{DD}/2$ rather than ground to minimize the maximum voltage across the capacitor and therefore the dielectric stress.

DRAM cells are arranged in arrays consisting of rows and columns, as shown in Figure 4.2. The gate terminals of all cell transistors of a given row are connected to the row decoder by a common *word line*. The drain terminals of all cell transistors of a given column are connected to a *sense amplifier* by a common *bit line*. The length

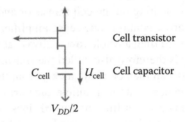

FIGURE 4.1 DRAM cell.

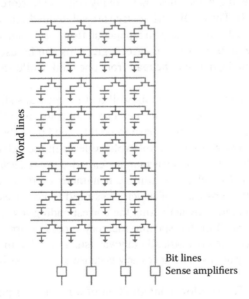

FIGURE 4.2 DRAM array.

of the bit line and word line, and therefore the size of the array, is limited by technological factors such as bit line capacity and noise sensitivity of the sense amplifiers. Larger memory devices are composed of multiple arrays internally.

To access a cell, the bit line is first precharged to half the operating voltage, $V_{DD}/2$. After that, the cell capacitor is connected to the bit line by switching on the cell transistor. Since the capacitance of the bit line is larger than the cell capacitance, most of the charge is transferred from the cell capacitor to the bit line. This is called *destructive read*, as the cell contents are destroyed in the process. The voltage of the bit line rises or falls slightly, depending on the stored data. After that, the sense amplifier is switched on. The sense amplifier basically consists of a latch formed by two cross-coupled inverters. When the sense amplifier is switched on, the small voltage difference between its inputs causes it to leave its metastable state and amplify the voltage difference until it settles in one of its stable states. The value can then be read from the bit line by conventional logic. Simultaneously, the cell capacitor voltage is restored as long as the cell transistor remains on.

The whole process of switching on the cell transistor and amplifying the voltage is called *activation*. The array contains one sense amplifier for each column. Since all cells of a row share a word line, a whole row is always activated at the same time. The set of sense amplifier is therefore also called the *row buffer*. While activation is a slow process [8], the data can then be read quickly from the row buffer.

A real-world DRAM device contains multiple such arrays: first, due to the technological limits for the length of bit lines and word lines, a logical array has to be subdivided into multiple physical arrays. Second, a DRAM device with an input/output (I/O) word length larger than 1 can have separate arrays for each word in the bit (as an alternative, the device designer can also choose to store the bits of a word in the same array) [8]. Third, DRAM devices are subdivided into a number of banks (typically eight), which are largely independent from each other, each containing its own set of sense amplifiers and control logic. This can be used to increase data throughput: for example, data can be read from one bank while another bank is busy activating a row.

Since the cell capacitor dielectric as well as the cell transistor (in its off state) have a finite resistance, charge from the cell capacitor is lost over time due to leakage current. To avoid data loss, all cells have to be refreshed periodically. As explained above, activating a row restores the full voltage level in the cells. Modern DRAM devices are typically refreshed in a manner called *auto refresh*: the controller is required to periodically submit a minimum number of refresh commands. The DRAM device itself keeps track of which rows need to be refreshed and performs the actual refresh operation by activating the corresponding rows. The details of the refresh operation depend on the specific memory type. For applications where the whole memory is accessed periodically anyway (such as video memory), an explicit refresh operation may not even be necessary because each access implicitly refreshes the row.

Some types of DRAM allow configuration of a number of parameters, such as the latency between a read command and the data becoming available on the output pins. For this purpose, they provide a number of registers, called *mode registers*, which can be written to by the controller.

Larger semiconductor devices have a higher probability for production defects, which reduces yield. To improve overall yield, SDRAM devices contain redundant rows and columns [9]. While this slightly increases the size of the die, it allows remapping nonfunctional rows and columns, thereby repairing a die with defective regions. The redundant row and column replacements are determined during post-production and are configured by means of fuses on the device. The replacement values are loaded from the fuses into latches during the startup procedure of the device and are used in the address decoding processes.

Furthermore, most DRAM devices use some form of address or data scrambling for various optimization reasons [9]. Both redundancy and scrambling mean that logically adjacent cells (i.e., cells with adjacent addresses) may not be physically adjacent. This makes interpretation of test results and error patterns challenging.

4.2.2 TYPES OF DRAMS

Early DRAM types had an asynchronous interface with an address bus, a data bus, and a number of control signals. The *row address strobe* (RAS) signal was used to latch the row address and activate a row. The *column address strobe* (CAS) signal was used to read and write data from and to the row buffer, depending on the *write enable* (WE) signal. Finally, the trailing edge of the RAS signal precharged the array.

Modern DRAM types use a synchronous interface (synchronous DRAM [SDRAM]). This facilitates integration with other synchronous logic (such as microprocessors) and allows performance-increasing techniques such as pipelining. SDRAM devices have a combined address/command bus. The signal names—RAS, CAS, and WE—are still present, but have lost much of their original meaning. These signals form a 3-bit bus capable of encoding $2^3 = 8$ commands. Furthermore, some commands that do not require the address bus (such as the precharge command) may reuse some of the address bits for additional command bits (e.g., for selecting whether to precharge a single bank or all banks).

Data throughput is further increased twofold by using a double-data-rate (DDR) interface, which transfers data on both the positive and the negative clock edge. Some SDRAM types (e.g., DDR3 [10]) only use a DDR interface for the data bus and a single-data-rate (SDR) interface for the address/command bus, while others (e.g., LPDDR3 [11]) use a DDR interface for both the data bus and the address/command bus.

In applications where power consumption is critical, specially designed low-power SDRAMs are used. These devices employ a variety of power-saving techniques, such as selectively disabling parts of the array or adapting the refresh rate to the actual die temperature instead of the worst-case die temperature.

The most prominent use for SDRAM is as main memory (RAM) in computers (desktop computers, notebooks, and servers) or mobile devices such as cell phones or tablets. As of 2014, the current generation for personal computer (PC) memory is DDR3 SDRAM, the successor to DDR SDRAM and DDR2 SDRAM. Mobile devices use LPDDR2 or LPDDR3 devices (note that LPDDR3 is the third generation of low-power DDR SDRAM, not a low-power version of DDR3). Graphics cards use GDDR5 devices (again, the name refers to the fifth generation of graphics DDR SDRAM).

4.3   RADIATION EFFECTS

DRAM devices are subject to both total-dose and single-event effects (SEEs). The large number of possible error patterns, memory standards, and manufacturers make exhaustive testing infeasible. Publications typically report results from SEE and/or TID tests (sometimes restricted to heavy ion or proton SEE tests) with a handful of related device types from different manufacturers of one memory generation. Different research groups typically focus on different aspects and error patterns, so the results presented in literature are necessarily incomplete. We can therefore only give an overview over different types of errors and error patterns observed by different groups.

4.3.1   SEEs

There are several different types of SEEs commonly found in DRAM devices:

- Single-event upsets (SEUs), which are isolated single-bit or multibit errors
- Stuck bits, which are isolated single-bit errors that cannot be rewritten
- Single-event functional interrupts (SEFIs), which are corrupted rows, columns, or whole address ranges
- Single-event latch-up
- Current increase

It is worth noting that the classification of errors depends on the application. For example, an application that cannot tolerate a temporary current increase may be classified as a SEFI (necessitating countermeasures), while other applications may treat it as a separate class of errors or even ignore it if its magnitude is low enough.

4.3.1.1   SEUs

SEUs originate from the array or the data read path and cause different types of data corruption:

- A strike in the array changes the stored data. The cell can either be written again afterward (SEU), or it is permanently stuck at a specific value (stuck bit).
- A strike in the read path can either cause a transient error (i.e., an error that disappears when the same cell is read again without an intermediate write operation) or a persistent data error. Depending on which part of the circuitry is upset, it is possible that the erroneous value is written back to the array by the sense amplifier.

Tables 4.1 and 4.2 show a compilation of cross sections for SEUs caused by heavy ions and protons, respectively, as reported by different groups. We only show the value for the saturation cross section. The threshold linear energy transfer (LET) is typically very small: usually, errors are observed even at the lowest LET available at a facility (e.g., 1.8 MeV cm²/mg for the RADEF facility [12]), indicating that the threshold LET is smaller than this LET.

TABLE 4.1

Reported SEU Cross Sections for Heavy Ion Irradiation

Device Type	Capacity	Manufacturer	σ_{sat} (cm²/bit)	Reference
DRAM	16 Mbit	Various	10^{-7} to 10^{-6}	Harboe-Sorensen et al. [13]
SDRAM	512 Mbit	Elpida	$\approx 10^{-9}$	Adell et al. [14]
SDRAM	512 Mbit	Elpida	10^{-9} to 10^{-8}	Guertin et al. [15]
SDRAM	512 Mbit	Unspecified	$\approx 10^{-9}$	Hafer et al. [16]
SDRAM	512 Mbit	Various	$\approx 10^{-9}$	Langley et al. [17]
SDRAM		Samsung	$\approx 10^{-8}$	Henson et al. [18]
DDR	1 Gbit	Samsung	10^{-11} to 10^{-10}	Ladbury et al. [6]
DDR2		Samsung	10^{-13} to 10^{-12}	Ladbury et al. [19]
DDR2	1 Gbit	Samsung	10^{-10} to 10^{-9}	Ladbury et al. [20]
DDR2	1 Gbit	Samsung	10^{-11} to 10^{-10}	Ladbury et al. [20]
DDR2	1 Gbit	Micron	10^{-11} to 10^{-10}	Li et al. [21]
DDR2	2 Gbit	Various	10^{-15} to 10^{-14}	Koga et al. [22]
DDR2	2 Gbit	Elpida	10^{-8}	Hoeffgen et al. [23]
DDR3	4 Gbit	Various	10^{-11} to 10^{-10}	Herrmann et al. [24]
DDR3		Samsung	10^{-10} to 10^{-9}	Grürmann et al. [25]
DDR3		Elpida	10^{-11} to 10^{-10}	Grürmann et al. [25]
DDR3		Samsung	$\approx 10^{-13}$	Ladbury et al. [19]
DDR3		Micron	$\approx 10^{-10}$	Koga et al. [26]

TABLE 4.2

Reported SEU Cross Sections for Proton Irradiation

Device Type	Capacity	Manufacturer	σ_{sat} (cm²/bit)	Reference
DRAM	16 Mbit	Various	10^{-14} to 10^{-13}	Harboe-Sorensen et al. [13]
SDRAM	512 Mbit	Elpida	10^{-19} to 10^{-17}	Guertin et al. [15]
SDRAM	512 Mbit	Unspecified	$\approx 10^{-18}$	Langley et al. [17]
SDRAM	512 Mbit	Samsung	$\approx 10^{-16}$	Langley et al. [17]
SDRAM	1 Gbit	Samsung	10^{-17} to 10^{-16}	Ladbury et al. [6]
SDRAM	2 Gbit	Various	10^{-20} to 10^{-19}	Quinn et al. [27]
SDRAM	2 Gbit	Various	10^{-19} to 10^{-18}	Koga et al. [22]
DDR3		Various	$\approx 10^{-20}$	Koga et al. [26]

In some cases, different groups obtain different results. This may be due to different die revisions, differences in the measurement setup (e.g., different temperature), or measurement errors. Generally, the cross section for protons is several orders of magnitude smaller than the cross section for heavy ions.

SEUs have one of two possible polarities: up errors are cells written as 0 and read as 1, and down errors are cells written as 1 and read as 0. The values 0 and 1 refer to the logic value presented at the interface. The internal structure of the device can cause the logic values of some cells to be stored in a cell as its inverse value (*data scrambling*). With a typical scrambling scheme, half of the bits are stored as inverse.

Harboe-Sorensen et al. [13] tested 15 16-Mbit DRAM devices from eight man-ufacturers. All devices had an equal share of "up" and "down" errors, except for a Texas Instruments device where more than 99% of the errors were in the down direction for heavy ions (but not for protons). Koga et al. [22] report that for DDR2 SDRAM devices from Micron, Hynix, and Elpida, all errors were in the down direc-tion for both heavy ions and protons, and for Samsung devices, this was the case for about 95% of errors. For DDR3 devices, they found about 70% down errors [26].

Harboe-Sorensen et al. [13] also considered the internal inverting scheme of the devices. They assume that memory cells can only be discharged, not charged, by an ion strike. All errors that affect a bit value corresponding to a discharged cell must therefore originate from the peripheral circuitry. They call the direction of cell fail-ure the "soft" and the other one the "hard" direction. For most parts and most ions, less than 1% of all errors are in the hard direction. Surprisingly, some parts show a substantial share of errors (between 2% and 88%) in the hard direction for some ions only. For proton tests, there is only a slight deviation from the typical behavior of bits failing in the soft direction.

The exact error mechanisms are hard to determine without detailed knowledge of the device internals. Common explanations are charge depletion [28], charge trap-ping [24], or displacement damage [29].

It is worth noting that some observed effects can easily be misinterpreted as SEUs, thereby overestimating the SEU cross section. For example, consider a run with 10 single-bit errors in the whole address space, four of which are in the same row. It is highly likely that these four errors are related and probably have been caused by one single event. Still, a naïve analysis might count them as SEUs. While this case is fairly easy to spot, it makes automated analysis difficult. There may also be more complex relationships between seemingly unrelated errors due to internal address scrambling, especially in conjunction with multibit upsets (MBUs). Bougerol et al. [30] note that it is easy to overestimate cell sensitivity and suggest laser testing to help identify different error patterns in order to later be able to properly interpret heavy-ion and proton test results.

Another pattern that illustrates this problem is shown in Figure 4.3. Apart from row SEFIs and SEUs, this error map shows two regions where the density of single-bit errors is almost 60 times higher than in the rest of the device. It is likely that these regions are due to a SEFI-like effect, such as a shifted reference voltage. Again, this case is easy to spot, but it is easy to imagine a situation where such errors would have been misinterpreted as bona fide SEUs, severely overestimating the SEU sensitivity.

Koga et al. [22] found that varying the refresh rate by a factor of 16 did not signifi-cantly affect the sensitivity for both heavy ions and protons.

4.3.1.2 Stuck Bits

Some single-bit errors cannot be resolved by any measure, not even by rewriting the cell or power cycling the device. The affected cells are permanently stuck at a fixed value. These errors are therefore called stuck bits. While it is conceivable that there could be other kinds of SEEs that survive a power cycle (e.g., errors caused by per-manent damage to a sense amplifier), to the knowledge of the authors, no such effects have been observed in DRAM devices.

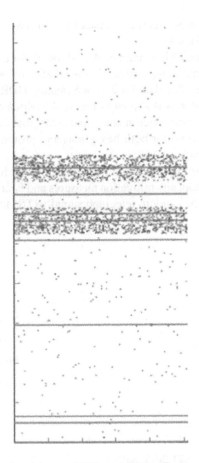

FIGURE 4.3 Error map showing row SEFIs and SEUs. The density of single-bit errors in the two pronounced regions is almost 60 times higher than in the rest of the device. 4-Gbit DDR3 SDRAM device from Elpida after heavy-ion irradiation with iron.

Stuck bits are also referred to as hard SEUs. However, the terms "hard" and "soft" are ambiguous as they are used in different meanings by different authors (e.g., for SEFIs that can only be removed by a power cycle [30] or for errors with a polarity that cannot be explained by the discharging of a cell capacitor [13]).

Stuck bits are a major concern: they cannot be removed by scrubbing, a common SEU mitigation technique in space applications where redundancy is used to periodically detect errors and correct erroneous data. They are also immune to reset or even to power cycling, both of which are common error mitigation techniques in terrestrial applications.

Stuck bits are attributed to a reduced cell retention time [31] (see also Section 3.2.1). This is supported by the fact that hard SEUs show a strong dependence on temperature [14,29]. For DDR3 SDRAM devices from Samsung and Elpida, Herrmann et al. [32] found that the cross section for stuck bits at room temperature is about an order magnitude below the cross section for SEUs; in other words, 1 in 10 SEUs causes a stuck bit. The corresponding cross sections versus LET for two 4-Gbit DDR3 SDRAM parts are shown in Figure 4.4.

Some hard SEUs anneal. Herrmann et al. [24] found, for a 2-Gbit DDR3 SDRAM device from Samsung, that most hard SEUs anneal after a few days and attribute the errors to charge trapping. The data for 2-Gbit Samsung DDR3 SDRAM parts under irradiation with different ions is shown in Figure 4.5. Koga et al. [26] tested 4-Gbit DDR3 SDRAM devices from Micron and found errors that did not appear until after the irradiation (latent errors) for both heavy ions and protons. None of these errors persisted past a power cycle.

Chugg et al. [29] examined stuck bits in more detail and found surprising results: stuck bits recovered without any mitigation measures and later spontaneously became stuck again, similar to the random telegraph signal (RTS) found in charge-coupled

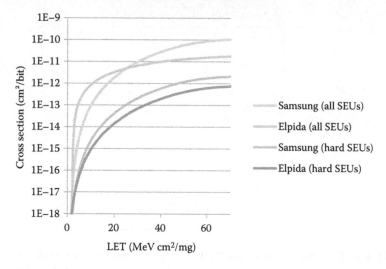

FIGURE 4.4 SEU and hard SEU cross sections for two 4-Gbit DDR3 SDRAM devices [32].

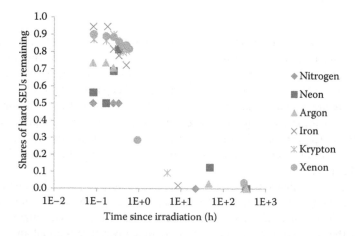

FIGURE 4.5 Annealing of hard SEUs [24].

devices (CCDs). From various experimental results, they conclude that this behavior is probably caused by multiple current leak paths. These paths, caused by displacement damage, spontaneously open and close under thermal influence, leading to charge loss when open. This model can be used to explain both the annealing of stuck bits and the latent errors observed by other groups.

4.3.1.3   SEFIs and Burst Errors

SEFIs are caused by heavy-ion strikes in the control circuitry. They typically manifest as one of the following:

- Row SEFIs, corrupting either a single row of the device address space, part of a row, or multiple rows (typically groups of 2, 4, or 8 rows)*
- Column SEFIs, corrupting either a single column of the device address space, part of a column, or multiple columns
- Device SEFIs, corrupting the whole device or extended regions of a device

Note that these terms are not used uniformly throughout literature. In particular, the term burst error (or error burst) may be used for row SEFIs or device SEFIs. Other researchers call a large run of errors a block error if the device recovered on its own [19] or a SEFI otherwise. The same distinction has been expressed by the terms transient SEFI and persistent SEFI as well [25].

Like stuck bits, SEFIs are a major issue for ground-based as well as space applications. Typical error correction mechanisms for space mass memories use $n = m + k$ memory devices to store a word consisting of m data bits and k redundancy bits. The error correction capabilities of such a system can be exceeded by a coincidence of two errors in the same word [33]. While a coincidence of two SEUs at the same location in different devices is highly unlikely, a device SEFI in one device will cause a coincidence with any SEUs currently existing in any of the other devices. Such coincidences may not be possible to correct. The strike of SEFIs on different kinds of storage applications has been discussed in more detail by Guertin et al. [15].

Like SEUs, SEFIs can be made persistent by the refresh operation that writes the erroneous data to the array, so the data corruption can remain even after the original source of the error (e.g., an upset latch) is removed.

A technique for mitigation of device SEFIs is called *software conditioning*. This term summarizes a variety of operations, such as rewriting the mode registers, resetting the DLL of the device, or calibrating the on-die termination (ODT) resistance [24]. These operations are usually performed during device initialization only, but can be repeated at arbitrary times during operation (subject to timing constraints). DDR3 SDRAM also provides a possibility to reset the device. The standard [34] does not guarantee data retention if the device is reset; however, it has been found that in practice, no data is lost until the reset condition is held for several seconds provided that the device is kept powered [35].

---

* Note that distinguishing between a single SEFI that corrupts multiple rows and multiple row SEFIs corrupting a single row each may not be possible experimentally due to insufficient temporal resolution. However, it is often possible to recognize the former by the related addresses of the affected rows.

Software conditioning has been shown to reduce the cross section of DDR2 SDRAM by different groups [15,21]. Herrmann et al. [24] found, for a 4-Gbit Elpida DDR3 device, that periodic software conditioning (during irradiation) decreases the sensitivity to device SEFIs by about one order of magnitude. For a 4-Gbit Samsung DDR3 device, software conditioning does not have any significant effect; the sensitivity with and without software conditioning is about the same as for the Elpida device with software conditioning. The cross sections versus LET are shown in Figure 4.6.

Software conditioning can also be used as a means for later device SEFI mitigation [15]. Grürmann et al. [25] found that a substantial share of device SEFIs can be removed by applying software conditioning measures or performing device reset (where no data loss was observed). For device reset, this is easily explained, for example, by an invalid state, which can be caused by an ion strike on a state machine register. Mode registers contain device configuration such as timing delays between a read command and the corresponding data becoming available on the data bus. An erroneous value in one of these registers can cause all read operations to fail but can be corrected by rewriting the correct value to the register. The same is true for undocumented, manufacturer-specific mode register bits that put the device into some kind of test mode.

Bougerol et al. [30] examined SEFIs in more detail by comparing heavy-ion and laser test results for a 256 Mbit DRAM from Micron. They found that a certain burst error pattern—in particular, multiples of 512 errors in the same column—could be traced to the fuse latches used for row/column redundancy. These errors could be removed by a mode register set (MRS) command even without writing to the memory. This indicates that this kind of error only affects the read path. They conclude that not only does the MRS command set the mode register value, but it also causes the fuse latches to be reloaded.

In the same work, Bougerol et al. [30] also observed error bursts with a length between 100 and 300 words. From detailed investigation, they concluded that the errors are due to single-event transients (SET) in a voltage buffer. They identified

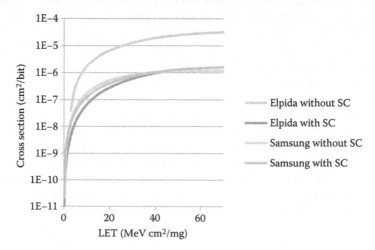

FIGURE 4.6 Effect of software conditioning on the device SEFI cross section for two DDR3 SDRAM part types [24].

an analog circuit on the die that may control the capacitor common voltage ($V_{DD}/2$). An SET in this circuit may cause capacitors to be discharged. Contrary to the burst errors due to fuse latch upset, the data stored in the affected cells is lost because the erroneous value is written back to the array by the following refresh operation. On the other hand, no particular action (like MRS) is required to recover from this kind of error.

Even more than for SEUs, the exact error mechanisms of SEFIs are hard to determine without detailed knowledge of the device internals. Possible explanations are upsets in configuration-register flip-flops, state machine latches, sense amplifiers, or address decoder circuits.

For space applications, error rates in modern SDRAM devices are mostly determined by SEFIs (originating from the periphery) rather than SEUs (originating from the array): device SEFIs are about as likely as SEUs [35] but corrupt several orders of magnitude more data. In a typical application, the bit error rate is dominated by coincidences between two device SEFIs [33].

4.3.1.4   Single-Event Latch-Up

Radiation can cause a latch-up (single-event latch-up [SEL]), which can in turn destroy the device if the system is not designed to incorporate latch-up protection measures.

Harboe-Sorensen et al. [13] tested 16-Mbit DRAM devices from four manufacturers and found SEL only for parts from Micron, but not from Fujitsu, Samsung, or Texas Instruments (TI). Additionally, they found some low-current latch-ups in devices from Micron and TI, where the devices failed partially and could only be recovered by a power cycle. They observed no SELs under proton irradiation.

Hafer et al. [16] found 512-Mbit SDRAM to be immune to SEL at an angle of incidence between $0°$ (normal) to $60°$ up to an LET of 111 MeV cm^2/mg at $105°$C.

No SEL was found for DDR2 SDRAM by Koga et al. [22], Harboe-Sørensen [7], and Li et al. [21], and for DDR3 SDRAM by Herrmann et al. [36].

4.3.1.5   Current Increase

While SEL is rare in SDRAMs, other instances of increased operating current have been observed. These cases are distinguished from latch-up in that the current does not increase to a level that destroys the part or that the current returns to its regular value without any corrective measures.

Hafer et al. [16] found, for 512-Mbit SDRAMs, that the idle current during heavy-ion irradiation was mostly stable; there were some spikes where the current increased to $\approx$ 46 mA from its idle value of $\approx$ 38 mA. They suspect that this is caused by the device temporarily erroneously performing a read operation.

Hoeffgen et al. [23] tested 2-Gbit DDR2 SDRAM and observed a sharp rise of the supply current from the usual 420 mA to over 600 mA.

Herrmann et al. [36] found, for 4-Gbit DDR3 SDRAM under heavy-ion irradiation, that under irradiation, the current (both idle and operating) is sometimes increased for some time (e.g., 1 minute). It then returned to a lower value or the baseline value. After the end of the irradiation, the idle current dropped to its baseline value within seconds. No permanent current increase was observed. In some cases, the current increase was severe: for example, an idle current of $\approx$ 200 mA has been observed compared to a baseline idle current of $\approx$ 20 mA.

4.3.2 TOTAL-DOSE EFFECTS

There are several different types of total-dose effects commonly found in DRAM devices:

- Decreased retention time, causing isolated errors
- Current increase
- Functional failure

4.3.2.1 Retention Time

The retention time is the time (after a cell has been written) until the charge has leaked off the cell capacitor to a point where the stored value can no longer be reliably read. Each cell must be refreshed before the retention time is reached. The refresh interval must therefore be shorter than the retention time or errors will occur. It is well-known that the retention time strongly depends on the device temperature [37].

The specification defines a minimum refresh interval designed to be longer than the worst-case cell retention time. For example, for DDR3 SDRAM, 8192 refresh operations are required in 64 ms (corresponding to a refresh interval of 7.8 µs*).

The per-cell retention time can be determined experimentally by writing a pattern to the memory, disabling refresh[†] for a given time t, and finally reading the memory contents and comparing them to the original pattern. If the cell contains an error (i.e., its read value is different from the original pattern), its retention time is smaller than t; otherwise, it is larger than t. By repeating this procedure for different t values, it is possible to determine the retention time for each cell.[‡] This allows determining the frequency distribution of retention times.

A simpler experiment is to determine the number of bit errors for different values of t. These bit errors occur in the weakest cells (i.e., the cells with the lowest retention time, either from the production variation or from radiation damage).

Lee et al. [38] found, for a Samsung 64-Mbit DRAM device, that after 70 krad, almost 15% of the bits showed an error. After annealing for 24 hours, some of the cells recovered, but the device was unreliable. After 24 hours at 100°C, the device did not show bit errors anymore.

Bacchini et al. [39] found, for a Promos 256-Mbit SDRAM, that ^{60}Co irradiation with 65.1 krad reduced the retention time by up to more than 90% and that for 11 cells, the retention time was lower than the 64-ms refresh interval defined in the specification. The observed errors did not anneal at either room temperature or 85°C.

---

* Refresh operations can be postponed or pulled in, subject to certain timing restrictions.
† This requires a controller that allows disabling refresh operation, which may not be the case for commercially available controllers.
‡ Note that it is necessary to repeat the whole wait time for each new value of t because reading a cell simultaneously refreshes it.

Bertazzoni et al. [40,41] found, for a Micron 256-Mbit SDRAM, that the retention time decreased proportionally with the dose for irradiation with up to 70 krad. They also found a degradation of the data retention time due to heavy ion irradiation, but at a much smaller total dose (0.84 krad). They hypothesize that, in contrast with ^{60}Co irradiation, the dose it is highly localized, so the local dose may be much higher than the global dose.

Scheick et al. [42] found, for a Toshiba 16-Mbit DRAM, that the median retention time decreases exponentially with absorbed dose.

4.3.2.2 Data Errors

When reading data from a device that has been exposed to photon radiation, errors occur. These errors are mostly randomly-distributed, single-bit errors. These errors may be caused by decreased retention time. In this case, the effect could be mitigated by decreasing the refresh interval.

Herrmann et al. [32] found, for a 4-Gbit Samsung DDR3 SDRAM device that was operated during irradiation, that no errors occurred up to a dose of 250 krad and the error density remains below 10^{-6} up to a dose of 300 krad, as shown in Figure 4.7. The error density showed a strong dependence on the temperature, suggesting that it may, in fact, be caused by decreased retention time. After annealing at room temperature for 480 hours, between 60% and 98% of the errors disappeared (depending on the sample). Other DDR3 SDRAM part types, such as 4-Gbit Hynix and 4-Gbit Elpida, showed an even lower error density around 10^{-8} at 85°C after 376 krad of unbiased irradiation [13].

As with SEUs, it is possible to distinguish between the two error directions, up and down. Harboe-Sorensen et al. [13] report, for 14 tested types, about 50% of errors in each direction.

Herrmann et al. [36] also found an error pattern that has not been explained so far: keeping a single row in the device active for an extended time (but within the

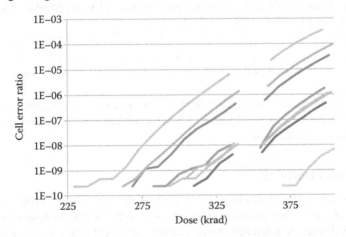

FIGURE 4.7 Errors in eight 4-Gbit Samsung DDR3 SDRAM devices during *in situ* ^{60}Co irradiation at room temperature [32]. *Cell error ratio* or *error density* is the number of bit errors in a device, divided by the total number of bits in the device.

constraints of the specification) appears to cause errors in several distinct regions of the device address space, with error densities up to 20%. This pattern was also observed for other part types. All of these errors were in the down direction.

4.3.2.3 Current Increase

Operating current typically increases with received dose.

Harboe-Sorensen et al. [13] defined parametric failure as a current increase of 20% over the baseline (unirradiated) current. They tested 15 devices from eight manufacturers and found parametric failure between 5 and 45 krad (Si).

Lee et al. [38] tested 64-Mbit DRAMs and found that, for a Samsung device, the standby current increased slightly up to 20 krad. At 30 krad, the current increased by a factor of almost 50 and stayed at that level until the end of the irradiation at 70 krad. In contrast, the standby current for a Mitsubishi device increased gradually.

Shaw et al. [43] report similar results for a Micron 64-Mbit DRAM device: the standby current increased lightly up to 15 krad, then increased rapidly by a factor of 8 between 15 and 19 krad, and stayed at this level until functional failure at 30 krad.

Herrmann et al. [32,44] found, for 4-Gbit DDR3 SDRAM, an increase of less than 25% after ≈420 krad for a Samsung part, an increase by a factor of 10 after ≈120 krad for a Micron part, and no increase at all after ≈420 krad for a Hynix part.

4.3.2.4 Functional Failure

TID damage consists of gradual degradation of the device functionality or parameters. If a critical part of the peripheral circuitry is damaged, the part may not be functional any more, even though most of the array is still intact (functional failure). However, for a realistic application and many device types, the device becomes unusable due to too many errors in the array or parametric failure (e.g., current requirement). Many researchers therefore specify criteria for functional failure in order to determine a maximum dose up to which a device is usable and to facilitate comparison between different devices regarding a specific application. These criteria are motivated by a specific—real or hypothetical—mission.

Harboe-Sorensen et al. [13] define functional failure in 16-Mbit DRAMs as 1024 errors or more in a single run. This corresponds to an error density of $6 \cdot 10^{-5}$. They tested 15 devices from eight manufacturers and found functional failure between 5 and 55 krad (Si).

Lee et al. [38] found functional failure at 70 krad for a Samsung 64-Mbit DRAM device and at 50 krad for a Mitsubishi 64-Mbit DRAM device without specifying a criterion for functional failure.

4.4 CONCLUSIONS

DRAM devices suffer from a large variety of radiation effects, both single-event and total-dose effects. The classification of effects is highly application-specific.

Many SEEs are of a dynamic nature, causing different types of errors depending on the activity the device was performing during the time of the strike. SEUs can easily be mitigated using error-correcting codes, but other errors are more severe. While modern DDR3 SDRAM devices appear to be immune to SEL, they still have failure modes that can necessitate a power cycle of the device, causing loss of stored data.

The main total-dose effects are (a) a reduction of retention time, which can be mitigated to some extent by an increased refresh rate, and (b) an increased operating current, which can be prohibitive for use of a specific device type in a given application. Overall total-dose robustness of modern SDRAM devices is good, making them suitable for high-dose applications.

REFERENCES

1. T. Perez and C. A. F. De Rose, Non-Volatile Memory: Emerging Technologies and Their Impacts on Memory Systems, Technical Report No. 60. Pontifícia Universidade Católica do Rio, Port Alegre, 2010.
2. T. Sasada, S. Ichikawa and M. Shirakura, mass data recorder with ultra-high-density stacked memory for spacecraft, in *IEEE Aerospace Conference*, 2005.
3. ESA, JUICE is Europe's next large science mission, online. Available: http://www.esa .int/Our_Activities/Space_Science/JUICE_is_Europe_s_next_large_science_mission.
4. A. Taber and E. Normand, Single event upsets in avionics, *IEEE Transactions on Nuclear Science,* Vol. 4, No. 2, pp. 120–126, 1993.
5. T. O'Gorman, The effect of cosmic rays on the soft error rate of a DRAM at ground level, *IEEE Transactions on Electron Devices,* Vol. 41, No. 4, 1994.
6. R. Ladbury, M. Berg, H. Kim, K. LaBel, M. Friendlich, R. Koga, J. George, S. Crain, P. Yu and R. Reed, Radiation performance of 1 Gbit DDR SDRAMs fabricated in the 90 nm CMOS technology node, in *IEEE Radiation Effects Data Workshop,* 2006.
7. R. Harboe-Sorensen, F.-X. Guerre and G. Lewis, Heavy-ion SEE test concept and results for DDR-II memories, *IEEE Transactions on Nuclear Science,* Vol. 54, No. 6, pp. 2125–2130, 2007.
8. B. Keeth, R. J. Baker, B. Johnson and F. Lin, *DRAM Circuit Design,* Piscataway, NJ: IEEE Press, 2008, pp. xv, 421.
9. A. van de Goor and I. Schanstra, Address and data scrambling: Causes and impact on memory tests, in *IEEE International Workshop on Electronic Design, Test and Applications,* 2002.
10. JEDEC, *The DDR3 SDRAM Specification,* 2009.
11. JEDEC, *JEDEC Standard No. 209-3—Low Power Double Data Rate 3,* 2012.
12. A. Virtanen, R. Harboe-Sorensen, A. Javanainen, H. Kettunen, H. Koivisto and I. Riihimaki, Upgrades for the RADEF facility, in *Radiation Effects Data Workshop (REDW),* 2007.
13. R. Harboe-Sorensen, R. Muller and S. Fraenkel, Heavy ion, proton and Co-60 radiation evaluation of 16 Mbit DRAM memories for space application, in *IEEE Radiation Effects Data Workshop,* 1995.
14. P. Adell, L. Edmonds, R. McPeak, L. Scheick and S. McClure, An approach to single event testing of SDRAMs, *IEEE Transactions on Nuclear Science,* Vol. 57, No. 5, 2009.
15. S. Guertin, G. Allen and D. Sheldon, Programmatic impact of SDRAM SEFI, in *IEEE Radiation Effects Data Workshop (REDW),* 2012.
16. C. Hafer, M. Von Thun, M. Leslie, F. Sievert and A. Jordan, Commercially designed and manufactured SDRAM SEE data, in *IEEE Radiation Effects Data Workshop,* 2010.
17. T. Langley, R. Koga and T. Morris, Single-event effects test results of 512MB SDRAMs, in *IEEE Radiation Effects Data Workshop,* 2003.

18. B. Henson, P. McDonald and W. Stapor, SDRAM space radiation effects measurements and analysis, in *IEEE Radiation Effects Data Workshop*, 1999.
19. R. L. Ladbury, K. A. LaBel, M. D. Berg, E. P. Wilcox, H. S. Kim, C. M. Seidleck and A. M. Phan, Use of commercial FPGA-based evaluation boards for single-event testing of DDR2 and DDR3 SDRAMs, in *NSREC*, Vol. 60, No. 6, 2013.
20. R. Ladbury, M. Berg, K. LaBel and M. Friendlich, Radiation performance of 1 Gbit DDR2 SDRAMs fabricated with 80–90 nm CMOS, in *IEEE Radiation Effects Data Workshop*, 2008.
21. L. Li, H. Schmidt, T. Fichna, D. Walter, K. Grürmann, H. W. Hoffmeister, S. Lamari, H. Michalik and F. Gliem, Heavy ion SEE test of an advanced DDR2 SDRAM, in *21. Workshop für Testmethoden und Zuverlässigkeit von Schaltungen und Systemen*, 2009.
22. R. Koga, P. Yu, J. George and S. Bielat, Sensitivity of 2 Gb DDR2 SDRAMs to protons and heavy ions, in *IEEE Radiation Effects Data Workshop (REDW)*, 2010.
23. S. Hoeffgen, M. Durante, V. Ferlet-Cavrois et al., Investigations of single event effects with heavy ions of energies up to 1.5 GeV/n, in *12th European Conference on Radiation and Its Effects on Components and Systems (RADECS)*, 2011.
24. M. Herrmann, K. Grürmann, F. Gliem, H. Schmidt, G. Leibeling, H. Kettunen and V. Ferlet-Cavrois, New SEE test results for 4 Gbit DDR3 SDRAM, in *RADECS Data Workshop*, 2012.
25. K. Grürmann, M. Herrmann, F. Gliem, H. Schmidt, G. Leibeling and H. Kettunen, Heavy ion sensitivity of 16/32-Gbit NAND-Flash and 4-Gbit DDR3 SDRAM, in *IEEE NSREC Data Workshop*, 2012.
26. R. Koga, J. George and S. Bielat, Single event effects sensitivity of DDR3 SDRAMs to protons and heavy ions, in *IEEE Radiation Effects Data Workshop*, 2012.
27. H. Quinn, P. Graham and T. Fairbanks, SEEs induced by high-energy protons and neutrons in SDRAM, in *IEEE Radiation Effects Data Workshop (REDW)*, 2011.
28. L. W. Massengill, Cosmic and terrestrial single-event radiation effects in dynamic random access memories, *IEEE Transactions on Nuclear Science,* Vol. 43, pp. 576–593, 1996.
29. A. Chugg, A. Burnell, P. Duncan and S. Parker, The random telegraph signal behavior of intermittently stuck bits in SDRAMs, *IEEE Transactions on Nuclear Science,* Vol. 56, No. 6, pp. 3057–3064, 2009.
30. A. Bougerol, F. Miller, N. Guibbaud, R. Gaillard, F. Moliere and N. Buard, Use of laser to explain heavy ion induced SEFIs in SDRAMs, *IEEE Transactions on Nuclear Science,* Vol. 57, No.1, 2010.
31. L. Scheick, S. Guertin and D. Nguyen, Investigation of the mechanism of stuck bits in high capacity SDRAMs, in *IEEE Radiation Effects Data Workshop*, 2008.
32. M. Herrmann, K. Grürmann, F. Gliem, H. Schmidt and V. Ferlet-Cavrois, In-situ TID test of 4-Gbit DDR3 SDRAM devices, in *Radiation Effects Data Workshop (REDW)*, 2013.
33. D. Walter, M. Herrmann, K. Grürmann and F. Gliem, From memory device cross section to data integrity figures of space mass memories, in *European Conference on Radiation and Its Effects on Components and Systems (RADECS)*, 2013.
34. JEDEC, *JESD79-3E–DDR3 SDRAM Specification.*
35. M. Herrmann, K. Grürmann, F. Gliem, H. Kettunen and V. Ferlet-Cavrois, Heavy ion SEE test of 2 Gbit DDR3 SDRAM, in *12th European Conference on Radiation and Its Effects on Components and Systems (RADECS)*, 2011.
36. M. Herrmann, K. Grürmann, F. Gliem, H. Schmidt, M. Muschitiello and V. Ferlet-Cavrois, New SEE and TID test results for 2-Gbit and 4-Gbit DDR3 SDRAM devices, in *RADECS Data Workshop*, 2013.

37. J. A. Halderman, S. D. Schoen, N. Heninger, W. Clarkson, W. Paul, J. A. Cal, A. J. Feldman and E. W. Felten, Lest we remember: Cold boot attacks on encryption keys, in *17th USENIX Security Symposium*, 2008.
38. C. Lee, D. Nguyen and A. Johnston, Total ionizing dose effects on 64 Mb 3.3 V DRAMs, in *IEEE Radiation Effects Data Workshop*, 1997.
39. A. Bacchini, M. Rovatti, G. Furano and M. Ottavi, Total ionizing dose effects on DRAM data retention time, *IEEE Transactions on Nuclear Science*, Vol. 61, No. 6, 2014.
40. S. Bertazzoni, D. Di Giovenale, M. Salmeri, A. Mencattini, A. Salsano and M. Florean, Monitoring methodology for TID damaging of SDRAM devices, in *19th IEEE International Symposium on Defect and Fault Tolerance in VLSI Systems*, 2004.
41. S. Bertazzoni, D. Di Giovenale, L. Mongiardo et al., TID and SEE characterization and damaging analysis of 256 Mbit COTS SDRAM for IEEM application, in *8th European Conference on Radiation and Its Effects on Components and Systems*, 2005.
42. L. Scheick, S. Guertin and G. Swift, Analysis of radiation effects on individual DRAM cells, *IEEE Transactions on Nuclear Science*, Vol. 47, No. 6, 2000.
43. D. Shaw, G. Swift and A. Johnston, Radiation evaluation of an advanced 64 Mb 3.3 V DRAM and insights into the effects of scaling on radiation hardness, *IEEE Transactions on Nuclear Science*, Vol. 42, No. 6, pp. 1674–1680, 1995.
44. M. Herrmann, K. Grürmann and F. Gliem, TN-IDA-RAD-14/3—In-situ and unbiased TID test of 4-Gbit DDR3 SDRAM devices, 2014.

5 Radiation Effects in Flash Memories

Marta Bagatin and Simone Gerardin

CONTENTS

5.1 INTRODUCTION

Nonvolatile memories (NVMs) represent a large part of the semiconductor market. They are used in a growing number of applications that demand very high capacity, and also in critical scenarios where failures, even of single bits, can lead to significant economic losses or life-threatening situations.

Flash comes in two different architectures: NOR and NAND. Traditionally, NOR flash devices have been used for code storage and in situations where random data access is important. The manufacturer guarantees that every single bit is functional and meets retention and endurance specifications without any corrective action by the user. NAND flashes, on the other hand, are selected when density and serial access are the key parameters. Bad blocks and bit fails are possible, so the user must implement an appropriate error correction code (ECC) meeting datasheet requirements.

NAND flash memories are the most aggressively scaled devices: memories with a feature size of 16 nm and storage density of 128 Gbit per die are available on the market. Three-dimensional (3-D) devices based on several layers of memory elements and a more relaxed feature size are also being mass-produced. NVMs close follow Moore's law, but fundamental scaling limits are rapidly being approached and semiconductor companies and research institutions are actively developing replacements based on different storage mechanisms.

Floating gate (FG) is the leading commercial flash technology. It is based on the inclusion of a floating polysilicon electrode between the gate and the channel of a metal-oxide-semiconductor field-effect transistor (MOSFET). By injecting and removing charge (electrons or holes) from the floating gate, the threshold voltage of the device can be altered and used to represent one or more bits of information.

Until about a decade ago, radiation effects in NVMs were analyzed only in the context of space applications and with a focus on the high-voltage peripheral circuitry, which was the only source of sensitivity back then. Later, following a reduction in the amount of stored charge, single-event effects induced by heavy ions were demonstrated in the FG cells and more recently also neutrons and alpha particles have been shown to induce upsets in NAND flash bits, although to an extent comparable to other bit-fail mechanisms, and therefore under the control of mandatory ECC.

This chapter will analyze radiation effects, separating the effects involving the FG cells from those due to the peripheral circuitry, both in the context of sea-level applications and in the more severe space environment.

5.2 FLOATING GATE TECHNOLOGY

Flash memories cells resemble MOSFETs [1,2] but they include an additional charge storage element, typically a polysilicon FG placed between the silicon bulk and the gate, called a control gate (see Figure 5.1). The charge storage element is isolated on one side by the tunnel oxide and on the other side by an oxide-nitride-oxide (ONO) stack.

By changing the amount of charge in the storage element, the threshold voltage (V_{th}) of the transistor can be altered. The state with positive or no charge in the storage element is called *erased* and the one with a net negative charge is called

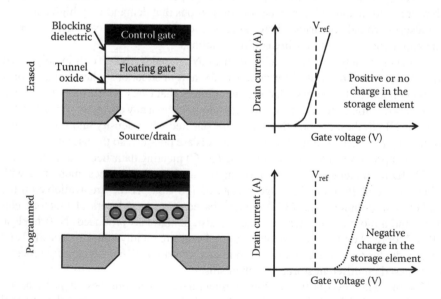

FIGURE 5.1 Structure (left) and I-V characteristics (right) of an erased (top) and a programmed (bottom) FG cell.

programmed (Figure 5.1). Read-out is carried out by biasing the control gate at a fixed voltage, intermediate between the V_{th} of the programmed and erased cells, and comparing the drawn current with one or more reference sources.

The I-V characteristics of a FG cell are close to those of a standard MOSFET, with some peculiar features due to the presence of the storage element, which is capacitively coupled to all the other terminals. As a result, the FG potential, which determines the density of carriers in the channel, is influenced not only by the control gate voltage, but also by the drain, and even worse, by neighboring cells [3].

To alter the amount of stored charge, electrons or holes can be injected over or through the tunnel-oxide potential barrier. Injection over the barrier is accomplished using channel hot electrons (holes). In this scheme, the cell is turned on by raising the gate voltage and at the same time a high voltage is applied to the drain, creating a large lateral field close to the drain terminal [3]. The efficiency of this process is poor because a lot of current must flow between source and drain and just a small part of it is injected into the FG.

Injection through the barrier exploits Fowler-Nordheim (FN) tunneling, which is more efficient but slower [3]. This quantum mechanical effect can be used for both program and erase and requires thin-enough oxide barriers, ~10 nm or less. However, thin tunnel oxides are critical for cell reliability, since carriers can tunnel in and out of the FG more easily, not only during program and erase, but also in storage mode.

Cell parameters (e.g., V_{th}) show a large variability in devices containing billions of cells. Tightening algorithms are employed during program and erase operations to diminish the V_{th} spread [4]. They work by introducing two additional reference levels, program verify and erase verify, and ensuring that the final V_{th} of a programmed (erased) cell is above the program verify level (lower than the erase verify level).

With tight control on the amount of charge injected in the FG, it is possible to store multiple bits per cell (multilevel cell [MLC]) [5] instead of a single bit per cell (single-level cell [SLC]). The peripheral circuitry of MLC devices is more complex due to the multiple reference levels and the much tighter read margins with respect to SLC devices. MLCs are slower and less robust, but they can offer almost twice the density and half the cost per bit using two bits per cell.

There are two major array organizations: NAND and NOR [6]. To avoid the use of a selector for each cell and increase density, erase is performed on whole blocks (a few Mbits) in both types. In the NOR array several cells are connected in parallel to a bit line with drain contacts shared between two adjacent cells. A fast random access is obtained (<100 ns), word program is much slower (~5 µs), and block erasure is even slower (~200 ms). The programming is carried out by channel hot carrier injection whose high current limits the operation parallelism to just a few cells. Erase is performed by FN tunneling. The manufacturer guarantees that all individual bits are functional and meet retention and endurance specifications. Thanks to these features, flash NOR is ideally suited as a read-mostly memory for code storage.

In the NAND architecture the cells are connected in strings of 16 or more elements with two selection devices. In turn, the strings are connected to bit lines. At least in the planar implementation, the physical bit size is the smallest in the semiconductor industry and considerably more compact than the NOR cell (hence much higher capacity is available) thanks to the FN program mechanism and the lack of a drain contact for each pair of cells. The FN-based program warrants a great level of parallelism. Page programming (which typically involves a few kBytes) is carried out in about 0.2 ms; block erase is carried out by FN tunneling and involves a few MBytes in parallel and takes about 2 ms; random access is poor, but serial access has very good performance. Because the manufacturer does not guarantee each single bit and each single block, the use of external ECC and bad-block management is compulsory. As a result of these features, flash NAND is best suited for data storage but with proper buffers can also be used for code.

Due to the high electric fields during program and erase, the reliability of FG devices is always critical [7]. Intrinsic phenomena affect all the cells in a uniform way, and single-bit failures occur in just a few cells due to extrinsic defects (e.g., contamination particles during processing) or to peculiar configurations of intrinsic point defects. The reliability of NVMs is measured by two main parameters: endurance and retention. The endurance is defined by the number of program/erase cycles that can be successfully carried out on a memory (10^5 is the typical specification for an SLC flash memory). It is limited by the generation of traps and charge trapping in the tunnel oxide during the high-field program and erase operations [8]. Retention is the period of time that information can be stored without corruption and is limited by stress-induced leakage current (SILC) [9]. SILC is a leakage current through the tunnel oxide due to trap-assisted tunneling.

Similar to other types of memories, flash devices include decoders and buffers, but also other peculiar functional blocks: a microcontroller is needed to execute the complex program/erase algorithms and charge pumps are required to generate the high voltages to inject and remove charge from the floating gates. As we will see in the following, charge pumps are one of the weakest points of flash memories as far as radiation sensitivity is concerned.

5.3   RADIATION EFFECTS ON FLOATING GATE CELLS

Whereas in the past only the peripheral circuitry was believed to be sensitive to radiation, FG cells in contemporary flash memories are also a source of errors [10–15]. We will start our discussion with FG cells, investigating total-dose and single-event effects and then move on to analyze effects in the peripheral circuitry.

In general, an error occurs in a FG cell when ionizing radiation induces a V_{th} shift large enough to move the cell V_{th} beyond the read voltage. This can be the result of both total-dose accumulation and single-event effects. Historically, many results were first collected on NOR devices. However, unless explicitly mentioned, the results presented here apply both to NOR and NAND devices with slight or no modifications.

5.3.1   Bit Errors Induced by Total Dose

Total dose causes the V_{th} of FG cells to move toward the neutral V_{th} (i.e., the state with no charge in the FG). In other words, total dose tends to remove the charge from FGs. This is shown in Figure 5.2 in the V_{th} distributions of a two-bit-per-cell NOR device irradiated with a total-dose source (protons) [14]. The V_{th} distributions

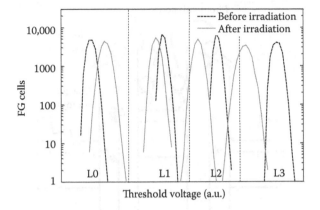

FIGURE 5.2 Threshold voltage shifts induced by total ionizing dose on a 90-nm MLC NOR for cells in the four levels (L0 is the erased level, L1, L2, and L3 the programmed levels, going from the lowest to the highest V_{th}). The V_{th} of all cells is impacted and moves toward the neutral distribution located between L0 and L1. (Adapted from M. Bagatin, S. Gerardin, G. Cellere, A. Paccagnella, A. Visconti, M. Bonanomi, S. Beltrami, *IEEE Trans. Nucl. Sci.*, Vol. 56, pp. 3267–3273, December 2009.)

globally shift: the programmed ones (L1, L2, and L3) to the left and the erased one (L0) to the right. All the distributions move toward the intrinsic distribution, which is typically found between the erased and the first programmed state. These threshold voltage shifts are responsible for the errors that will be shown later when we will also include the impact of the periphery.

The three mechanisms responsible for V_{th} shifts due to total dose are depicted in Figure 5.3 [16–22]: charge induced by radiation in the surrounding oxides (tunnel and ONO) is injected into the FG, (a typically small amount of) charge is trapped in the tunnel oxide, and the carriers in the FG obtain enough energy from the radiation to escape from the potential well (photoemission). Except for charge trapping, these mechanisms tend to reduce the amount of charge in the FG. Due to reliability issues, the tunnel oxide thickness has not been scaled with the cell size. As a result, total-dose effects tend not to be impacted too much by scaling [19]. The use of different sources of total dose (gamma rays, x-rays, protons, etc.) leads to small variations that are discussed in [23]. Thanks to their response to total dose, dosimeters based on FG have been developed [24,25].

Error annealing has been reported after total dose [14]. After the exposure, the number of errors decreases over time, but in some cases it can also increase (in particular for erased cells and for cells programmed at the lowest V_{th} levels [14]). Annealing has been attributed to the removal of positive trapped charge from the oxides, which, in case of erased cells, has the effect of shifting V_{th} distributions to higher V_{th} values.

Retention and endurance after total-dose levels not leading to functional failures have been investigated in a few publications [26–28]. No significant effect on the endurance of NAND flash memories was observed for doses between 50 and

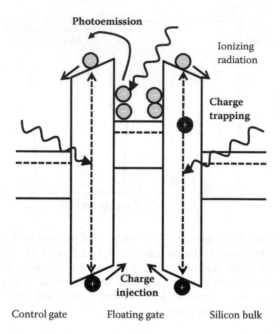

FIGURE 5.3 The three mechanisms responsible for threshold voltage shifts in FG arrays exposed to total dose: charge injection, charge trapping, and photoemission. (After E. Snyder, P. McWhorter, T. Dellin, J. Sweetman, *IEEE Trans. Nucl. Sci.*, Vol. 36, pp. 2131–2139, December 1989.)

200 krad (silicon oxide [SiO_2]), depending on the manufacturer. On the contrary, in [27], retention issues after total ionizing dose (TID) exposure up to 50 krad (SiO_2) were recorded, although the number of affected bits was quite small and was not expected to pose a significant challenge to ECCs.

5.3.2 BIT ERRORS INDUCED BY SINGLE-EVENT EFFECTS

Bit errors due to FG discharge can also be produced by sources of single-event effects, such as heavy ions [29], protons [30], neutrons, and alpha particles [31]. The V_{th} distributions are affected by heavy ions as illustrated in Figure 5.4 for a NOR MLC array irradiated with a large number of Si ions [29] (for NAND, see for instance [32]). A secondary peak appears after particle strikes [33], especially for the higher-V_{th} distributions. Similar to the total-dose case, the affected cells, both erased and programmed, which are now a subset of the total number of cells, move toward the intrinsic distribution. The number of cells in the secondary peak is related to the irradiation fluence. The distance between the peaks (i.e., the average ΔV_{th}) is related to the ion linear energy transfer (LET) and to the electric field in the tunnel oxide [34]. An almost linear relationship between the average V_{th} shift and the electric field and also between the average V_{th} shift and the LET of the impinging particle is typically reported [34]. The cells with the largest V_{th} shifts give rise to digital errors

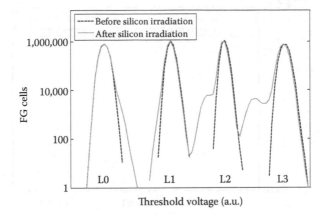

FIGURE 5.4 Threshold voltage shifts induced by heavy-ion irradiation (silicon ions with LET = 9.8 MeV cm²/mg) on 90-nm MLC NOR flash for cells in the four levels (L0 is the erased level and L1, L2, and L3 the programmed levels, going from the lowest to the highest V_{th}). A secondary peak is visible, especially on the two higher V_{th} levels. (Adapted from M. Bagatin, S. Gerardin, A. Paccagnella, G. Cellere, A. Visconti, M. Bonanomi, *IEEE Trans. Nucl. Sci.*, Vol. 57, pp. 3407–3413, December 2010.)

at the output. Statistical properties of the array (variations between cells) and of the energy deposition processes (straggling) are of extreme importance in the determining the error rate [33].

Besides the secondary peak, a distribution of hit cells is often visible between the primary and secondary peaks, especially in scaled devices [33,35]. Geant4-based simulations have shown that the secondary peak is linked to the ions that go through the FGs, whereas the transition region is related to cells struck by energetic delta-electrons emitted even quite far from the primary track [33].

Ion energy plays a role as well, although to a much smaller extent than LET [36]. For a given LET, the number of digital errors, the threshold voltage shift, and the amount of charge loss depend on the energy of the impinging particle. Due to the smaller radius and higher density of the track, lower-energy ions [36] are believed to be more effective in discharging FG cells.

The dependence of heavy-ion-induced upsets on the angle of incidence has been studied in several works to gain insight into the underlying mechanisms and improve error predictions. In [37,38] a thorough experimental investigation of the single-event upset (SEU) sensitivity was performed rotating and tilting the samples, showing a very complex dependence and a preferential direction for the generation of multiple-bit upsets (MBUs), which occur much more frequently along one direction [39]. In old SLC memories with feature size larger than 50 nm, MBUs account only for a few percent. In more modern devices, at high LET the majority of SEUs are clustered MBUs [40].

A significant effort has been spent to identify the sensitive volume [41]. Measurements have shown that especially at high LET and for NAND devices, the sensitive volume is larger than the tunnel oxide but smaller than the whole FG. No

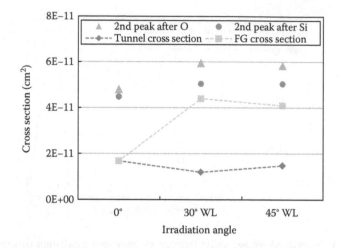

FIGURE 5.5 Cross sections for errors and secondary peak for the highest program level in a 41-nm NAND flash irradiated with silicon (Si) (LET = 9.8 MeV cm²/mg) as a function of the ion incidence angle along the WL. The apparent cross section of the tunnel oxide and the FG are superimposed. (Adapted from S. Gerardin, M. Bagatin, A. Paccagnella, A. Visconti, M. Bonanomi, S. Beltrami, *IEEE Trans. Nucl. Sci.*, Vol. 58, pp. 2621–2627, December 2011.)

conclusive correlation has been found either with the tunnel oxide or with the FG. This is illustrated in Figure 5.5, where the error cross section is plotted versus the angle of incidence and compared with the apparent tunnel-oxide cross section for 41-nm NAND devices. As seen, the secondary peak dependence on the angle shows an increase from normal to tilted exposures along the wordline (WL), and this is compatible, although not fully, with a thick sensitive volume (large part of the FG) [41].

The physical mechanisms responsible for the discharge of FG cells struck by heavy ions have not been completely identified yet. The seemingly linear dependence between ΔV_{th} and the LET of the impinging particle was used to support the idea that the FG discharge does not depend on charge yield and so must occur before charge recombination [34]. This led to the assumption that a transient leakage path is created across the tunnel oxide by the heavy ions. This path is active as soon as the particle hits the cell and is believed to be resistive in nature, thus explaining the linear relation between the discharge and the electric field. Good agreement with experimental data has been found using this model [34]. However, the exact physical origin of the path is still debated. The initial proposal was a collapse of the band structure due to the high density of radiation-generated carriers. More recently it has been suggested that the leakage path consists of temporary defects that favor the escape of stored carriers through trap-assisted tunneling [42]. The leakage path model leads to the conclusion that the sensitive volume is the tunnel oxide, which is not entirely consistent with the angular data reported in [41].

Another model presented in [43] starts from the generation of energetic carriers by the striking particles and the carrier fluxes in and out the FG due to tunneling currents. According to the authors in [43], the unbalance between the tunneling currents generates the discharge of the FG. By carefully modeling the energy relaxation of the carriers, agreement has been obtained with experimental data [43] without assuming difficult-to-explain conductive paths.

Earlier models [44] attributed the upsets and the shifts in V_{th} entirely to charge trapping and interface trap generation in the oxides surrounding the FG due to microdose effects [45]. However, even though this interpretation has been refuted by clear evidence of charge loss, a small charge trapping component with heavy ions has been identified [15,46], the neutralization of which due to thermal or tunneling annealing is responsible for the decrease over time of the number of heavy-ion-induced FG errors [47].

Whereas upset bits can be generally rewritten without issues, some cells may be permanently damaged by the irradiation in a quite subtle way. In [48,49] some FG errors were observed after exposure and corrected by a new program operation. However, 1-1/2 hours after programming, a tail in the V_{th} distribution was observed and grew over time, eventually leading to errors. This suggested that the irradiation can induce some permanent damage [48,49], although only with very high-LET ions. It has been attributed to the generation of radiation-induced leakage current (RILC). Radiation can create neutral defects in the tunnel oxide, giving rise to a permanent leakage path that compromises the retention capability of the affected cell. On the other hand, the ONO layer appears to be more robust to permanent damage due to the larger thickness and lower electric field [50].

Scaling has a deep impact on SEUs in FG cells. Figure 5.6 shows the most important technological trend [31,51]: the threshold LET greatly decreases as a function of

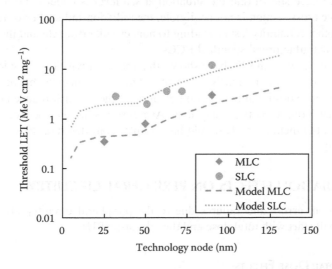

FIGURE 5.6 Dependence on the cell feature size of threshold LET (experimental and modeled) for FG upsets for MLC and SLC NAND cells. (Adapted from A. Gasperin, A. Paccagnella, G. Ghidini, A. Sebastiani, *IEEE Trans. Nucl. Sci.*, Vol. 56, pp. 2218–2224, August 2009.)

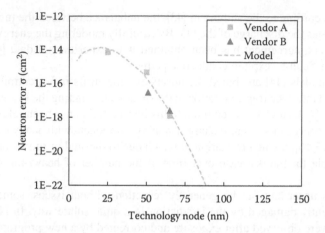

FIGURE 5.7 Neutron-induced cross section as a function of feature size for NAND flash cells. Experimental data is compared with a model based on the transient conductive path. (Adapted from A. Gasperin, A. Paccagnella, G. Ghidini, A. Sebastiani, *IEEE Trans. Nucl. Sci.*, Vol. 56, pp. 2218–2224, August 2009.)

the feature size. Indeed, shrinking of the cell size is accompanied with a reduction of the FG charge. As a consequence, the deposition of the same amount of charge on the FG or in its vicinity causes a comparatively larger shift of the transistor V_{th} toward the reference level, and as a result, a larger count of SEUs. Data taken with radiation sources reproducing the terrestrial neutron spectrum (Figure 5.7) and radioactive alpha source have shown that the situation at sea level is under control [31,52,53]. Although the cross section is not negligible, the radiation-induced error rate is comparable to other reliability issues leading to nonzero bit error rate and therefore correctable by manufacturers' specified ECCs.

Finally, the authors in [29] have shown that the sensitivity to heavy-ion-induced SEUs is increased with previously accumulated total dose in cells that have not been erased after the exposure due to the combination of the V_{th} shift induced by the total dose and that induced by the heavy ions. As a result, synergies between total dose and SEUs in FG memory cells should be carefully considered in the error rate predictions [29].

5.4 RADIATION EFFECTS ON PERIPHERAL CIRCUITRY

In this section, errors and failures due to the peripheral circuitry are presented. Again we will start with total dose and then discuss SEUs.

5.4.1 TOTAL DOSE EFFECTS

When also taking into account the peripheral circuitry, total dose leads to corruption of stored information due to the discharge of FGs and to functional errors (e.g., all cells stuck at 0 or 1) due to failures in charge pumps and/or decoders. It is possible to

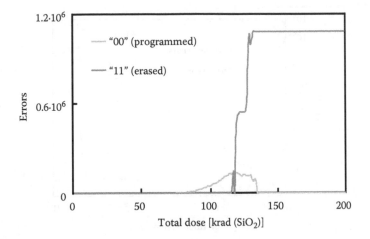

FIGURE 5.8 Read errors in programmed cells and in erased cells of a 90-nm NAND flash memory as a function of dose during x-ray irradiation. FG cells and charge pumps were exposed, whereas the row decoder was shielded during the irradiation. (Adapted from M. Bagatin, G. Cellere, S. Gerardin, A. Paccagnella, A. Visconti, S. Beltrami, *IEEE Trans. Nucl. Sci.*, Vol. 56, pp. 1909–1913, August 2009.)

study the sensitivity of each building block by irradiating a small part of a memory. Figure 5.8 shows an example of such a study [54], illustrating the number of errors in a NAND 90-nm SLC memory exposed to a source of x-rays. Typically, errors in the FGs occur at doses from few krad(Si), especially in MLC devices, to a few tens of krad(Si). At higher doses, which vary considerably depending on manufacturer, memory type, and so forth, failures in some of the building blocks occur [55–58]. For instance, in Figure 5.8, the errors at low doses are due to threshold voltage shifts in FG cells, whereas the behavior at high dose is determined by the failure of charge pump circuitry.

Charge pumps have long been known as the most sensitive block from the point of view of total-dose damage [12,55]. Their purpose is to provide the high voltages needed for program and erase and also for read operations in the most recent devices.

Charge pumps are realized using multiple stages in series. Each stage works by charging a capacitor and putting it in series to the next stage. The number of stages determines the final output voltage. The output of a charge pump must be very precise for program and erase operations to work correctly. Because of the thick oxides used to sustain the high voltage and the need for an accurate output voltage, this block has always been one of the most critical from an ionizing radiation standpoint. Charge pumps are typically the first to fail in total-dose tests. An example of degradation in the output voltage of a charge pump as a function of total dose is presented in [54].

5.4.2 SINGLE-EVENT EFFECTS

Figure 5.9 shows the number of errors during a read loop in a NAND memory exposed to heavy ions (bromine (Br), LET = 41 MeV cm²/mg) [15]. As we can see, in

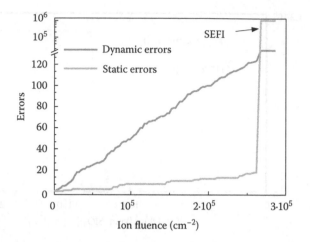

FIGURE 5.9 Dynamic errors (due to the corruption of PB latches) and static errors (due to the corruption of FG cells) in a flash NAND 90-nm memory irradiated with Br ion as a function of the accumulated fluence. After a fluence of about $2.7 \cdot 10^5$ ions/cm^2, a SEFI takes place. (Adapted from M. Bagatin, S. Gerardin, G. Cellere, A. Paccagnella, A. Visconti, S. Beltrami, R. Harboe-Sorensen, A. Virtanen, *IEEE Trans. Nucl. Sci.*, Vol. 55, pp. 3302–3308, December 2008.)

the first part of the exposure, two types of errors occur: some (called static errors) are persistent until an erase/program operation is performed and are related to strikes in the FG cells; others (called dynamic errors) are related instead to the peripheral circuitry and are present only in one read cycle and disappear in the next one. With a lower probability, single-event functional interrupt (SEFI) may occur, causing for instance a whole block to fail read as result of heavy-ion strikes in the onboard microcontroller causing bursts of errors (see Figure 5.9).

Dynamic errors can originate from the corruption of the page buffer (PB) latches [15], a temporary storage area where data stay before being output at the device pins or before being programmed into the array. While in the PB, the data can be corrupted by radiation, as with SRAMs [15]. Since data in the PB is rewritten each time a page is accessed in the FG array, dynamic errors persist for only one read cycle.

SEFIs are often observed during radiation tests [59] (block erase SEFI, partial erase SEFI, write SEFI, read SEFI, etc.). These are almost invariably related to strikes in the on-chip microcontroller. They can be cured in different ways: simple repeat, reset, or, in the most serious cases, power cycle. The nonvolatility of flash memories is a significant advantage in this area, since no data loss occurs after a power cycle.

Some groups have reported the occurrence of high-current spikes and destructive events (DEs) during irradiation of NAND flash memories under heavy-ion beams at high LET [60–62]. These events lead to the inability to erase or program significant parts of the affected memory (or even the whole array).

Some associate current spikes to DE [60], whereas others speculate that the two are separate phenomena [61]. Some other researchers believe these are artifacts of the measurement conditions and may not be an issue in the space environment [62].

No agreement has been found on the underlying mechanisms, and certainly more research is needed in this sense.

DE cross-section curves from several test campaigns can be found in [40,62].

5.5 CONCLUSIONS

The NVM market is dominated by NAND FG devices, which have the most aggressively scaled feature size and are also the first to commercially pursue 3-D integration. They are attractive for a wide variety of applications, including critical ones, both on Earth and in harsh environments, such as space. No rad-hard device matches the capacity offered by state-of-the-art NAND flash. As a result, radiation sensitivity of commercial devices has been carefully evaluated.

Flash FG cells are sensitive to total-dose and single-event effects. Reported effects comprise the corruption of the information stored in FG cells after exposure to sources of total-dose or single-event effects, including terrestrial neutrons and alphas emitted by radioactive contaminants. The underlying mechanisms are well established for total-dose degradation but are still under investigation for single-event effects. The reduction in stored charge leads to a continuous reduction in the threshold LET, which becomes lower and lower with each new generation. Total-dose-induced corruption is less dependent on scaling, but varies widely between parts and manufacturers. Charge pumps have been early recognized as one of the most sensitive parts as far as total-dose degradation is concerned. Because of the high voltages required and the impossibility of scaling the gate oxide, total-dose tolerance is not improving as in standard low-voltage CMOS circuits. Finally, recent observations of destructive events during heavy-ion irradiations may pose a threat in space but are still controversial among researchers.

REFERENCES

1. R. Bez, E. Camerlenghi, A. Modelli, A. Visconti, Introduction to Flash Memory, *Proc. IEEE*, Vol. 91, pp. 489–502, April 2003.
2. J. V. Houdt, R. Degraeve, G. Groeseneken, H. E. Maes, Physics of Flash Memories, in *Nonvolatile Memory Technologies with Emphasis on Flash*, J. Brewer, M. Gill, eds. Hoboken, NJ: John Wiley & Sons, pp. 129–177, 2008.
3. P. Pavan, R. Bez, P. Olivo, E. Zanoni, Flash Memory Cells—An Overview, *Proc. IEEE*, Vol. 85, pp. 1248–1271, August 1997.
4. G. G. Marotta, G. Naso, G. Savarese, Memory Circuit Technologies, in *Nonvolatile Memory Technologies with Emphasis on Flash*, J. Brewer, M. Gill, eds. Hoboken, NJ: John Wiley & Sons, pp. 63–128, 2008.
5. A. Fazio, M. Bauer, Multilevel Cell Digital Memories, in *Nonvolatile Memory Technologies with Emphasis on Flash*, J. Brewer, M. Gill, eds. Hoboken, NJ: John Wiley & Sons, pp. 591–616, 2008.
6. G. Forni, C. Ong, C. Rice, K. McKee, R. J. Bauer, Flash Memory Applications, in *Nonvolatile Memory Technologies with Emphasis on Flash*, J. Brewer, M. Gill, eds. Hoboken, NJ: John Wiley & Sons, pp. 19–62, 2008.
7. N. Mielke, T. Marquart, N. Wu, J. Kessenich, H. Belgal, E. Schares, F. Trivedi, E. Goodness, L. Nevill, Bit Error Rate in NAND Flash Memories, in *Reliability Physics Symposium*, 2008. *IRPS 2008*. April 27–May 1, 2008, pp. 9–19.

8. N. Mielke, H. Belgal, I. Kalastirsky, P. Kalavade, A. Kurtz, Q. Meng, N. Righos, J. Wu, Flash EEPROM Threshold Instabilities due to Charge Trapping during Program/Erase Cycling, *IEEE Trans. Device Mater. Rel.*, Vol. 4, pp. 335–344, September 2004.

9. K. Naruke, S. Taguchi, M. Wada, Stress Induced Leakage Current Limiting to Scale Down EEPROM Tunnel Oxide Thickness, *Electron Devices Meeting, 1988. IEDM '88. Technical Digest., International*, pp. 424–427, 1988.

10. S. Gerardin, A. Paccagnella, Present and Future Non-Volatile Memories for Space, *IEEE Trans. Nucl. Sci.*, Vol. 57, pp. 3016–3039, December 2010.

11. S. Gerardin, M. Bagatin, A. Paccagnella, K. Grürmann, F. Gliem, T. R. Oldham, F. Irom, and D. N. Nguyen, Radiation Effects in Flash Memories, Nuclear Science, *IEEE Trans. Nucl. Sci.*, Vol. 60, No. 3, pp. 1953–1969, June 2013.

12. T. R. Oldham, R. L. Ladbury, M. Friendlich, H. S. Kim, M. D. Berg, T. L. Irwin, C. Seidleck, K. A. LaBel, SEE and TID Characterization of an Advanced Commercial 2Gbit NAND Flash Nonvolatile Memory, *IEEE Trans. Nucl. Sci.*, Vol. 53, pp. 3217–3222, December 2006.

13. H. Schmidt, K. Grürmann, B. Nickson, F. Gliem, R. Harboe-Sørensen, TID Test of an 8-Gbit NAND Flash Memory, *IEEE Trans. Nucl. Sci.*, Vol. 56, No. 4, pp. 1937–1940, August 2009.

14. M. Bagatin, S. Gerardin, G. Cellere, A. Paccagnella, A. Visconti, M. Bonanomi, S. Beltrami, Error Instability in Floating Gate Flash Memories Exposed to TID, *IEEE Trans. Nucl. Sci.*, Vol. 56, pp. 3267–3273, December 2009.

15. M. Bagatin, S. Gerardin, G. Cellere, A. Paccagnella, A. Visconti, S. Beltrami, R. Harboe-Sorensen, A. Virtanen, Key Contributions to the Cross Section of NAND Flash Memories Irradiated with Heavy Ions, *IEEE Trans. Nucl. Sci.*, Vol. 55, pp. 3302–3308, December 2008.

16. J. Caywood, B. Prickett, Radiation-Induced Soft Errors and Floating Gate Memories, *21st Annual Reliability Physics Symposium, 1983*, pp. 167–172, April 1983.

17. E. Snyder, P. McWhorter, T. Dellin, J. Sweetman, Radiation Response of Floating Gate EEPROM Memory Cells, *IEEE Trans. Nucl. Sci.*, Vol. 36, pp. 2131–2139, December 1989.

18. P. McNulty, S. Yow, L. Scheick, W. Abdel-Kader, Charge Removal from FGMOS Floating Gates, *IEEE Trans. Nucl. Sci.*, Vol. 49, pp. 3016–3021, December 2002.

19. G. Cellere, A. Paccagnella, A. Visconti, M. Bonanomi, P. Caprara, S. Lora, A Model for TID Effects on Floating Gate Memory Cells, *IEEE Trans. Nucl. Sci.*, Vol. 51, pp. 3753–3758, December 2004.

20. J. Wang, S. Samiee, H.-S. Chen, C.-K. Huang, M. Cheung, J. Borillo, S.-N. Sun, B. Cronquist, J. McCollum, Total Ionizing Dose Effects on Flash-based Field Programmable Gate Array, *IEEE Trans. Nucl. Sci.*, Vol. 51, pp. 3759–3766, December 2004.

21. J. Wang, G. Kuganesan, N. Charest, B. Cronquist, Biased-Irradiation Characteristics of the Floating Gate Switch in FPGA, *Radiation Effects Data Workshop, 2006 IEEE*, pp. 101–104, July 2006.

22. G. Cellere, A. Paccagnella, A. Visconti, M. Bonanomi, S. Beltrami, J. Schwank, M. Shaneyfelt, P. Paillet, Total Ionizing Dose Effects in NOR and NAND Flash Memories, *IEEE Trans. Nucl. Sci.*, Vol. 54, pp. 1066–1070, August 2007.

23. G. Cellere, A. Paccagnella, A. Visconti, M. Bonanomi, A. Candelori, S. Lora, Effect of Different Total Ionizing Dose Sources on Charge Loss from Programmed Floating Gate Cells, *IEEE Trans. Nucl. Sci.*, Vol. 52, pp. 2372–2377, December 2005.

24. L. Scheick, P. McNulty, D. Roth, Dosimetry Based on the Erasure of Floating Gates in the Natural Radiation Environments in Space, *IEEE Trans. Nucl. Sci.*, Vol. 45, pp. 2681–2688, December 1998.

25. N. Tarr, G. Mackay, K. Shortt, I. Thomson, A Floating Gate MOSFET Dosimeter Requiring No External Bias Supply, *IEEE Trans. Nucl. Sci.*, Vol. 45, pp. 1470–1474, June 1998.

26. T. R. Oldham, M. Friendlich, M. A. Carts, C. M. Seidleck, K. A. LaBel, Effect of Radiation Exposure on the Endurance of Commercial NAND Flash Memory, *IEEE Trans. Nucl. Sci.*, Vol. 56, No. 6, pp. 3280–3284, December 2009.

27. T. R. Oldham, D. Chen, M. Friendlich, M. A. Carts, C. M. Seidleck, K. A. LaBel, Effect of Radiation Exposure on the Retention of Commercial NAND Flash Memory, *IEEE Trans. Nucl. Sci.*, Vol. 58, No. 6, pp. 2904–2910, December 2011.

28. M. Bagatin, S. Gerardin, A. Paccagnella, A. Visconti, S. Beltrami, M. Bertuccio, L. T. Czeppel, Effect of Total Ionizing Dose on the Retention of 41 nm NAND Flash Cells, *IEEE Trans. Nucl. Sci.*, Vol. 58, No. 6, pp. 2824–2829, December 2011.

29. M. Bagatin, S. Gerardin, A. Paccagnella, G. Cellere, A. Visconti, M. Bonanomi, Increase in the Heavy–Ion Upset Cross Section of Floating Gate Cells Previously Exposed to TID, *IEEE Trans. Nucl. Sci.*, Vol. 57, pp. 3407–3413, December 2010.

30. M. Bagatin, S. Gerardin, A. Paccagnella, V. Ferlet-Cavrois, J. R. Schwank, M. R. Shaneyfelt, A. Visconti, Proton-Induced Upsets in SLC and MLC NAND Flash Memories, *IEEE Trans. Nucl. Sci.*, Vol. 60, pp. 4130–4135, December 2013.

31. S. Gerardin, M. Bagatin, A. Paccagnella, V. Ferlet-Cavrois, A. Visconti, C. Frost, Neutron and Alpha Single Event Upsets in Advanced NAND Flash Memories, *IEEE Trans. Nucl. Sci.*, Vol. 61, pp. 1799–1805, August 2014.

32. G. Cellere, A. Paccagnella, A. Visconti, M. Bonanomi, S. Beltrami, Single Event Effects in NAND Flash Memory Arrays, *IEEE Trans. Nucl. Sci.*, Vol. 53, pp. 1813–1818, August 2006.

33. S. Gerardin, M. Bagatin, A. Paccagnella, G. Cellere, A. Visconti, M. Bonanomi, A. Hjalmarsson, A. Prokofiev, Heavy-Ion Induced Threshold Voltage Tails in Floating Gate Arrays, *IEEE Trans. Nucl. Sci.*, Vol. 57, pp. 3199–3205, December 2010.

34. G. Cellere, A. Paccagnella, A. Visconti, M. Bonanomi, A. Candelori, Transient Conductive Path Induced by a Single Ion in 10 nm SiO2 Layers, *IEEE Trans. Nucl. Sci.*, Vol. 51, pp. 3304–3311, December 2004.

35. G. Cellere, A. Paccagnella, A. Visconti, M. Bonanomi, Secondary Effects of Single Ions on Floating Gate Memory Cells, *IEEE Trans. Nucl. Sci.*, Vol. 53, pp. 3291–3297, December 2006.

36. G. Cellere, A. Paccagnella, A. Visconti, M. Bonanomi, S. Beltrami, R. Harboe-Sorensen, A. Virtanen, Effect of Ion Energy on Charge Loss from Floating Gate Memories, *IEEE Trans. Nucl. Sci.*, Vol. 55, pp. 2042–2047, August 2008.

37. K. Grürmann, D. Walter, M. Herrmann, F. Gliem, H. Kettunen, V. Ferlet-Cavrois, SEU and MBU Angular Dependence of Samsung and Micron 8-Gbit SLC NAND-Flash Memories under Heavy-Ion Irradiation, *Radiation Effects Data Workshop, 2011 IEEE*, pp. 1–5.

38. K. Grürmann, D. Walter, M. Herrmann, F. Gliem, H. Kettunen, V. Ferlet-Cavrois, MBU Characterization of NAND-Flash Memories under Heavy-Ion Irradiation, *RADECS 2011 Proceedings*, pp. 207–212.

39. M. Bagatin, S. Gerardin, A. Paccagnella, V. Ferlet-Cavrois, Single and Multiple Cell Upsets in 25-nm NAND Flash Memories, *IEEE Trans. Nucl. Sci.*, Vol. 60, No. 4, pp. 2675–2681, August 2013.

40. K. Grürmann, M. Herrmann, F. Gliem, H. Schmidt, G. Leibeling, H. Kettunen, V. Ferlet-Cavrois, Heavy Ion sensitivity of 16/32-Gbit NAND-Flash and 4-Gbit DDR3 SDRAM, *Radiation Effects Data Workshop, 2012 IEEE*, pp. 114–119.

41. S. Gerardin, M. Bagatin, A. Paccagnella, A. Visconti, M. Bonanomi, S. Beltrami, Angular Dependence of Heavy-Ion Induced Errors in Floating Gate Memories, *IEEE Trans. Nucl. Sci.*, Vol. 58, pp. 2621–2627, December 2011.

42. M. Beck, Y. Puzyrev, N. Sergueev, K. Varga, R. Schrimpf, D. Fleetwood, S. Pantelides, The Role of Atomic Displacements in Ion-Induced Dielectric Breakdown, *IEEE Trans. Nucl. Sci.*, Vol. 56, pp. 3210–3217, December 2009.
43. N. Butt, M. Alam, Modeling Single Event Upsets in Floating Gate Memory Cells, *IRPS 2008*, pp. 547–555, 2008.
44. S. M. Guertin, D. M. Nguyen, J. D. Patterson, Microdose Induced Data Loss on Floating Gate Memories, *IEEE Trans. Nucl. Sci.*, Vol. 53, pp. 3518–3524, December 2006.
45. S. Gerardin, M. Bagatin, A. Cester, A. Paccagnella, B. Kaczer, Impact of Heavy-Ion Strikes on Minimum-Size MOSFETs with Ultra-Thin Gate Oxide, *IEEE Trans. Nucl. Sci.*, Vol. 53, pp. 3675–3680, December 2006.
46. H. Schmidt, D. Walter, M. Bruggemann, F. Gliem, R. Harboe-Sorensen, P. Roos, Annealing of Static Data Errors in NAND-Flash Memories, *RADECS 2007 Proceedings*, pp. 1–5, September 2007.
47. M. Bagatin, S. Gerardin, G. Cellere, A. Paccagnella, A. Visconti, S. Beltrami, M. Bonanomi, R. Harboe-Sørensen, Annealing of Heavy-Ion Induced Floating Gate Errors: LET and Feature Size Dependence, *IEEE Trans. Nucl. Sci.*, Vol. 57, pp. 1835–1841, December 2010.
48. G. Cellere, L. Larcher, A. Paccagnella, A. Visconti, M. Bonanomi, Radiation Induced Leakage Current in Floating Gate Memory Cells, *IEEE Trans. Nucl. Sci.*, Vol. 52, pp. 2144–2152, December 2005.
49. M. Bagatin, S. Gerardin, A. Paccagnella, Retention Errors in 65-nm Floating Gate Cells after Exposure to Heavy Ions, *IEEE Trans. Nucl. Sci.*, Vol. 59, pp. 2785–2790, December 2012.
50. A. Gasperin, A. Paccagnella, G. Ghidini, A. Sebastiani, Heavy Ion Irradiation Effects on Capacitors With and ONO as Dielectrics, *IEEE Trans. Nucl. Sci.*, Vol. 56, pp. 2218–2224, August 2009.
51. M. Bagatin, S. Gerardin, A. Paccagnella, A. Visconti, Impact of Technology Scaling on the Heavy-ion Upset Cross Section of Multi-Level Floating Gate Cells *IEEE Trans. Nucl. Sci.*, Vol. 58, pp. 969–974, August 2011.
52. S. Gerardin, M. Bagatin, A. Ferrario, A. Paccagnella, A. Visconti, S. Beltrami, C. Andreani, G. Gorini, C. Frost, Neutron-Induced Upsets in NAND Floating Gate Memories, *IEEE Trans. Device Mater. Rel.*, Vol. 12, pp. 437–444, June 2012.
53. M. Bagatin, S. Gerardin, Soft Errors in Floating Gate Memory Cells: A Review, *Microelectron. Reliab.*, Vol. 55, pp. 24–30, 2015.
54. M. Bagatin, G. Cellere, S. Gerardin, A. Paccagnella, A. Visconti, S. Beltrami, TID Sensitivity of NAND Flash Memory Building Blocks, *IEEE Trans. Nucl. Sci.*, Vol. 56, pp. 1909–1913, August 2009.
55. D. Nguyen, L. Scheick, TID, SEE and Radiation Induced Failures in Advanced Flash Memories, *Radiation Effects Data Workshop, 2003 IEEE*, pp. 18–23, July 2003.
56. D. Nguyen, S. Guertin, G. Swift, A. Johnston, Radiation Effects on Advanced Flash Memories, *IEEE Trans. Nucl. Sci.*, Vol. 46, pp. 1744–1750, December 1999.
57. T. Langley, P. Murray, SEE and TID Test Results of 1 Gb Flash Memories, *Radiation Effects Data Workshop, 2004 IEEE*, pp. 58–61, July 2004.
58. D. Nguyen, C. Lee, A. Johnston, Total Ionizing Dose Effects on Flash Memories, *Radiation Effects Data Workshop, 1998. IEEE*, pp. 100–103, July 1998.
59. H. Schmidt, D. Walter, F. Gliem, B. Nickson, R. Harboe-Sorensen, A. Virtanen, TID and SEE Tests of an Advanced 8 Gbit NAND-Flash Memory, *Radiation Effects Data Workshop, 2008 IEEE*, pp. 38–41, July 2008.
60. F. Irom, D. N. Nguyen, G. Cellere, M. Bagatin, S. Gerardin, A. Paccagnella, Catastrophic Failure in Highly Scaled Commercial NAND Flash Memories, *IEEE Trans. Nucl. Sci.*, Vol. 57, No. 1, pp. 266–271, February 2010.

61. M. Bagatin, S. Gerardin, A. Paccagnella, G. Cellere, F. Irom, D. N. Nguyen, Destructive Events in NAND Flash Memories Irradiated with Heavy Ions, *Microelectron. Reliab.*, Vol. 50, Nos. 9–11, pp. 1832–1836, 2010.
62. T. R. Oldham. M. Berg, M. Friendlich et al., Investigation of Current Spike Phenomena During Heavy Ion Irradiation of NAND Flash Memories, *Radiation Effects Data Workshop, 2011 IEEE*, pp. 152–160, July 2011.

6 Microprocessor Radiation Effects

Steven M. Guertin and Lawrence T. Clark

CONTENTS

6.1   INTRODUCTION

Microprocessors are critical components in computing systems. In this chapter, we discuss radiation effects in microprocessors. Reduced to their constituent circuits, microprocessors are finite state machines comprised of logic gates, state elements such as flip-flops or latches, and basic memory elements such as static random access memory (SRAM) cells. In practice, the combination of these elements can create an extremely complicated device. This is especially so when viewed by their behavior after a multitude of possible soft errors in the machine.

A standard microprocessor block diagram is shown in Figure 6.1. This is a simple five-pipeline stage, in-order, scalar machine. However, many modern microprocessors are deeply pipelined (i.e., eight or more stages) superscalar devices with out-of-order instruction execution with additional complications such as register renaming. Many modern integrated circuits (ICs) also increase in complexity by integrating multiple microprocessors on a single IC and include memory and high-speed IO interfaces, making a system-on-a-chip (SOC) device. Nevertheless, many microprocessor effects can be understood in terms of the basic structure of Figure 6.1.

6.1.1   Fundamental Soft-Error Mechanisms and Circuits

The primary soft-error effects in simple structures are illustrated in Figure 6.2. A single-event upset (SEU) is due to ionization (as a result of the passage of an ion) near a storage (latch) node. Referring to Figure 6.2a when node A collects charge, its voltage is driven low. The latch feedback in turn causes node AN to transition and the new value is stored. A single-event transient (SET) is a finite-duration voltage upset of the logic state due to ionizing radiation. In the SET the voltage change is transient since the input of the gate is static and unaffected as there is no feedback in the circuit. Figure 6.2b shows a negative SET at node E. After some finite time, node E transitions back to its initial state. As shown, SETs may propagate through logic fed by the generating gate, potentially fanning out to affect many gates. However, they are important only when they propagate to, and are captured by, a state element such as a latch or flip-flop circuit where they affect the machine state.

Essentially all of the subsystems in a microprocessor are constructed out of a relatively small set of low-level circuits [1] that represent good points for the presentation and exploration of radiation effects. We present the specific circuits here and expand on their radiation effects in the next section. Pipeline data and most configuration information is stored in flip-flops (FFs) [2] while latches may be used for some information. Referring to Figure 6.3, latches are the primary nonmemory array storage since flip-flops are usually comprised of two back-to-back latches, providing clock-edge-triggered operation. The feedback-configured inverters in the latches make them primarily sensitive to SEU. When opaque, any latch will upset if the radiation deposited charge exceeds the cell Q_{CRIT} [3]. If erroneous data due to a SET in the logic path (not shown) feeding the D input in Figure 6.3 reaches the storage nodes H and S and meets the setup and hold timings as the clock closes the latch (i.e., turns off the complementary metal-oxide semiconductor [CMOS] pass gate controlled by EN) then the incorrect state is captured. However, the timing window where the

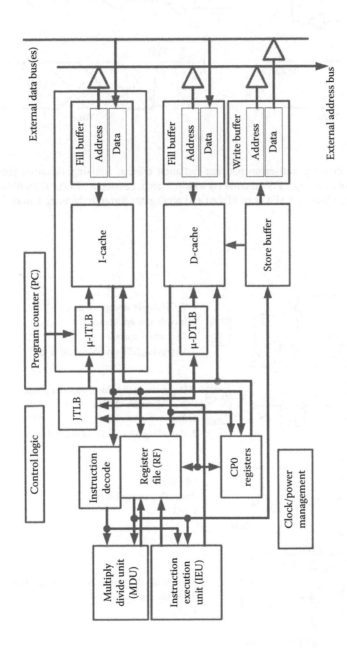

FIGURE 6.1 Typical microprocessor general structure.

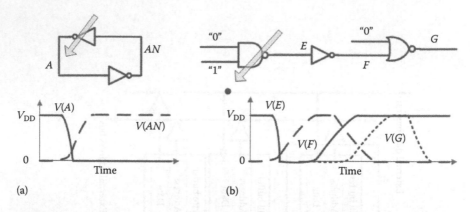

FIGURE 6.2 Primary soft-error-inducing radiation effects. Ionizing radiation produces charge that is collected by a PN junction (a) within a latch, causing a SEU or (b) within combinational logic, whereby a SET is produced and propagates through the logic fanout.

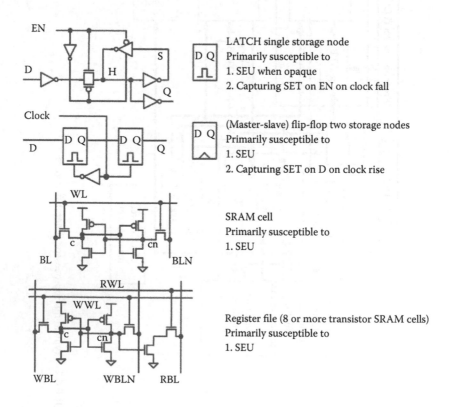

LATCH single storage node
Primarily susceptible to
1. SEU when opaque
2. Capturing SET on EN on clock fall

(Master-slave) flip-flop two storage nodes
Primarily susceptible to
1. SEU
2. Capturing SET on D on clock rise

SRAM cell
Primarily susceptible to
1. SEU

Register file (8 or more transistor SRAM cells)
Primarily susceptible to
1. SEU

FIGURE 6.3 Typical microprocessor memory circuits (state elements).

data upset must occur naturally makes the cross section of such single-event effects (SEEs) smaller. Additionally, if the clock is inadvertently activated due to a SET, the incorrect state may be captured. Caches are generally implemented using SRAM [4]. These cells use six transistors, allowing very high density. Some SOCs include noncache-embedded SRAM arrays or cache configurable as SRAM for application use [5]. The standard two-port register file (RF) cell is shown at the bottom of Figure 6.3. It is used in RFs, translation lookaside buffers (TLBs), and other small memory arrays. Note that extra ports can be added, which is key to allowing multiple accesses in superscalar designs. Some more modern microprocessors have used RF cells in the caches [6] for speed, or more recently, for low-voltage operation.

6.1.2 CHAPTER OUTLINE AND ORGANIZATION

This chapter discusses the general impact of radiation on microprocessors focused on soft errors. The examination begins with the structure of microprocessors and evaluates the most common responses to radiation. The SEEs we focus on are soft errors, but due to their nature as electronic components, other SEE types discussed in this book may apply to microprocessors as well.

Some detail about microprocessor architecture is necessary to guide the discussion, expanding on the brief introduction already presented. With this information it will be possible to explain how the rather simple SEE radiation effects can result in very complex device response. The various levels of abstraction in microprocessor structure discussed here are the following. Standard or simple circuits include SRAM cells, FFs, and logic gates. These are combined to create blocks such as SRAM arrays and register files. These blocks, with added logic and pipeline registers, combine to create subsystems such as an arithmetic logic unit (ALU), multiplier, instruction and data translation lookaside buffer (ITLB and DTLB respectively), queues, cache memory, and the instruction pipeline. Finally, subsystems combine to form system-level structures such as central processing unit (CPU) cores, cache systems, or memory management units (MMUs).

The Highly Efficient Radiation Hardened by Design (RHBD) Microprocessor for Enabling Spacecraft (HERMES) under development at Arizona State University (ASU) is a microprocessor whose architecture has been modified to enable correction of essentially all forms of SEE [7]. It thus provides a simple example design for discussing microprocessor errors that can be caused by SEE. HERMES uses the basic structure in Figure 6.1 but adds extensive redundancy. Using dual modular redundancy (DMR) execution pipelines, HERMES compares the data in one (upset) DMR copy with the other (correct) copy primarily at write-back to prevent committing incorrect speculative state to architectural state. The checking response and recovery to correct state may be disabled. Thus, by observing HERMES' erroneous pipeline state with an injected error, its impact can be understood as it propagates through the pipeline.

Radiation effects on the various structures that make up a microprocessor can be most easily understood at the individual circuit/structure-level element and subsystem levels, and this chapter will work through those levels to build up through subsystem responses, highlighting how individual SEEs manifest system errors.

The impact of radiation effects seen in device testing is the system response. This response often obscures the root causes because the actual SEE is occurring in a simple structure that is several levels of abstraction away from the system response; the SEE then causes incorrect operation of a building block that in turn results in an incorrect subsystem operation, and finally a subsystem error is manifest as a system-level error. Although the topic is quite broad, a fair amount can be understood through a detailed examination of the most important radiation effects and their most common structure-level and subsystem-level impacts to a running microprocessor. We briefly discuss the full system-level impact later in the chapter, but it is largely outside of the primary goals of this chapter because it treats the microprocessor as a black box and depends on the user application.

This chapter is organized as follows. First, we present microprocessor structure, focused on examining the fundamental radiation-sensitive circuits and building up to modern microprocessor structure. We then detail the ways that SEEs affect the elements that make up a microprocessor. Next, we go into depth on SEEs on microprocessors, looking at many of the relevant topics in SEE. Finally, we present a discussion of some RHBD microprocessor considerations. These cover issues encountered in testing, quantifying difficulties in evaluating communication upset rates, a brief exploration of software fault tolerance, and some discussion of how we can develop system response from understanding of low-level structure data.

6.2 MICROPROCESSOR STRUCTURE

A microprocessor is fundamentally just a state machine that executes instructions. The instructions can be very complex such as in a complex instruction set computer (CISC) (e.g., ×86 instructions, including looping modifiers). However, in modern microprocessors, even these tend to be very simple, as they are decoded into reduced instruction set computer (RISC)-like micro-operations (micro-ops) (i.e., simple register-to-register and load/store operations) before being executed. The instructions control the microprocessor to set its internal logic state configuration. The portion of the state configuration that is used in subsequent instructions, as well as the state of IO connections to other devices, is known as its architectural state [8,9]. There is considerable stored state in the microprocessor that may or may not commit to architectural state. This is speculative state. Dismissing interactions between programs, the operation is normally completely deterministic.

When a radiation effect is manifest on a microprocessor, the system becomes nondeterministic. Sometimes this loss of determinism is benign. For example, it may perform exactly the right operation but may pick up a timing delay, increasing execution latency, or execute otherwise unneeded operations, which have no impact on the architectural state. At other times, a soft error produces incorrect architectural state. If this incorrect state is detected, it may be corrected; otherwise, the program operation may be incorrect. The latter is referred to as silent data corruption (SDC). This may cause slightly erroneous calculations at the least, or complete failure of the device. Almost any response between these extremes is possible. These behaviors can also be temporary, intermittent, or permanent. In order to understand this wide range of responses it is necessary to look deeper.

6.2.1 Pipelining and Speculative and Architectural State

In Figure 6.4 we show a basic microprocessor pipeline. Any particular microprocessor may vary considerably from what is shown, but the general ideas in the context here can be used to understand the implications for the majority of specific cases. The instruction fetch (I) stage contains the instruction fill buffer, ITLB, and cache, as well as the program counter (PC) registers and logic. Assuming a cache hit, instructions are fetched from the cache and delivered to the execution (E) pipeline stage. Transparent latches drive the register file decoders, labeled S0 and S1 decode. The register file starts operation in the I stage and completes it with delivery of the register contents chosen by the instructions. In the E stage, instructions are fully decoded to control the ALU and subsequent operations, including destination register decoding information. Multiply and divide instructions, which can take multiple cycles, begin operation in this stage. This decoded control state is carried through the pipeline. Load and store operations access the data cache in the memory (M) stage. The DTLB, fill buffer, and data cache are augmented with a store buffer. The latter is essential since a write may be to a location that is not in the cache. Otherwise, these circuits are very similar to those in the I-cache, and in many designs they are identical. The A stage provides time to align data returning from the data cache for subsequent write-back to the register file in the W stage.

One of the first operations performed by any board support package (BSP) boot loader is to set up the configuration of the device to work properly on its board. Further changes of configuration are also possible during later execution. In the

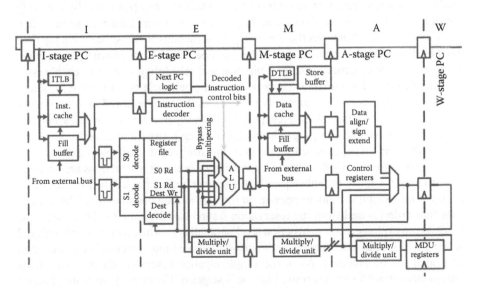

FIGURE 6.4 A microprocessor pipeline enables continuous use of each stage of execution logic. Modern microprocessors also have multiple pipelines with a scheduler. The destination write register (chosen by the Dest decode) is accessed through the destination port (Dest Wr) at the write back (W) stage.

MIPS architecture this storage is in the CP0 and other configuration registers (see Figure 6.1) [10]. A radiation event in the configuration registers can easily result in a device that no longer operates correctly, since its basic operating mode has been altered. One example is the registers that control the SRAM redundancy configuration. These bits control which portions of the memory arrays are bypassed since they are defective. A soft error that dynamically changes the redundancy control bits at minimum changes the memory configuration and may expose defective circuits to the machine. Even commercial microprocessors typically harden some of the registers controlling such key functions [11]. The connection to the outside circuitry is provided through input/output (IO) connections. Radiation events here can result in erroneous signals to and from the microprocessor.

Each clock cycle in the pipeline requires pipeline registers that store partial completion of an instruction. Referring to Figure 6.4, this includes the pipeline registers at the boundary of each of the I, E, M, A, and W stages. Even in normal (no soft-error) operation, the data in these registers may or may not be correct. This is speculative state, which commits to architectural state when the instruction is retired [12], in Figure 6.4, and writes to the register file, data cache, or external memory (via the write buffer). Not all speculative state commits to become architectural state. As a simple example, the microprocessor predicts a jump as either taken or not taken, but that prediction may not be the case. If the prediction is incorrect as determined by actually finishing execution of the conditional instruction, then the subsequent speculative pipeline state (in the earlier stages) is discarded (flushed) and execution resumes from the correct instruction. The speculative state that does not commit to architectural state but has been affected by a SEE constitutes benign error cases [9]. Branch prediction is a key feature in microprocessors that have deep pipelines. By correctly predicting the branch most of the time, the pipeline can be kept busier. However, when a branch is predicted incorrectly, the in-flight instructions in the speculative pipeline must be discarded. Additionally, instead of just operating on one instruction at a time, many modern microprocessors use a pipeline and scheduling system to enable issuing and tracking of multiple instructions on each clock cycle (see Figure 6.5). These superscalar systems can be extremely complex as they are required to issue four or more instructions on each clock cycle [13]. Initially, superscalar microprocessors executed instructions solely in program order, executing instructions concurrently only when there were no dependencies. However, this limited the available parallelism.

In the out-of-order (OOO) ×86 microprocessors, complex instructions are decoded into streams of micro-ops. Subsequently, the reorder buffers track dependencies while enabling the execution units to operate on the data as data becomes available (i.e., in data flow order) issued from the reservation station (see Figure 6.6). This allows the hardware to exploit greater instruction level parallelism—while a micro-op is delayed waiting for its input data, subsequent micro-ops whose input data has been calculated or loaded can be executed. As in the simple pipeline described above, instructions are speculative until they are retired (in the R stages of Figure 6.6). To ensure precise exceptions, instructions retire in order—exceptions are never issued on an instruction while it is speculative, since in the architecturally correct version, it may not have occurred. The original Pentium Pro (P6) microarchitecture featured a 13-stage pipeline. The P6 could decode and retire three instructions per cycle. The Pentium-4 (Intel

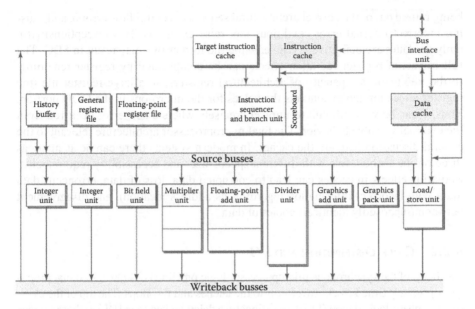

FIGURE 6.5 Microprocessor architecture with superscalar execution (Motorola 88110). (After J. Shen and M. Lipasti, *Modern Processor Design: Fundamentals of Superscalar Processors*, McGraw-Hill, New York, 2005.)

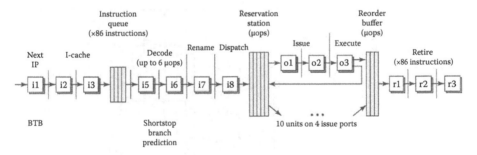

FIGURE 6.6 Intel Pentium Pro pipeline. The o1 through o3 stages are the out-of-order engine. Instructions may reside there for many more than three clocks. (After J. Shen and M. Lipasti, *Modern Processor Design: Fundamentals of Superscalar Processors*, McGraw-Hill, New York, 2005.)

Netburst architecture) replaced the L1 instruction cache with a trace cache that stores decoded micro-ops. The original version had 20 pipeline stages [14]. At the Prescott iteration, the pipeline had expanded to 31 stages [15]. The cache holds streams (traces) of decoded micro-ops in the predicted path of program execution. The instructions are stored in the trace cache in the predicted branch order.

The core has internal registers that hold critical state information that can be upset by radiation. The pipeline also issues control signals to specific execution units in the core, such as the ALUs, branch handling units, floating point units (FPUs), and any other resources provided. Between the instruction pipelines and execution units, radiation effects can result in errors in executed instructions (data errors), instructions being retired out of the correct architectural sequence (control flow errors), and causing incorrect internal error conditions that raise exception flags (exceptions) [16]. Only the latter are definitively visible, while the former two may result in SDC. The full impact of register radiation effects can be complicated by register renaming. In the ×86 there is a paucity of architectural registers, so a large register file (the register allocation table) contains the values for the micro-ops in the OOO portion. Thus, it can be very difficult to unravel precisely where a soft error occurred from the externally visible behavior. The final microprocessor architecture relevant to the general discussion here are the caches. In modern systems there can be as many as three on-die cache levels with different speed, size, and accessibility requirements. Radiation effects in caches can lead to corrupted data, loss of data, erroneous data (data with incorrect information regarding its address location), and disagreement between (supposedly identical) copies of data.

6.2.2 Clock Distribution and I/O

The heart of the synchronous microprocessor logic pipeline is its clock tree that propagates precisely timed clock waveforms to the latches and flip-flops. The top of the clock tree is a phase-locked loop (PLL) (occasionally a delay-locked loop [DLL]) that provides a frequency locked to the external reference clock (RefClk in Figure 6.7) and skew to the external circuitry [1]. The clock tree allows the drive of the PLL to be increased

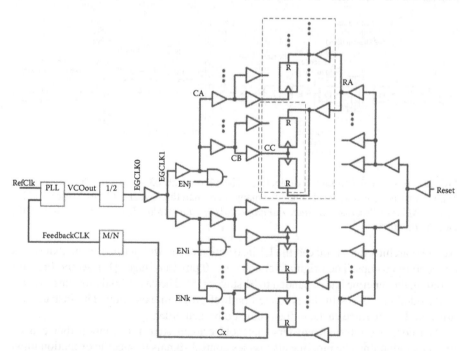

FIGURE 6.7 Processor clock (left) and reset trees (right). The structures are similar due to the need for very high fanout but the timing constraints on the reset tree are less stringent. SEUs or SETs can cause clock gate enables (ENx) to be misasserted, also causing clock misassertions.

as it fans out in layers. Its latency may approach the clock period in high-performance designs. The upper topologies are frequently gridded instead of trees, while the stages closer to the flip-flops are usually trees to allow clock gating. The final clock stages typically drive about five to 50 flip-flops. Clock gating (not shown) allows clocks driving unused logic to be left unasserted to save power. A SEE in the logic controlling these gated clocks can cause a misassertion, propagating to a large number of state elements, which then capture their state at the wrong time. It is highly likely that these FFs have inputs in unintended states when this occurs, resulting in an error. The skew in the clock arrival time varies from as little as +/– 15 to as much as +/– 50 ps for embedded designs. The clock signal fans out to many individual clock signals driven by a buffer tree to have large final drive strength. Because it connects to hundreds or thousands of flip-flops, a SET on the tree may affect many but not all of the state elements at the leaves of the tree.

The reset tree has less, but still stringent skew requirements, as the reset signal must deassert to all flip-flops in the same clock cycle. As shown in Figure 6.7, it fans out from one to many reset signals similar to the clock tree. Analogously, a hit in the tree can affect many state elements depending on the point in the tree where the SET occurs and the SET duration, which depends on the drive strength at that node.

The entire microprocessor is connected to the outside world through IO drivers. These provide level shifting between the core voltage and the often higher I/O voltages, as well as buffering to provide the high drive strength needed to drive high board-level capacitances. The IO circuits also include electrostatic discharge (ESD) protection features and often configuration information to sculpt the output waveforms (buffer strength and slew rates).

6.2.3 SOC CIRCUITS

Before moving on, it is relevant to note how the new structures in modern devices are made of the same low-level structures as general microprocessors. The general discussion areas break down roughly as SOC devices and multicore devices. SOCs are more general as they can include systems with multiple or many microprocessors. A good example of a SOC with multiple cores is the Qualcomm Snapdragon microprocessor [17]. The Snapdragon has up to four ARM-based Krait cores, each similar to the ARM Cortex-A15. It also includes integrated multimedia and graphics cores and supports wireless communications and other peripheral services directly on the chip. The predominant digital portions (e.g., the modem DSP and graphics engines) are also composed of the logic gates and data storage seen in Figures 6.2 and 6.3. Other services, such as on-chip accelerometers, radio frequency devices, and analog or mixed-signal portions of the device may have significantly different radiation responses. SOC device-type specific radiation effects include upset of their data transfer to the microprocessor via direct memory access (DMA). These data transfer radiation effects are also covered by the types of radiation effects we are considering here.

6.3 GENERAL RADIATION EFFECTS IN MICROPROCESSORS

Given the subsystem architecture of a microprocessor and the low-level structures that are used to build them, as presented above, we are now in position to discuss radiation effects and how they manifest in the low-level structures. In the next section we will follow up on this discussion by looking at how the SEEs produce microprocessor errors.

Total ionizing dose (TID) is the alteration of circuit elements by the biasing of oxides due to radiation-deposited fixed charge. The alteration can be permanent or semipermanent, where device parameters change but may recover somewhat due to annealing processes. The details of TID are discussed in Chapter 1. For microprocessors, the most important TID effects are changes in leakage current and modification of transistor slew rates potentially manifesting as increased gate delay [18–20]. Thus, TID effects generally result in a loss of device operating margin. Consequently, it is recommended that margin is added to the mission environment specification in order to ensure circuits will work after TID degradation.

Single-event latch-up (SEL) occurs when a silicon-controlled rectifier (SCR), or PNPN device, is turned on by the charge collected from an ion track [19]. Such structures are inherently present in CMOS due to the parasitic NPN-PNP bipolar transistors intrinsic to bulk CMOS. SEL thus results in a high-current power-to-ground connection, which can lead to an overcurrent condition locally or in a feeder circuit. Typically, a resistive load somewhere in series with the SEL dissipates significant heat and may be permanently damaged via thermal stress. SEL is also problematic because it may require special consideration during SEE evaluation of a microprocessor in order to ensure SEL sensitivity is properly captured [21,22]. Devices fabricated on silicon-on-insulator (SOI) lack the PNPN series devices and thus have intrinsic resistance to SEL [23].

The microprocessor state is a combination of long-term data and currently used sequential state in latches and/or flip-flops as well as memories that determine the program execution results, and all are susceptible to SEU. The basic latch charge collection and upset mechanism was recalled in Section 6.1.1 and in detail in Section 1.5.1. In a conventional (unhardened) microprocessor, all of the FFs and many of the memory arrays are not protected or checked for soft errors. Regarding the latter, the Pentium II has 39 separate memory structures, including write buffers, TLBs, queues, and register files [24]. Caches are usually constructed using SRAM (e.g., [4]). Small memories are generally made of register file (see Figure 6.3) or FF cells and sometimes latches. Frequently their SEE performance is similar to that of cache bits, but since the storage node capacitance is greater, the SEE sensitivity may be reduced.

SETs can affect microprocessors in any combinational logic path. However, SETs in the clock and reset trees are among the easiest to diagnose. The clock tree is the primary means of data synchronization. The strict skew requirements make the clock tree susceptible to corruption due to soft-error induced jitter [25–27]. Any clock tree SET may result in unpredictable behavior by adding a clock edge at the wrong time. Additionally, SETs in reset signals (global or otherwise) can result in partial reset affecting portions of the machine. The high fanout means that a SET at any point in the clock or reset tree may affect the state of many flip-flops or latches. Referring to Figure 6.7, a SET at node CC will inadvertently clock flip-flops in its fanout (indicated by the small box). A SET on reset node RA will reset all of the flip-flops in the larger box. A clock SET at a node high in the tree (e.g., CA) may affect a significant portion of the machine state, as is evident. Experiments with fully hardened (DICE) FFs have shown vulnerabilities to such clock and reset SETs [28]. Finally, the PLL and clock dividers, while low cross section, if corrupted, affect the entire design. This is particularly important in broad-beam testing where high flux can cause loss of phase lock. Since clock gating and reset signals are controlled by bits in the microprocessor state, SEUs can also trigger a partial reset or cause erroneous clock edges. One specific example of a partial reset behavior has been observed in [29] where 16-bit portions of the register data were found to be set to all zeros under SET events. SETs can cause direct responses in asynchronous portions of a device. To that end, SETs can raise machine conditions that erroneously trigger interrupts or exceptions.

6.4 SEEs IN MICROPROCESSORS

In this section we discuss the specific responses of microprocessors to SEEs by covering the following specific topics. Cache SEE response is covered in detail. We present information regarding register upsets and pipeline or execution events. We also discuss environmental difficulties regarding SEE mechanisms and the impact on testing, focusing on operating frequency and temperature. We begin this section by discussing cache soft errors, since cache size and capacity naturally gives them a large overall error cross section.

Essentially, the program flow can be thought of as a directed graph where instructions are nodes and the graph edges describe the program flow [30,31]. Control flow errors (i.e., changes in the program flow) violate the graph describing correct operation. These include changes in branch destinations where a branch is supposed to exist, but also branch insertions. A branch insertion can be due to the PC being corrupted or calculated incorrectly. Branch target buffer (BTB) destination errors are another root cause of such a control flow error. The data portion of the BTB, which is a cache for jump destinations, contains the destination addresses. A SEU can change this value. Additionally, data errors can cause a branch error. A simple example of this is an upset of a loop counter causing a loop to continue when it should have stopped or to end prematurely. Oh et al. [32] investigated a MIPS 4400 scalar pipeline with control flow errors injected for a variety of programs. The error behavior varied widely, but SDC results ranged from 1.8% for a Tower of Hanoi solver to 55% for a fast Fourier transform (FFT) calculation. The control flow error was caught by the operating system (OS) 55% of the time for both programs, presumably due to protection violations by the destination addresses. Correct program operation resulted only 10% to 24% of the time across numerous programs.

This is in contrast to the results of (complete, not just control flow) error injection campaigns on a superscalar OOO pipeline [16] where the faults were benign 72% to 98% of the time depending on the number of valid in-flight instructions, which ranged from zero to 100 in the tests. The large amount of speculative state in the deep OOO pipeline results in large portions of the queues and register files containing dead state (i.e., state that will not be subsequently used by the running programs).

6.4.1 SEEs IN CACHES

SEEs in caches can manifest in many ways. How the caches respond to the events depends on how the caches are constructed and where the SEE occurs. In the previous section we specifically discussed SRAM single-bit upset (SBU); however, there are ways in which caches are sensitive to SET. In addition, cache SEEs are dependent on the tag architecture. An example two-way set associative cache architecture is shown in Figure 6.8. At its most fundamental level, this cache is comprised of two memory arrays, the tag and data arrays, respectively, with added logic for address comparison and data steering. Microprocessor caches store data in lines that contain between 16 and 64 bytes. Each line can contain a section of the main memory [33]. The primary tag purpose is to store the physical memory address of the line, while the data portion stores the corresponding data. The cache tag also includes which lines are valid (indicated by the valid bit) and their state (indicated by the dirty bit) among other architecture-specific information, such as whether it is locked (i.e., cannot be replaced without OS intervention). Cache memory is further subdivided into ways according to the associativity. Each way can be accessed simultaneously, with the ways of a single index referred to as a set. The example of Figure 6.8 has a set associativity of two, since two ways are accessed at a time. Set associativity provides added flexibility in the placement of data in the cache, avoiding lines that have the same set address from replacing each other.

Referring to Figure 6.8, when the cache is accessed the set is decoded and one entry per way is read out of each of the tag and data arrays. The portion of the address that is not decoded to determine the set of each of these entries is compared to the address requested. If one matches, the result is a cache hit, and the match selects the correct word from the cache line via the way multiplexer. In the event of a miss, the data is not resident in the cache and must be fetched from main memory or higher-level cache. The design in Figure 6.8 is a late way select architecture. It is fast, since the tag and data are accessed simultaneously, but dissipates substantially more

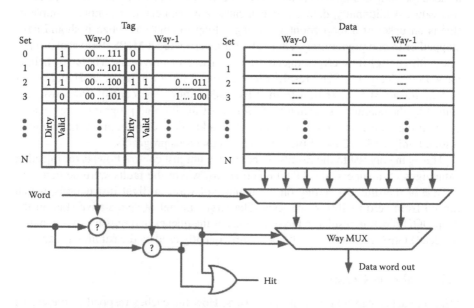

FIGURE 6.8 Typical microprocessor cache structure.

power since the data array ways that are not selected need not have been accessed. Nonetheless, this is the most commonly used first-level cache organization due to its low latency [4]. Level 2 (and higher-level caches) generally look up the tag first and then only access the data array that is selected. This saves considerable power, particularly since higher-level caches also require higher associativity. The key point is that except for the comparators and some control circuitry, the cache is physically comprised of multiple SRAMs—smaller-tag SRAMs where each word is the width of the tag portion of the address, usually on the order of 20 bits, and large data SRAMs, with widths equal to the line size.

Most modern microprocessors use a write-back policy, where the caches are the primary data storage locations. With a write-back policy, modified data is only written back to main memory when it is evicted from the cache. This reduces the memory bus traffic as compared to a write-through policy, where the main memory is written in parallel with the cache [33]. From a soft-error perspective the latter is ideal—there is a redundant copy of the data that is stored in the cache. In the event an error is detected in the cache, it can simply be invalidated and the natural cache operation will refetch the data as the lines are needed. Write-through-only caches also thus lack a dirty bit because data never differs from that in memory (except while the write to the bus is pending). In the write-back case, there is no up-to-date copy of the data stored in the main memory. Generally, a write-back cache can support either policy. Consequently, write-through is often selected by the OS in systems where reliability is critical, at the expense of some performance.

Bit upsets in the caches have been observed to be dependent on two mechanisms. The first is the fundamental sensitivity of the SRAM cells (i.e., SEU). Examples of this are provided in [34–37]. In these tests, the cache is essentially used as SRAM. It is first prepared with a known data pattern, then it is disabled, and SEUs are allowed to accumulate. The state of the cache is checked after exposure for bit errors, which in such a test may accumulate from many separate particle strikes. This mechanism is in contrast to the active operating mode, where the SRAM array may be more or less sensitive to SEU than when it is actively exercised, such as in a real application or operational cache. It is important to note that this type of sensitivity cannot be determined with debugging tools such as cache readers because they do not operate the cache at normal application speed. Caches are constructed from many arrays (hundreds in the case of large L3 caches). To save power, all arrays are inactive when not being accessed in a given clock cycle. The distinction between different cache operating modes is becoming less important as most microprocessors manufacturers are increasingly interested in reducing power consumption, and leaving unused cache lines in a low-power state is increasingly common [38]. Inactive arrays have reduced sensitivity to some sources of SEU, such as SET. However, if the low-power state includes reduced supply voltages, the SEU rate may increase due to lower Q_{CRIT}. More power can be saved by flushing the cache and shutting down (gating) its power, but from a soft-error perspective in this case the cache does not exist.

When incorrect data is fetched from a cache there are three options for how the system handles it. First, it can pass the incorrect data to the user application with no indication of error. Second, the cache can indicate that incorrect data has been encountered. Third, it can attempt to correct it (discussed later in this section).

Which of these possibilities is used is a function of the microarchitecture and the software? While first-level caches have until recently often been protected only by parity, larger, higher-level caches are usually error-correcting code (ECC)-protected. ECC is also commonly referred to as error detection and correction (EDAC). The second case above (error reporting) is the only option with parity. EDAC allows error correction and generally provides reporting in cases where correction fails.

Each cache line's control bits can contribute to another class of errors. The control bits for cache lines indicate specific information (e.g., dirty or clean state) and a bit flip can result in this information being altered. Changes in address can result in incorrect cache misses and loss of dirty data (i.e., data that is altered but not yet written to external memory) or the dirty data may be flushed from the cache to the wrong location. For example, referring to Figure 6.8, a SEU that alters the valid bit in set 1, way 0 from 1 to 0 is benign—the data in way 0 is not dirty and thus will be refetched when the next access misses because the valid bit is a 0. Other errors can be more subtle (and virtually impossible to test for, so they are not usually reported), such as changing the authorization level or lock status of a cache line. The most commonly reported SEE type is loss of a cache line, as above, which may indicate the address changed, the valid bit changed, or for some other reason the cache line had to be refetched. However, the results may not be benign. This type of SEE is discussed in [39], where loss of data in cache lines is observed by putting the microprocessor into a state where main memory has a recognizably different data pattern than what is in the cache. These events are SEE-induced cache-line invalidations. An example is changing the valid bit in set 2, way 0 from 1 to 0. In this case, the associated data is dirty (as indicated by a dirty bit of one). Consequently, the data has been altered in the cache but will not be transferred back to the main memory. When this cache line is stranded due to the valid bit flip, the updated data is lost. Consequently, the microprocessor architectural state is disrupted—a subsequent access to that memory location will return the preupdated data.

Finally, there is the possibility of cache control errors, particularly as SETs become a more prevalent contributor to the SEEs. If the wrong location is written, it may get the correct parity or EDAC, and a SDC can occur. SETs in SRAM decoders and control logic were observed by McDonald et al. to cause upsets that could not be mitigated using EDAC [40]. These errors affected the SRAM addressing and control circuits, causing inadvertent writes, or worse, writes of data with correct ECC but to the wrong locations. Since such control errors can allow the correct ECC, they may cause SDC. Mavis et. al. reported local word-line (WL) misassertions in modern RHBD SRAMs with hierarchical WLs [41]. Dynamic errors have also been observed in hardened by process SRAMs [42]. In these designs, the resistor-hardened SRAM cells exhibited no static errors, but did exhibit dynamic (operating) errors at low linear energy transfer (LET). In this case, the errors appeared to affect individual bit lines during writes, rather than control circuits, as entire words were not erroneous.

Other SET-induced errors can include writes to multiple cells or writes from read-out cells to others. The latter can occur in the case where a WL glitches on after the bit lines (BLs) are fully driven but before BL precharging occurs. A SET may affect the set chosen in the instruction cache. This is likely to cause a cache miss, but a subsequent fill operation will access a line that is already in the cache since the PC was not affected, only the logic it controls. The resulting multiple cache entry type of error is rarely comprehended in the design. As mentioned above, a SEU in the I-cache tag or data arrays will result in a miss or incorrect instruction being fetched, respectively. In most microprocessors, the latter will cause a parity exception, whereupon the line can be refetched so execution can continue—recall that I-cache lines are never dirty and so may be invalidated on an error. The TLBs are prone to the same types of errors but can also remap the address during physical translation.

6.4.1.1 Cache Hardening and Errors Seen on Silicon

The HERMES design adds myriad error detectors to the cache [36]. The error detection circuits can detect SRAM periphery as well as input errors to the cache. The latter include SET-induced short clocks or nonmatching DMR addresses. Broad-beam experiments on this design have included both the static and dynamic operational modes described above. As mentioned above, when tested in the static fashion, the cache blocks are basically SRAMs, and multiple ionizing particle hits can build up. The dynamic testing is more interesting. Here reads and writes occur to the caches and the error detection mechanisms are activated. When an error is detected, the entire cache is read out to observe any latent errors. Since such latent errors could be due to multiple particle strikes, there could be more than one per word. As is common practice, in HERMES the columns are interleaved (by four cells in the tag arrays and eight cells in the data arrays) to separate bits used in the same parity calculation. Multiple strikes hitting the same row could cause two independent bit upsets, thwarting the parity. HERMES caches also have dual-redundant valid bits—a mismatch between them constitutes an error. Note however, that normal cache operations will replace lines when their storage space is needed by other lines. This automatically scrubs (i.e., flushes) out some upsets.

In broad-beam heavy-ion testing the HERMES caches never had an entry with correct parity but incorrect data, indicating the bit spacing is sufficient. The tag response was dominated by incorrect hit versus miss responses, as expected. Most often, when a tag bit is upset, the resulting address will be outside the program space. A subsequent access will indicate a miss. Referring to Figure 6.9, the tag storage holding address 00 ... 010 in set two is upset to 00 ... 011. Subsequent accesses to the original address will miss. It is possible that the soft-error-induced address is in the program space (e.g., when a least significant bit is struck). The access then produces a tag parity error but not a miss because in HERMES tag parity is only checked on a hit. Again, this is common practice—the added routing and the power such routing adds makes checking parity on the missing ways expensive. HERMES uses dual redundant tag comparators, each comparing one of the valid bits (to 1). Consequently, mismatching hit/miss responses can also be produced when the valid bits mismatch.

The HERMES cache peripheral error detection provides some insights into the relative cross section of such errors versus memory array bit upsets. In general, periphery upsets are much rarer, as shown in Figure 6.10. At LET = 20 MeV-cm²/mg, incorrect WL assertions are about three times more likely than a tag compare mismatch, which in turn is about three times more likely than a write-enable error. Word-line misassertions have a cross section about one-tenth of the SRAM arrays. This is partly due to

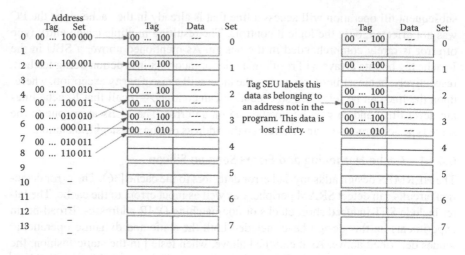

FIGURE 6.9 SBU in a cache tag can result in moving cache data to an incorrect address.

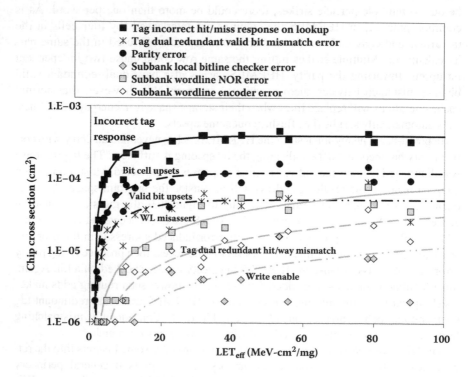

FIGURE 6.10 Hermes cacheF testchip heavy-ion broad-beam cross sections. The array SEUs dominate, but control logic, tag address compare, and WL misassertion errors do occur. (After X. Yao, D. Patterson, K. Holbert, and L. Clark, *IEEE Trans. Nucl. Science*, Vol. 57, No. 4, pp. 2089–2097, August 2010.)

the small tag array size, which makes the physical area of the periphery circuits substantial compared to the storage area. At high LET (e.g., greater than 50 MeV-cm^2/mg) a WL misassertion is nearly as likely as a valid bit upset. Note that a WL misassertion could be caused by a SET in the circuitry that drives the address into the cache as well as generated within the WL decode itself. The response of the data arrays is similar but on a different scale. Due to the much larger arrays (but nearly identical decode circuitry) the cross section of WL misassertions is approximately two orders of magnitude below the array bit upset cross section. Finally, it is worth noting that when a timing error occurred, there was not always a data upset. In broad-beam testing of the HERMES caches, all errors reported or found were correctable by the microarchitecture and associated recovery operations.

6.4.1.2 Error Detection and Correction

Since caches are known to have relatively high sensitivity to modified data bits (from SEE or otherwise), they are nearly all protected with parity or EDAC. Parity allows error detection but not correction. Thus, in the case of a write-through cache, the affected cache line can be invalidated and its data will be naturally refetched by regular cache operation. In contrast, a parity error in a dirty line of a write-back cache cannot be alleviated. The HERMES results above, using a write-through cache with parity protection, show that the combination of a write-through cache and parity can be highly effective. The most common approach is EDAC using single-error correction, double-error detection (SECDED). EDAC is inefficient for small data widths because the error-correcting bits scale with the log of the number of protected bits, and most microprocessor architectures support byte writes. Consequently, a read-modify-write operation is usually required to recalculate EDAC [43]. As a result, for first-level caches, parity-only protection is often chosen due to its simplicity.

Verification of EDAC system operation is important. An example of backing out the per-bit sensitivity using knowledge of the EDAC architecture of a cache can be found in [44]. The EDAC verification is performed by letting SEEs accumulate so that it is possible to observe if the error protection works properly (as the observed errors will often exhibit multiple-bit upsets [MBUs] directly in the readout, with a characteristic ratio of SBUs to MBUs, implying that SBUs were corrected). As indicated above, cache data can also be lost due to parity errors that result in discarding the cache line with the error. An example of lost cache data due to parity is available in [39].

6.4.1.3 Cache SEE Example

As an example of a cache upset's impact on the microprocessor behavior, a SEE is injected into the HERMES RTL model with error checking disabled. The upset is injected into the I-cache tag array that causes the lookup of the cache line corresponding to virtual addresses 0x80000450/4/8/C (physical addresses 0x00000450/4/8/C) to miss when it should have hit. This is the most common cache error due to a tag upset, since most upsets will not map a cache resident line to another cache resident line. The tag is stored in location 8 of the way 0, bank 2, top subbank. It has a value of 0x000029B. The last two bytes are the parity, least recently filled flag (LRF), lock, and valid bits. The bit upset creates a value of 0x800029B. As a result, the next lookup of this tag misses and the line is refilled from memory. Note that the parity is now incorrect. However, like most designs, the parity is only checked on a hit. Since the original corrupted line is still in way 0, the new line just brought in from memory is now put into way 1 (location 8 of the way 1, bank 2, top subbank). As long as the corrupted tag is not accessed before it gets evicted, the error is benign and even if accessed will trigger a parity error.

6.4.2 REGISTER UPSETS

Most microprocessor operations involve at least one scratch register. Other architectural registers are also important, such as the program counter, the link register, and the stack pointer. Bits in these registers can experience SBU [35,45]. When upset, these registers can cause many types of errors, partly due to the cascade of faults that fans out from the initial error. On the other hand, sometimes the data they hold is no longer needed, making the error benign. Thus, the result of a SBU in a register can range through errors in calculations, no event if the register contents are not needed, and can even include latent errors such as when the bit error occurs during copying of a user application.

Another type of register upset occurs when the wrong register is used for a given operation. For example, when a result is ready to be stored back to a register, the target register number or address may be affected by a SEE. If the result is written to another register (due to a SBU in an instruction FF or a SET during data storage), then the contents of the intended target and/or the erroneous target may be incorrect. In this case the event is referred to as register clobber. A clobbered register, in general, is about as problematic to the microprocessor as a SBU. If the register is written before its next use, then no error is encountered. However, if the upset value is retrieved for use, errors will likely occur.

Some modern microprocessors use register relabeling to enable acceleration of operations [13,14]. In order to do this, copies, and sometimes the live values of registers, are put in an array with an index that is used to indicate what register is contained in the given entry in the array. If any of the labeling information becomes corrupt due to SEE, the result is essentially the register clobber case.

6.4.2.1 RF Upset Example

In this example, we assume an RF that is not protected by parity. A SEU is injected into RF read value modifying the value in register 1 from 0x0000FFFF to 0x0000FFFE. This value is subsequently added to 0x00000001 by the add immediate unsigned (ADDIU) at PC = 0x80000510, resulting in many erroneous bits in the result register, since the carry does not propagate in the corrupted case. The ASM code is

```
PC = 0x800004E4   --> LUI    $1,  0x0000
PC = 0x800004E8   --> ORI    $1,  $1, 0xFFFF
            # SEU injected into R1A modifies the value from
            #    0x0000FFFF to 0x0000FFFE
    . . .
PC = 0x80000510   --> ADDIU  $1,  $1, 0x0001
PC = 0x80000514   --> SW     $1,  $2, 0x0099
.   PC = 0x80000518   --> BNE    $1,  $3, 0xFF56
```

After the ADDIU there are many erroneous bits in the speculative microprocessor state. It commits to the architectural state in the RF as register 1 is written as the result of the ADDIU instruction. It commits to the D-cache (and subsequently memory at the subsequent store word [SW] instruction). Finally, in this (contrived) example, it manifests as a control flow error at the branch not equal (BNE) instruction, assuming the new value fails to compare the same way. This is one of the factors that makes soft-error root cause diagnosis difficult. On a dump of the register contents, both registers will show an error, but if the dump occurs after many more operations, determining the root-cause upset is problematic. Generally register files are parity-protected. Thus, the initial RF SBU would be detected at the readout and at commit to architectural state (at the W stage) the microprocessor will take an exception. However, a SET on the readout path or the in the execution unit logic that upset the same bit after the parity is checked would have the same result as the case where there is no RF parity protection.

Experiments on the HERMES RF test chips showed that the cross section of the RF storage (using 8-T cells as in Figure 6.2, but with three read ports) is similar to that of the pipeline, which has nearly the same number of FFs as the RF has latches [46]. The WB error bars include the accumulated SEUs, and it is thus likely, based on the latch count, that an approximately equal number of pipeline upsets are due to FF SEUs. This indicates that while SETs are prevalent, SEU dominates SET at the testing frequencies, which were at or below 200 MHz (see Figure 6.11). That is, the data in Figure 6.11 includes a significant proportion of SEUs in the WB errors; thus SETs are not the dominant cause of the errors at testing frequencies.

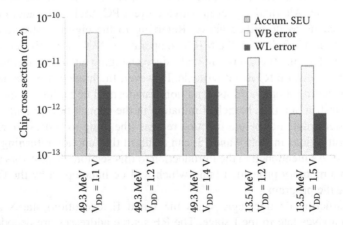

FIGURE 6.11 Hermes DMR pipeline RF testchip RF proton upset results. The WB errors (caught at write-back to the RF) dominate. About one-third of those errors were originally read from the RF (labeled accumulated SEU). WL errors arise from an incorrect register address or pipeline SET. (After L. Clark, D. Patterson, N. Hindman, K. Holbert, and S. Guertin, *IEEE Trans. Nucl. Sci.*, Vol. 58, No. 6, pp. 3018–3025, © 2011 IEEE.)

6.4.3   PIPELINE AND EXECUTION UNIT SEEs

Another microprocessor area with potential for many types of SEEs are the execution units. In this case, the error is most likely caused by incorrect data due to a FF SEU, but can also be the result of SETs that get captured. This error type is generally difficult to capture and separate from other SEE types that have a similar signature. As noted above, the wrong result value can simply be stored in the RF. One case where rotate operations (bitwise shifts where the bits shifted out of the register are inserted on the other side) were observed to have SEE is discussed in [29]. We provide a comparison of the cross section for execution errors relative to the effective sensitivity for the testing (effective sensitivity is discussed in Section 6.5). For most devices, the cross section of errors in the pipeline and execution units are well below the effective sensitivity of most tests and thus cannot be observed in limited test time at usable flux rates. The microprocessor is more likely to manifest almost exactly the same behavior as a pipeline or execution unit error due to upsets of register files or caches during accelerated ground testing. Moreover, since in testing the result of a pipeline soft error may not manifest in the external behavior for thousands of cycles, it is difficult to definitively resolve the root cause of pipeline errors in testing even with special test vectors. When full-speed operation can be conducted through such external test vectors, only a handful of these events will be detected.

6.4.3.1   Pipeline Errors

Table 6.1 presents a small portion (proc_2) of the Dhrystone benchmark compiled into MIPS assembly language as it executes through the scalar pipeline illustrated in Figure 6.4. Each pipeline stage exposes the executing instruction to errors from different portions of the microprocessor. In stage I (instruction fetch) the PC may be incorrect due to a SEU in the pipeline PC storage or a SET in the next address calculation logic. Alternatively, even with a correct PC, SETs can cause errors that impact the cache as described above. Referring to the eighth instruction (in the branch delay slot) in Table 6.1, the MIPS no operation (NOP) is a shift left logical of R0 (SLL R0). This instruction is quite safe since changing either source register will have no effect because R0 is not writable. However, the branch if not equal to zero (BNEZ) preceding it can have its condition transformed or be turned into another nonbranch instruction (i.e., branch elimination in the control flow graph). A change in the register pointed to by the function returns (the jump to address in register A [JR RA] instructions in slots 9 and 13) can result in the function returning to a random address. As mentioned in the introduction to this section, these errors are likely to result in a memory protection fault, which will be intercepted by the OS, likely terminating the program.

The instruction decode logic spans the I and E (execution) stages since the instruction arrives late in the I stage. The RF source addresses are decoded in the I stage to allow the RF access at the clock beginning the E stage. If these are corrupted, the wrong source data will be selected for the instruction. Note that load/store instructions use these values as base and offset address values, so if the instruction is a load (see LW at clock 6 in Table 6.1), the wrong addresses will be used, most likely corrupting the RF or returning data to the wrong location for a load or store,

TABLE 6.1

Instruction Flow for a Small Basic Block in Dhrystone

Pipeline stage

Clock	I			E	M	A	W
	PC			PC	PC	PC	PC
1	814:	lw	v0, 0 (a0)				
2	818:	li	v1, 65	814:			
3	81c:	addiu	a1, v0, 10	818:	814:		
4	820:	lbu	a3, −32632 (gp)	81c:	818:	814:	
5	825:	beql	a3, v1, 8000083c	820:	81c:	818:	814:
6	828:	lw	v1, −32636 (gp)	824:	820:	81c:	818:
7	82c:	bnez	a2, 80000824	828:	824:	820:	81c:
8	830:	nop		82c:	828:	824:	820:
9	834:	jr	ra	830:	82c:	828:	824:
10	838:	nop		834:	830:	82c:	828:
11	83c:	addiu	a2, a1, −1	838:	834:	830:	82c:
12	840:	addiu	a2, a1, −1	83c:	838:	834:	830:
13	844:	jr	ra	840:	83c:	838:	834:
14	848:	sw	v0, 0 (a0)	844:	840:	83c:	838:
15				848:	844:	840:	83c:
16					848:	844:	840:
17						848:	844:
18							848:

Note: Many errors can occur in the speculative state of pipeline stages I through A. Only the 12-PC MSBs are shown for brevity.

respectively. Similarly, in simple RISC pipelines, the adder calculates the final load or store address, so a SET within it or an incorrect input value due to an RF SEU can radically affect the result. In the E stage, the logic has varying degrees of soft-error vulnerability, particularly to SETs, based on size and timing criticality. In this stage the instruction is fully decoded into (frequently one-hot) control signals that propagate down the rest of the speculative portion of the pipeline. Small errors here can cause significant impact, changing instruction behavior radically. The adder, with high fanout in the carry chain, can affect many data bits due to an error. Logic operations, since they generally do not fan out (i.e., are bitwise) have low susceptibility. Shifts and rotates fall between the logic and add operations.

The M (data memory) stage is susceptible to DTLB and D-cache errors similar to those for instructions, but affecting data. In the event of a write-back cache, recovery is less likely. Most addressing errors in this stage will be due to E stage upsets, although cache control and addressing errors due to SETs may occur as described in Section 6.4.1. The A (memory alignment) stage has less logic but overlaps with the continued execution of multicycle multiply-and-divide instructions. These have a larger error cross section than the adder, since they have larger adders as well as large combinational logic blocks (i.e., the Wallace tree and Booth encoder). Finally, the W (write-back) stage writes the RF in the first (high) clock phase. Results that are upset by SET or pipeline latch SEU at this stage will be stored to the RF and may be forwarded to the ALU via the bypass multiplexers. A SET might affect only the latter, again complicating root-cause analysis of specific errors.

6.4.3.2 Pipeline Error Mitigation in HERMES

The large number of possible errors and their broad range of manifested behaviors makes complete pipeline error detection difficult. To address this, the entire speculative pipeline of Figure 6.4 is DMR in HERMES. The DMR operation was experimentally verified on a test chip, showing the efficacy at avoiding the commission of incorrect speculative state to the architectural state [46]. The two pipeline copies are completely independent and spatially separated to mitigate the possibility of a single error upsetting both copies. At the write-back to the RF, which is the point where speculative state commits to architectural state, both copies are compared. The RF write is aborted and a soft-error exception is thrown if the DMR copies mismatch, whereupon the exception handler cleans the RF (and other machine) state. Using this approach, no incorrect RF state has been committed in broad-beam testing. Since the PC is critical to determining the instruction to restart in the pipeline after a soft-error exception, it (and other critical architectural state) is protected using self-correcting triple-mode redundant (TMR) logic.

6.4.4 FREQUENCY DEPENDENCE

SEUs exhibit no frequency dependence, although their correctability can be impacted by the operating frequency. Specifically, an EDAC-protected memory may fail if run at too low a frequency or too high a flux rate. In this case, multiple errors from multiple strikes can accumulate within a codeword defeating the correction scheme. This is why scrubbing is important to avoid multiple-hit-induced defeats when ECC is used. SETs have the opposite frequency dependency. Recall that SETs manifest as errors when they are captured into FFs or latch on the edge of the clock or enable signals. These SEE types thus produce failures with a probability proportional to the operating clock period, saturating at some value.

This is shown in Figure 6.12. The maximum operating frequency $f_{CLK} = 1/P_{CLK}$ where P_{CLK} is the clock period, determined by the total delay of the circuits in the worst-case pipeline stage

$$\text{path delay} = T_{CLK2Q} + T_{CL} + T_{SETUP} \tag{6.1}$$

where T_{CLK2Q} is the delay from clock to the driving FF Q output, T_{CL} is the delay of the combinational logic in the stage, and T_{SETUP} is the FF setup time. A path is critical if it requires the entire time. Paths that complete sooner (i.e., have a smaller T_{CL}), are noncritical. For correct operation

$$\text{path delay} < P_{CLK}. \tag{6.2}$$

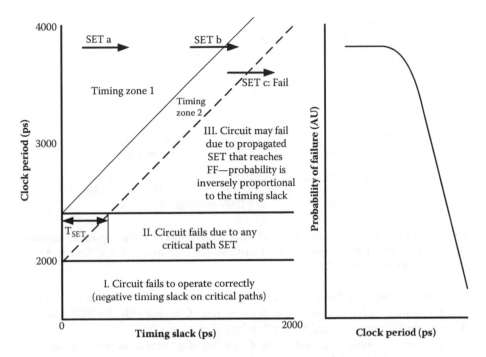

FIGURE 6.12 SETs are more likely to be captured into processor architectural state when they occur near a clock edge. Higher operating frequencies are more likely to result in these being captured, as there is less time window for them to occur and then dissipate before capture.

Designers refer to circuit delays by their slack (i.e., how much margin is left when evaluating Equation 6.2). Violations of Equation 6.2 (i.e., paths with negative slack) place the design in clock period slice I of Figure 6.12, where the circuit always fails.

However, robustness to SETs requires positive slack. When operating in a radiation environment, a SET of duration T_{SET} adds to the path delay. However, it may or may not cause a failure. If the SET occurs in clock period slice II on a critical path a failure always occurs, since the period is less than $T_{SET} + T_{CLK2Q} + T_{CL} + T_{SETUP}$. SETs occur asynchronously. A SET on a critical path will always fail within one T_{SET} of P_{CLK}, as the result will be pushed out past P_{CLK}. Assuming a P_{CLK} of 2 ns (500 MHz maximum frequency), T_{SET} of 400 ps, until that amount is added to P_{CLK}, all SETs on critical timing paths (as well as a fraction that are captured on nontiming critical paths) are captured, as shown in Figure 6.12. Eventually, increasing P_{CLK} makes all paths noncritical, even with a SET. SETs that occur on noncritical paths (in timing zone 1 of Figure 6.12) do not cause failures if they dissipate before the clock capturing edge (the dashed line) like SET a and SET b. Only SETs that occur late in the clock cycle (e.g., SET c) where their upset is captured by crossing the dashed sampling timing cause upsets. Because SEUs are not timing-sensitive (they track fluence), they add to (and probably dominate) the failure rate, creating a maximum failure rate as shown. Being asynchronous, a SET has an equal probability of

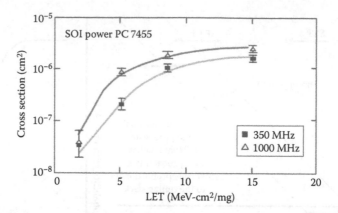

FIGURE 6.13 SETs are clearly responsible for more errors at higher operating frequencies in a PPC test. (From F. Irom, F. H. Farmanesh, *IEEE Trans. Nucl. Sci.,* Vol. 51, No. 6, pp. 3505–3509, © 2004 IEEE.)

occurrence at any time, but finite duration. Thus, the longer the clock period, the less likelihood that a SET is captured by a linear rate. Consequently, for SET-sensitive devices, the failure rate is reduced with a longer clock period.

There are a couple of reasons why frequency dependence is not often studied when examining the SEE sensitivity of a microprocessor. The first is that at low frequency the probability is already at about 20%–30% of maximum so it cannot increase much before saturating. Additionally, frequency-dependent upsets are mixed with those that are not frequency-dependent and possibly much more likely to occur. For example, frequency dependence does not affect static elements, such as cache bits, so it can also be difficult to observe frequency-dependent events in a sea of upsets in cache and register elements. This weak frequency dependence has been explored in PowerPC microprocessors, as seen in Figure 6.13 [47]. Referring back to Figure 6.12, the dominance of SEU makes the failure rate sensitivity to frequency much less pronounced than the above analysis, where SET dominates, would indicate. Frequency dependence becomes more important in RHBD microprocessors, where cache and register bits are hardened against SEE directly or via EDAC. Moreover, SET mitigation via latch hardening requires an increased setup time. The circuit will operate correctly but with increased probability of failure as this hardened FF T_{SETUP} is violated.

6.4.5 TEMPERATURE EFFECTS

Soft errors are often, but not always, weakly dependent on temperature (resistors are very temperature-sensitive, so hardening by design [HBD] designs may show strong dependence). SEL is strongly dependent on temperature, increasing in cross section by perhaps as much as an order of magnitude, and lowering the threshold LET [48]. The primary way that temperature is important in microprocessor SEE evaluation is that modern microprocessors often require a heat sink to operate normally (i.e., at full capacity and clock rate). This significantly constrains efforts to test for SEE. In

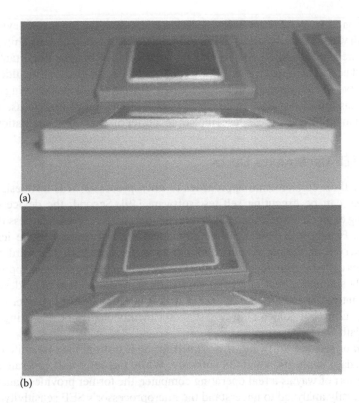

(a)

(b)

FIGURE 6.14 Maestro ITC microprocessors before (a) and after (b) thinning for heavy-ion testing. (Reprinted with permission from S. M. Guertin, B. Wie, M. K. Plante, A. Berkley, L. S. Walling, and M. Cabanas-Holmen, Radiation and Its Effects on Components and Systems Data Workshop, Biarritz, France, 2012.)

broad-beam testing it is normally necessary to expose the device under test (DUT) die so that short-range ions can penetrate to the sensitive (transistor) regions of the test device. This requires removing heat sinks and in some cases thinning the device, as shown in Figure 6.14. This type of preparation is in direct opposition to the need to provide appropriate thermal control. One solution is to blow cold air over the device during testing. As a consequence, testing in vacuum is often not possible with modern microprocessors. Another solution is to run the device at a lower frequency, which reduces the power dissipation and the cooling required. Thus, temperature-related test problems make obtaining full-spectrum SEE test data difficult.

6.5 SPECIAL TOPICS

This chapter has focused on microprocessor radiation response. So far, we have focused on the main concepts. Except for the cases of frequency and temperature, we have avoided most specific issues that complicate data collection or testing of modern, complex devices. In this section we address some of these issues that drive the difficulties in predicting, understanding, and observing radiation effects, again

focusing on SEE. We discuss SEE test design and look at the impact of device structures with weak SEE robustness. Difficulties related to on-chip and off-chip communication are also described. MBU and angular effects can be very important and are
discussed next. The impact of RHBD and FT methods can create difficulties across
the entirety of SEE impact on microprocessors. Testing situations involving complex
test systems are discussed. We conclude with a discussion of how to use the data
collected to build system response, including a brief discussion of simulation tools.

6.5.1   SEE Test Stimulus Design

There are three fundamental approaches to microprocessor SEE test design. First,
the device can be executing self-test software [49]. Second, the device external
debugging equipment interfacing to debugging ports can be used to access the internals [50]. Finally, the device can be stimulated via test vectors where its behavior is observed on the output pins, looking for differences [51]. The third option is
sometimes called the golden chip method. Generally, only the self-test approach can
accurately simulate a real operating environment. With modern SOC devices, the
complex interaction of on-chip resources makes microprocessor behavior nondeterministic, ruling out golden chip or test vector methodologies or relegating them to
nonstandard DUT operation.

In situ or self-test software, if designed for SEE testing, may be quite different
from standard application and OS software. While the latter will generate SEEs in
the same sort of way as a real operating computer, the former provides data that can
be more easily analyzed to understand the microprocessor's SEE sensitivity. That is,
it is very difficult to determine the root cause of a SEE in a real operating computer
environment. The self-test software can be designed to be fault-tolerant using many
different techniques [32,49], but it should be understood that real upset response may
require device-specific test experience to understand and mitigate SEE responses.
Note that one form of self-test is to actually run an application on a complex OS, but
this usually results in very complex error modes that are difficult to interpret (see
Section 6.5.6). An important point to make regarding this structure, however, is that
some typical FT techniques may not be viable for SEE testing. One notable problem
area is the use of trap, exception, or interrupt handlers. In a terrestrial or space SEE
environment, upsets are uncommon. However, in accelerated beam-testing SEEs can
be much more common and using trap, exception, or interrupt handlers to perform
SEE test operations can result in situations such as trap in trap, where the microprocessor cannot proceed because upsets occur too quickly. Moreover, SEE-specific test
software may differ considerably from a normal operating environment such as with
an OS. It is usually necessary to extrapolate from results obtained in SEE testing to
a full-system response. This is discussed further in Section 6.5.7.

6.5.2   SEEs Are Usually Detected in the Weakest (Softest) Elements

When exposed to ionizing radiation, all of the elements discussed earlier in this
chapter can become upset or trigger a SET. There is usually one type of structure
that, through a combination of its sensitivity and number of instances, will contribute

the majority of SEEs that are detected. Since this may create difficulty observing other lower cross-section SEEs during testing, it is very important to identify such elements. Caches are the most common culprit because they are the single largest collection of sensitive circuits (SRAM cell latches).

A specific example is found in the Freescale P2020 microprocessor sensitivity and is discussed in [52], where it is seen that caches are about 100 times more likely to show an upset than any other part of the microprocessor and execution crashes are about as common as calculation mistakes. We use this to highlight the key details that must be understood to appropriately handle the soft elements. First, it is important to note that other error modes, such as calculation errors, are not significantly more common than register upsets. As noted above, register upsets are the second most likely upset mechanism after cache upsets. Second, once this is established, it is necessary to ensure that there is sufficient information about how the microprocessor will be used to determine if an alternate approach should be taken. In particular, it is possible to operate the P2020 without the L1 and L2 caches enabled. Although this results in a significant performance loss, if a user application will use it this way then it is necessary to understand the upsets that will occur when the caches are disabled. Additionally, most OSs use some noncacheable operations.

Testing can easily identify the cross section of the most sensitive targets but for general SEE understanding it is necessary to obtain data on the less sensitive targets. This can be accomplished by creating tests where the problematic sensitive targets are not active—for example, by disabling the caches. However, in some cases this is not possible, so it is necessary to establish the effective cross section below which the problematic structures lead to events that cannot be separated from the events that testing is targeting. An example is the Maestro ITC microprocessor, which cannot have its level 2 cache disabled [44]. This leads to incorrect counts when the level 2 cache upsets build up and overwhelm the EDAC in accelerated test environments. This background of errors that are not related to the desired test SEE limit the effective cross section that can be measured is referred to as the limiting cross section. For the Maestro ITC, the limiting cross section is shown in Figure 6.15. It should be understood that sometimes the limiting cross section depends on flux-dependent events, as is the case here. Conceptually it is possible to lower the flux, reducing the unwanted error types. However, there is a practical limit because an individual measurement of a microprocessor SEE sensitivity (i.e., the cross section at a specific LET) usually must be obtained in a period of time less than an hour. Thus, the beam flux must be high enough to obtain a statistically significant number of counts in an hour or less, but must be as low as reasonable to avoid unwanted events.

6.5.3 ON-CHIP NETWORKS AND COMMUNICATIONS

Modern devices include multiple cores and communication buses that may interface directly to peripherals. These include memory interfaces, high-speed IO, and cache coherency systems. These systems have their own SEE sensitivities. Normalization of these event types can be very confusing because the data can be corrupted at different rates in different parts of the communication systems. The most straightforward upset sensitivity is that data being transferred from one part of a device to

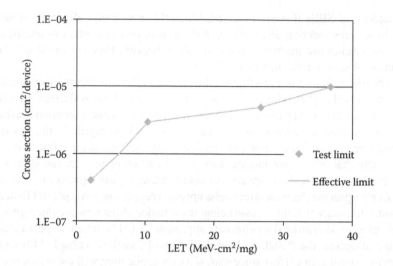

FIGURE 6.15 Example of a limiting cross section in testing of the Maestro ITC. (Reprinted with permission from S. M. Guertin, B. Wie, M. K. Plante, A. Berkley, L. S. Walling, and M. Cabanas-Holmen, Radiation and Its Effects on Components and Systems Data Workshop, Biarritz, France, 2012.)

another may become upset in transit. One way to treat this is with a traditional SEE cross section normalized to the number of data bits being transferred, as

$$\sigma = \frac{N}{\Phi \cdot B} \tag{6.3}$$

where N is the number of observed events, Φ is the number of particles incident on the test device, and B is the total number of bits transferred.

Another type of problem that can be caused by upsets in communication systems is deadlock. This is most likely to occur in systems that are designed to not have deadlock conditions through design rules that are violated by SEE. A deadlock will most likely result in complete communication bus failure.

When testing for communication errors it is important to realize that the data being transferred is stored in SEE-sensitive buffers (memory arrays) before and after transfer. Thus, it is important to have an estimate of the bit sensitivity when data is not being transferred. In this way it is possible to compare the transfer bit cross section to the intrinsic cross section for stored bit upset since the data storage effectively limits the sensitivity of a communications test. Efforts to test for communications errors in the P2020 were unsuccessful because the test software did not observe communications SEEs above the limiting cross section [52] (see Section 6.5.2).

6.5.4 MBU AND ANGULAR EFFECTS IN MICROPROCESSORS

At modern device feature sizes, single-ion strikes often create clusters of upsets [53]. These multicell upsets (MCUs) may or may not manifest as MBUs in specific memory

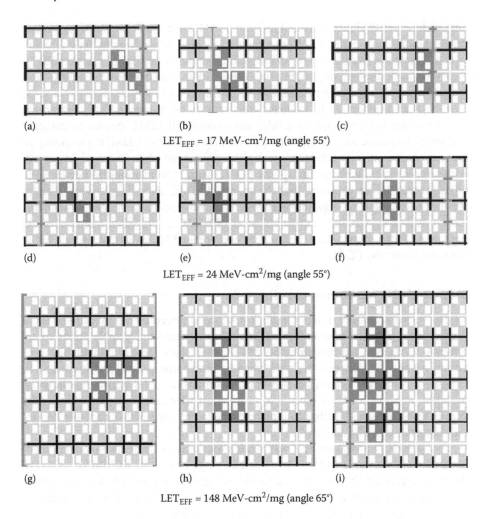

(a)   (b)   (c)

LET$_{EFF}$ = 17 MeV-cm^2/mg (angle 55°)

(d)   (e)   (f)

LET$_{EFF}$ = 24 MeV-cm^2/mg (angle 55°)

(g)   (h)   (i)

LET$_{EFF}$ = 148 MeV-cm^2/mg (angle 65°)

FIGURE 6.16 Examples of clusters of bit upsets (MCUs) as a result of a single ion strike. When ions strike at a grazing angle the cluster of upsets will follow the direction of the ion's passage. (a,d,g) Checkerboard data pattern, (b,e,h) all zeros data pattern, and (c,f,i) stripe data pattern.

(or ECC code) words depending on their physical organization. Figure 6.16 shows MCUs occurring in ion testing of a 90-nm SRAM. The exact layout has a strong impact on the MCUs, since the logical pattern does not necessarily match the physical layout. In this SRAM, every other column is flipped over the Y- (column) axis, so logical solid 0s actually produces vertical stripes. The horizontal dark separators indicate the N-well position (shared by adjacent cell rows). The vertical dark separators indicate the location of the substrate and well tap columns. MCU is exacerbated when ions strike at grazing angles, where charge collection can occur at multiple nodes. These upset clusters can also affect register and configuration bits that are stored in blocks. For EDAC-protected systems it is important that the bits from an MCU do not produce an MBU in a protected memory EDAC code word. Large MCUs are not prevalent until very high LET. Consequently, cell row interleaving by four or eight, which is common [2], allows very effective ECC or parity protection against MBUs.

MCUs have another important consequence. Rate calculations for EDAC protected systems require understanding how the EDAC will fail when it does. In some cases it is also necessary to evaluate sensitivity of bit storage cells by observing errors that are not corrected by EDAC (for example, if EDAC cannot be disabled and no performance metric registers are available to observe EDAC corrections). In these cases, the intrinsic cell cross section is compounded when an error is observed, with two or more bit errors in a given EDAC word. The relationship between the cross section, the ion exposure, and the observed EDAC errors is relatively straightforward to calculate but depends on EDAC implementation details. MCUs make this more complex by tending to create clusters of EDAC errors when one occurs. As mentioned above, the physical design must comprehend MCUs so that a single MCU does not thwart the EDAC.

6.5.5   RHBD MICROPROCESSOR BEHAVIOR

RHBD microprocessors deserve a quick discussion targeted on a mixture of event types and test issues. These microprocessors are often targeted at applications where power is a concern, but hardening always requires some form of redundancy, providing a trade-off between the performance, power dissipation. and hardness. In Chapter 7, hardening latches and flip-flops, as well as the delay impact, is discussed in detail. At the most basic level, it appears straightforward to harden a design by simply using hardened latches to replace every latch in the microprocessor design, including those in FFs and SRAM cells.

On older HBD processes, this was straightforward—resistors in each latch feedback loop allowed time to remove collected charge before the RC time constant of the resistor-gate capacitance allowed the following gate to transition. Moreover, most HBD processes use SOI substrates to significantly limit the amount of charge collected, thus speeding its removal. Since the familiar power equation

$$P = CV_{DD}^2 F + I_{LEAK} V_{DD} \qquad (6.4)$$

is unaffected by R and the total C was essentially unchanged, the power impact was small. Dense resistances (typically poly [54], but later constructed as a via or contact layer [55]) allowed compact hardened designs. However, the cost of developing and maintaining fabrication facilities for specialized processes, as well as difficulty in producing sufficiently high resistances in small areas, has made HBD processes increasingly rare and expensive. The industry has thus moved toward the alternative, RHBD, where conventional processes are used and hardening is achieved by clever circuit design and layout. Since it is still commercially available, SOI is very helpful in RHBD.

In RHBD designs, hardened latches add significantly to the power and area, and hardening against SETs naturally impacts the delay (by at least the SET duration),

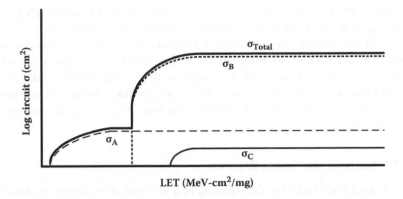

FIGURE 6.17 Multiple cross sections combine to produce more complicated device response.

since a SET adds to the timing critical path as pointed out in Section 6.4.4. Moreover, as explained in Chapters 3 and 7, multiple node charge collection in modern dense processes defeats many topologies that worked well at large geometries, where a single diffusion node tended to collect the charge. Additionally, any hardened latch topology has a Q_{CRIT} or SET duration above which it will fail. For this reason, hardened microprocessors often employ a combination of radiation-soft SRAM and other soft structures that are then protected by EDAC, parity, or other mechanisms that allow correction of the errors. They may include intrinsically hard structures, such as self-correcting triple-mode redundant circuits and latches. For example, HERMES relies on write-through caches protected by parity, dual redundant pipelines which restart instructions that finish the speculative pipeline with errors, and keeping key architectural state (and its control) in TMR [7].

Devices with mixes of RHBD and soft elements have complex SEE response where different mechanisms turn on at various LETs. Further, the mechanism with the largest cross section may change with LET. An example of this is shown in Figure 6.17. As mentioned, some SEE mechanisms, such as cache upsets, may be protected by EDAC that becomes overwhelmed during accelerated testing (i.e., errors accumulate from multiple strikes within one ECC code word between accesses). Consequently, RHBD device response in a given environment will be a combination of soft elements, RHBD elements that fail as the LET increases, and subsequent overwhelming of the fault tolerance. Combining these responses is described next.

6.5.6 COMPLEX SYSTEM TESTING

Section 6.5.1 hinted at the difficulty of testing a microprocessor that has an operating system running. This highlights a more general problem of testing microprocessors. Test systems for microprocessors are typically very complicated. The simpler the test methods and hardware, the more straightforward it is to determine how an observed error relates to the actual microprocessor SEE sensitivity. Thus, the microprocessor complexity pushes harder on the need for simple test methods and additional hardware. The other side of this situation is intentionally testing a very complex system. Sometimes this is the most expedient, or perhaps only reasonable, way to perform the test. Early testing of Intel microprocessors was conducted using the Windows NT OS and observing system crashes [56]. This test approach produces results that are very difficult to isolate to specific SEE sensitivities but does provide information about general SEE sensitivity of the operating configuration of the test system. Other researchers have simulated similar tests on Linux kernels and on other microprocessors [57].

6.5.7 BUILDING THE SYSTEM RESPONSE

In Section 6.5.6 we looked at the meaning of a SEE test of a complex system. This approach provides the end-result data for a specific configuration, but the majority of this chapter has focused on low-level structures' SEE sensitivities. The general approach we have taken is to understand these sensitivities and combine them to form the system SEE sensitivity. In the case of Windows NT, from Section 6.5.6, it is likely that most of the system crashes were due to cache parity or SBU events. We examine the general case as the final special topic of this chapter by looking at the microprocessor's operating mode, and then by looking at the duty cycle and sensitivity of the various microprocessor resources.

The operating mode of the microprocessor is perhaps the single most important determinant of the system response. Primarily, this is because for many microprocessors the caches and on-chip memories are the most sensitive structures. Thus, knowing how these resources will be utilized specifies the most common SEE impact. First, if error handling is not enabled correctly, then mechanisms like parity protection may actually guarantee interrupts when upsets occur, which may require termination of the offending thread (even in critical OS threads). Caches need to operate in the appropriate mode (usually write-through) to enable refetch of corrupt data. In SOCs, the actual peripheral resource usage is also important. Each resource must be considered, even though many have very low SEE rates relative to other structures, such as the microprocessor cores. One resource type that may be quite difficult to evaluate for SEE sensitivity is hidden or shadow resources. These can include branch prediction tables and other scratch memory areas that are not directly observable from a running program but may be integral to efficient instruction execution.

Once all of the resources used by the system are identified, the duty cycle for each should be taken into account. This may not be a simple calculation. For example, cache duty cycles may effectively be low if the information in the cache is actively updated when the working set is much larger than the cache. Another contributor to the actual duty cycle is the effect of optimized code. Modern microprocessor cores can issue many instructions per clock cycle to the various execution units. However, poorly optimized code will severely limit this, leaving many execution units idle, which greatly reduces the apparent upset rate [16]. Equation 6.5 provides a very rough way of including all the subsystems in a device in a comprehensive calculation of the system cross section for a given event type.

$$\sigma_i = \sum_j D_j \sigma_{ij} \qquad (6.5)$$

Here, σ_i is the cross section for system event type i, the sum is over all device subsystems j, D_j is the duty cycle for device subsystem j, and σ_{ij} is the cross section for subsystem j to exhibit system event type i. The latter is what gives rise to the limiting cross section (σ) in Section 6.5.2. For example, referring to Figure 6.17, there are three primary SEE contributing circuits with three cross sections, namely σ_A, σ_B, and σ_C. Since σ_A and σ_B have distinct threshold LETs and saturation cross sections, they are easily discerned. However, σ_C is obscured by the dominant σ_B, particularly with a log-scale Y-axis. As mentioned above, this situation is problematic when the source of the large dominant cross section (e.g., caches) cannot be removed from the test. In practice the situation is much more complex, and an example of some aspects of determining some of these factors can be found in the FT-UNSHADESuP microprocessor simulation tool [58] and in the use of debugging ports on hardware [59].

6.6 CONCLUSION

This chapter has discussed how SEEs develop from soft errors in low-level circuit elements into errors in microprocessor subsystems. Real-world examples have been discussed. Examples have ranged from the RHBD HERMES microprocessor to commercial communications devices such as the Freescale P2020.

The approach we have taken is to focus on how the fundamental circuit SEE responses build up into system errors. By taking this approach, it is expected that this material will be applicable to a large array of existing and future devices. In contrast, however, our approach has not looked closely at the issue of exactly what happens to a running system when a SEE occurs. This approach was chosen because, at the system level, designers can dramatically impact the SEE handling of a computer through the chosen design practices. Since this chapter is focused on SEEs in microprocessors, it is beyond our scope to discuss how the external circuit and support software will handle the SEEs manifest in a microprocessor.

We have highlighted the known limitations of SEE impact in some cases where it is common for SEEs to actually result in no observable error. As an important element to build understanding that will help the reader apply knowledge to microprocessors of interest, we have discussed specific issues regarding testing and construction of tests, in part because these tests and issues provide background on how the upsets can be observed. In real systems, however, it may be very difficult to predict the most important SEEs to try to mitigate. Some types of SEE are obviously important for mitigation. We have discussed cache and register bits being the two most likely elements of an unhardened device to upset.

The material here should be helpful for general understanding and it should be helpful for engineers trying to improve the SEE performance of a microprocessor design. This has been accomplished by exploring how SEEs impact and propagate at various levels of abstraction in a microprocessor architecture. We have also included material regarding SEE testing in an effort to provide more tools to understand these events and penetrate possible conceptual barriers. It is also our hope that engineers tasked with evaluating SEE sensitivity will be able to use the information presented here to develop well-designed tests to enable characterization of subject microprocessors.

REFERENCES

1. A. Chandrakasan, W. Bowhill, and F. Fox, eds., *Design of High-performance Microprocessor Circuits*, IEEE Press, New York, 2001.
2. L. T. Clark, Microprocessors and SRAMs for Space: Basics, Radiation Effects, and Design, Short Course, IEEE Nuclear and Space Radiation Effects Conference, Denver, CO, July 2010.
3. P. Dodd and F. Sexton, Critical Charge Concepts for CMOS SRAMs, *IEEE Trans. Nucl. Sci.*, Vol. 42, pp. 1764–1771, 1995.
4. J. Haigh, J. Miller, M. Wilkerson, T. Beatty, S. Strazdus, and L. Clark, A Low-Power 2.5-GHz 90-nm Level 1 Cache and Memory Management Unit, *IEEE J. Solid-State Circuits*, Vol. 40, No. 5, pp. 1190–1199, May 2005.
5. Freescale Semiconductor P2020EC Data Sheet, Rev. 2, 2013.
6. L. Chang et al., An 8T-SRAM for Variability Tolerance and Low-Voltage Operation in High-Performance Caches, *IEEE J. Solid-State Circuits*, Vol. 43, No. 4, April 2008.
7. L. Clark, D. Patterson, C. Ramamurthy, and K. Holbert, An Embedded Microprocessor Hardened by Microarchitecture and Circuits, *IEEE Trans. Comput.* (in press), 2015.
8. A. Biswas et al., Computing Architectural Vulnerability Factors for Address-Based Structures, *Proc. ISCA*, 2005.
9. S. Mukherjee, J. Emer, and S. Reinhardt, The Soft Error Problem: An Architectural Perspective, *Proc. HPCA*, pp. 243–247, 2005.
10. MIPS32 Architecture for Programmers, Vol. 1, 2001.
11. J. Chang et al., A 45nm 24MB On-Die L3 Cache for the 8-Core Multi-Threaded Xeon Processor, *VSLI Symp. Dig. Tech. Papers*, pp. 152–153, 2009.
12. J. Hennessy and D. Patterson, *Computer Organization and Design: The Hardware/ Software Interface*, Morgan Kaufmann, San Francisco, 1998.
13. J. Shen and M. Lipasti, *Modern Processor Design: Fundamentals of Superscalar Processors*, McGraw-Hill, New York, 2005.
14. G. Hinton et al., A 0.18-μm CMOS IA-32 Processor with a 4-GHz Integer Execution Unit, *IEEE J. Solid-state Circuits*, Vol. 36, No. 11, pp. 1617–1627, November 2001.
15. K. Diefendorff, Prescott Pushes Pipelining Limits, *Microprocessor Report*, February 2004.
16. N. Wang, J. Quek, T. Rafacz, and S. Patel, Characterizing the Effects of Transient Faults on a High-Performance Processor Pipeline, *Proc. Int. Conf. on Dependable Systems and Networks*, pp. 61–70, 2004.
17. Available at http://PDAdb.net/index.php?m=cpu&id=18255&c=qualcomm_snapdragon _msm8255.
18. S. Buchner, M. Sibley, P. Eaton, D. Mavis, and D. McMorrow, Total Dose Effect on the Propagation of Single Event Transients in a CMOS Inverter String, *Proc. RADECS*, pp. 79–82, 2009.
19. T. P. Ma and P. V. Dressendorfer, *Ionizing Radiation Effects in MOS Devices and Circuits*, Wiley-Interscience, New York, 1989.
20. I. S. Esqueda, H. J. Barnaby, and M. L. Alles, Two-Dimensional Methodology for Modeling Radiation-Induced Off-State Leakage in CMOS Technologies, *IEEE Trans. Nucl. Sci.*, Vol. 52, No. 6, pp. 2259–2264, December 2005.

21. W. A. Kolasinski, J. B. Blake, J. K. Anthony, W. E. Price, and E. C. Smith, Simulation of Cosmic-Ray Induced Soft Errors and Latchup in Integrated-Circuit Computer Memories, *IEEE Trans. Nucl. Sci.*, Vol. 26, pp. 5087–5091, 1979.
22. P. E. Dodd, J. R. Schwank, M. R. Shaneyfelt et al., Impact of Heavy Ion Energy and Nuclear Interactions on Single-Event Upset and Latchup in Integrated Circuits, *IEEE Trans. Nucl. Sci.*, Vol. 54, No. 6, p. 2303, December 2007.
23. K. Iniewski, ed., *Radiation Effects in Semiconductors*, CRC Press, Boca Raton, FL, 2012.
24. A. Carbine and D. Feltham, Pentium Pro Processor Design for Test and Debug, *IEEE Design and Test of Computers*, pp. 77–82, July–September 1998.
25. J. Leavy, L. Hoffman, R. Shovan, and M. Johnson, Upset Due to Single Particle Caused Propagated Transient in a Bulk CMOS Processor, *IEEE Trans. Nucl. Sci.*, Vol. 38, No. 6, December 1991.
26. N. Seifert et al., Radiation Induced Clock Jitter and Race, *IRPS Proc.*, Apr. 2005, pp. 215–222.
27. S. Chellappa, L. Clark, and K. Holbert, A 90-nm Radiation Hardened Clock Spine, *IEEE Trans. Nucl. Sci.*, Vol. 59, No. 4, pp. 1020–1026, 2012.
28. K. Warren et al., Heavy Ion Testing and Single Event Upset Rate Prediction Considerations for a DICE Flip-Flop, *IEEE Trans. Nucl. Sci.*, Vol. 56, No. 6, pp. 3130–3137, December 2009.
29. S. M. Guertin, C. Hafer, and S. Griffith, Investigation of Low Cross Section Events in the RHBD/FT UT699 Leon 3FT, *Proc. REDW*, 2011, pp. 1–8, July 2011.
30. A. Aho, M. Lam, R. Sethi, and J. Ullman, *Compilers: Principles, Techniques, and Tools*, Pearson/Addison-Wesley, Boston, 2007.
31. G. Saggese, A. Vetteth, Z. Kalbarczyk, and I. Ravishankar, Microprocessor Sensitivity to Failures: Control vs. Execution and Combinational vs. Sequential Logic, *Proc. Int. Conf. on Dependable Systems and Networks*, pp. 760–769, 2005.
32. N. Oh, P. Shirvani, and E. McCluskey, Control-Flow Checking by Software Signatures, *IEEE Trans. Reliability*, Vol. 51, No. 2, pp. 111–122, March 2002.
33. D. Patterson and J. Hennessy, *Computer Architecture: A Quantitative Approach*, 2nd Ed., Morgan Kaufmann, San Francisco, 1990.
34. F. Bezerra and J. Kuitunen, Analysis of the SEU Behavior of PowerPC 603R under Heavy Ions, Radiation and Its Effects on Components and Systems, *Proceedings of the 7th European Conference on European Space Agency SP-536*, IEEE 03TH8776, pp. 289–293, 2003.
35. G. M. Swift, F. H. Farmanesh, S. M. Guertin, F. Irom, and D. G. Millward, Single-Event Upset in the Power PC750 Microprocessor, *IEEE Trans. Nucl. Sci.*, Vol. 48, No. 6, pp. 1822–1827, 2001.
36. X. Yao, D. Patterson, K. Holbert, and L. Clark, A 90 nm Bulk CMOS Radiation Hardened by Design Cache Memory, *IEEE Trans. Nuc. Science*, Vol. 57, No. 4, pp. 2089–2097, August 2010.
37. F. Irom, *Guideline for Ground Radiation Testing of Microprocessors in the Space Radiation Environment*, JPL Publication 8–13, Jet Propulsion Laboratory, California Institute of Technology, Pasadena, CA, April 2008.
38. S. Narendra and A. Chandrakasan, *Leakage in Nanometer Technologies*, Springer, New York, 2006.
39. S. M. Guertin, P2020 Proton Test Report March 24, 2011, NASA Electronic Parts and Packaging Program (Internal Document), Goddard Space Flight Center, Greenbelt, MD, 2012.
40. P. McDonald, W. Stapor, A. Campbell, and L. Massengill, Non-Random Single Event Upset Trends, *IEEE Trans. Nucl. Sci.*, Vol. 36, No. 6, pp. 2324–2329, December 1989.
41. D. Mavis et al., Multiple Bit Upsets and Error Mitigation in Ultra-Deep Submicron SRAMs, *IEEE Trans. Nucl. Sci.*, Vol. 55, No. 6, pp. 3288–3294, December 2008.

42. L. Jacunski et al., SEU Immunity: The Effects of Scaling on the Peripheral Circuits of SRAMs," *IEEE Trans. Nucl. Sci.*, Vol. 41, No. 6, pp. 2324–2329, December 1989.
43. K. Mohr and L. Clark, Delay and Area Efficient First-Level Cache Soft Error Detection and Correction, *ICCD Proc.*, pp. 88–92, October 2006.
44. S. M. Guertin, B. Wie, M. K. Plante, A. Berkley, L. S. Walling, and M. Cabanas-Holmen, SEE Test Results for Maestro Microprocessor, Radiation and Its Effects on Components and Systems Data Workshop, Biarritz, France, 2012.
45. F. Bezerra et al., Commercial Processor Single Event Tests, in *Proc. RADECS Conf. Data Workshop Record*, 1997, pp. 41–46.
46. L. Clark, D. Patterson, N. Hindman, K. Holbert, and S. Guertin, A Dual Mode Redundant Approach for Microprocessor Soft Error Hardness, *IEEE Trans. Nucl. Sci.*, Vol. 58, No. 6, pp. 3018–3025, 2011.
47. F. Irom, F. H. Farmanesh, Frequency Dependence of Single-Event Upset in Advanced Commercial PowerPC Microprocessors, *IEEE Trans. Nucl. Sci.*, Vol. 51, No. 6, pp. 3505–3509, 2004.
48. A. H. Johnston, B. W. Hughlock, M. P. Baze, and R. E. Plaag, The Effect of Temperature on Single Particle Latchup, *IEEE Trans. Nucl. Sci.*, Vol. 38, No. 6, pp. 1435–1441, December 1991.
49. S. M. Guertin and F. Irom, Processor SEE Test Design, presented at Single Event Effects Symposium, La Jolla, CA, 2009.
50. Freescale, CodeWarrior™ Development Studio for Microcontrollers V6.3, Freescale Document 950-00087, 2009.
51. R. Koga, W. A. Kolasinski, M. T. Marra, and W. A. Hanna, Techniques of Microprocessor Testing and SEU Rate Prediction, *IEEE Trans. Nucl. Sci.*, Vol. 32, No. 6, pp. 4219–4224, 1985.
52. S. M. Guertin and M. Amrbar, SEE Test Results for P2020 and P5020 Freescale Processors, *Proc. REDW*, pp. 1–7, 2014.
53. G. Gasiot, D. Giot, and P. Roche, Multiple Cell Upsets as the Key Contribution to the Total SER of 65 nm CMOS SRAMs and Its Dependence on Well Engineering, *IEEE Trans. Nucl. Sci.*, Vol. 54, No. 6, December 2007.
54. H. Weaver et al., An SEU Tolerant Memory Cell Derived from Fundamental Studies of SEU Mechanisms in SRAM, *IEEE Trans. Nucl. Sci.*, Vol. 34, No. 6, pp. 1281–1286, December 1987.
55. T. Hoang et al., A Radiation Hardened 16-Mb SRAM for Space Applications, *Proc. IEEE Aerospace Conf.*, pp. 1–6, 2006.
56. J. W. Howard, Jr., M. A. Carts, R. Stattel, C. E. Rogers, T. L. Irwin, C. Dunsmore, J. A. Sciarini, and K. A. LaBel, Total Dose and Single Event Effects Testing of the Intel Pentium III (P3) and AMD K7 Microprocessors, Radiation Effects Data Workshop, Vancouver, Canada, 2001 IEEE, pp. 38–47, 2001.
57. W. Gu, Z. Kalbraczyk, and R. Iyer, Error Sensitivity of the Linux Kernel Executing on PowerPC G4 and Pentium 4 Processors, *Proc. Int. Conf. on Dependable Systems and Networks*, pp. 887–896, July 2004.
58. H. Guzman-Miranda, J. N. Tombs, and M. A. Aguirre, FT-UNSHADES-uP: A Platform for the Analysis and Optimal Hardening of Embedded Systems in Radiation Environments, *IEEE International Symp. on Ind. Electronics*, pp. 2276–2281, Cambridge, UK, June/July 2008.
59. M. Portela-García, C. López-Ongil, M. G. Valderas, and L. Entrena, Fault Injection in Modern Microprocessors Using On-Chip Debugging Infrastructures, *IEEE Trans. Dependable Secure Comput.*, Vol. 8, No. 2, March/April 2011.

7 Soft-Error Hardened Latch and Flip-Flop Design

Lawrence T. Clark

CONTENTS

7.1   INTRODUCTION

Using hardened latches is the most straightforward method to harden a logic design against soft-errors. By hardening the sequential circuits (i.e., the flip-flops [FFs] and latches) as well as hardening the memories appropriately, the rest of a digital design need not be altered. There is a significant timing impact, depending on the choice of hardened FF circuit and the level of hardness required.

This chapter describes the most commonly used designs as well their advantages and disadvantages from both the circuit quality and hardness perspectives. The types of hardening approach and other points are also discussed. Finally, appropriate circuit-level validation approaches, amenable to circuit rather than device level simulation are discussed. This is important since validation via testing requires a very slow feedback loop that requires design, fabrication, broad-beam or laser testing, and then redesign (and so forth). Consequently, design-only methodologies are useful so that the hardened FF and latch design process can be expeditious and successful.

The focus of this chapter is on the general concepts and designs that work on conventional bulk complementary metal-oxide semiconductor (CMOS) technologies. The approaches are also useful in silicon-on-insulator (SOI) technologies, but since SOI greatly attenuates the charge collection, hardening SOI digital circuits is considerably easier.

7.1.1   UNHARDENED LATCHES AND FLIP-FLOPS

7.1.1.1   Latch Circuits and Timing

We begin by reviewing unhardened latch and FF circuits as they comprise the baseline against which hardened circuits can be compared so that the relative impact of their hardening on the circuit performance (delay) can be quantified. Figure 7.1 shows the most commonly used D-type CMOS latch design. The output Q obtains the same logic state as the D input when the latch is transparent (i.e., when clock is high). The different circuit and logical views (Figure 7.1a through c) can be used interchangeably. On older processes, there is a layout density advantage to using the tristate gates in version (b). The logical view shows that the clock input CLK merely chooses between the feedback and the feed-forward path (this view is helpful in simplifying hardened circuit topologies). Note that the output and input polarity need not match. Generally, whether a latch of FF output is Q or the inverted value QN is chosen based on what is the fastest and lowest power. All modern computer-aided design (CAD) tools can deal with such inversion effortlessly.

The latch timing is shown in Figure 7.1d. The latch has four key timings: The setup time (t_{SETUP}) is the time required from the D pin to allow the setup node to achieve the correct state so that when the latch closes on the falling edge of the

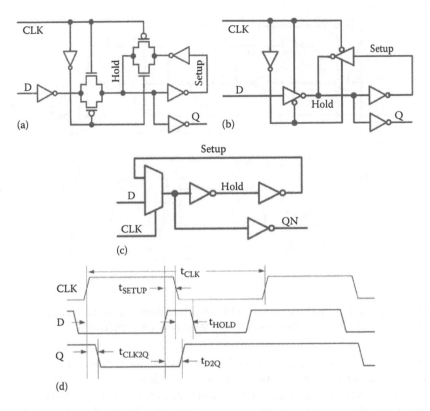

FIGURE 7.1 Basic CMOS (unhardened) latch design and timing. (a) Basic D type latch with CMOS transmission gates, (b) the same latch with tristate gates, (c) a logical equivalent, and (d) the timings required for both latches and flip-flops.

clock, the feedback maintains the state on setup and hold. The hold time (t_{HOLD}) is the time that D must be held before the falling edge of the clock so that the hold node does not have the wrong state, again so that Q = D in the feedback state. When the latch transitions from the opaque to the transparent state, Q transitions from the stored value to that of D. This delay is t_{CLK2Q}, which is minimized by connecting the inverter driving Q to the hold node rather than the setup node. When D transitions while the latch is in the transparent state, the time it takes to propagate to the output Q is t_{D2Q}. The timings can be easily estimated by eye simply by counting inversion delays between the pins. This makes easy an comparison between designs and is the approach taken in this chapter.

7.1.1.2 FF Circuits and Timing

The standard D FF circuit and timing is shown in Figure 7.2. The FF edge-triggered operation is obtained by placing two latches in series, the master and slave, respectively. This eliminates the transparency in the latch timing (i.e., there is no t_{D2Q}). Since the master latch holds during clock high, the setup and hold timings are measured to the master internal nodes. The slave timings are almost irrelevant since there is an entire clock phase to meet the setup time to node SS and the slave guarantees

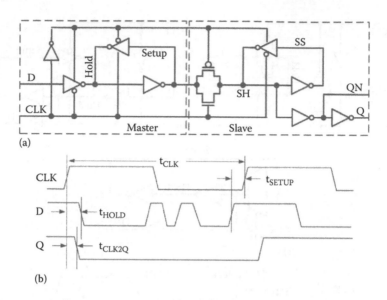

(a)

(b)

FIGURE 7.2 Basic CMOS (unhardened) FF design and timing. (a) A master and slave latch are put in series with minor circuit changes for layout efficiency. (b) Timing.

the hold timing to node SH. The CMOS transmission gate is required at the master input to avoid introducing an unnecessary inversion delay.

There are some basic circuit quality rules for obtaining reliable circuit operation. These are worth discussing as they are sometimes not followed in proposed latch and FF designs, but nearly always are in commercially successful designs. First, the D input should be buffered (i.e., drive a gate) and not diffusion inputs. When noise couples via wire cross capacitance, a diffusion input can couple below or above the power rail, discharging the hold node and altering the latch state. This is avoided by buffering. Consequently, standard cell designs rarely use unbuffered (i.e., bare) pass gate inputs (and only in environments with sophisticated noise analysis capabilities). Second, storage nodes should not be cell pins, since noise coupled to a wire can upset the latch state if they are. This requires a buffering inverter or gate between the storage nodes SS and SH and the Q or QN output. Appropriate input and output buffering within the latch circuit (cell) guarantees that the coupling will be low and uniform across usage. Finally, in some circuits, particularly some hardened designs, back-writing from the master to the slave latch can be an issue. Referring to Figure 7.2a, back-writing is the case where when CLK transitions high, node SH in the slave overpowers the master node setup so the master is written by the slave rather than the correct operation with the master writing the slave. Back-writing is generally avoided by appropriate circuit sizing and internal clock timing.

7.1.2 Upset Mechanisms

Single-event effect (SEE) is the general term describing device failures due to impinging radiation that deposits charge in the circuit upsetting the circuit state.

Depending on the incident particle, devices, and circuit types, there can be different failure modes from the collected charge. These range from benign to catastrophic, but the focus of this chapter is on those that cause soft errors (i.e., circuit state upsets that do not permanently damage the circuit). These upset mechanisms are reviewed in Chapter 1. In silicon, one linear energy transfer (LET) generates approximately 10 fC of charge per micron of track length per MeV-cm^2/mg of LET. A 32-nm technology generation transistor has less than 1 fC of total capacitance per micron of gate width, explaining the increasing importance of soft errors in scaled technologies [1]. Charge collection is at diodes, so electrons are collected by N-type diffusions and holes are collected by P-type diffusions. In a bulk CMOS process, the latter are in N-type wells of limited depth, reducing the charge due to collected holes considerably compared to that due to collected electrons in bulk CMOS technologies. Integrated circuits (ICs) are generally ground tested in neutron, proton, or ion beams to determine their susceptibility to SEE. The probability of upsetting a circuit is measured by its apparent target size (i.e., cross-section) in units of area, given by

$$\sigma = \frac{\text{Errors}}{\text{Fluence}}, \tag{7.1}$$

where fluence is measured in particles/cm^2. The primary goal of SEE mitigation is to limit the errors and hence the cross section.

7.1.2.1 Single-Event Upsets and Transients

The actual waveform is highly dependent on the amount of charge deposited as well as the driving or restoring circuit, which must remove the deposited charge. In the case of a latch, the deposited charge may be sufficient to upset the bistable circuit state, a charge amount termed the Q_{crit} of the latch, causing a single-event upset (SEU) [2]. In this case, the restoring circuit is turned off by the feedback, allowing the latch state to flip. For combinational circuits, the SEE is termed a single-event transient (SET) that can inadvertently assert signals or be captured as machine state when sampled by a latch. During an SET, as the collecting node is driven to the supply rail, the diode is no longer at a favorable voltage to collect charge until sufficient charge has been removed by the driver. Some charge may recombine, but due to long lifetimes in modern silicon (Si) substrates, it tends to linger in a manner analogous to a diode's diffusion capacitance. Consequently, the SET may be prolonged. Assuming N-type (substrate) collection, the PMOS transistor provides its maximum current, $I_{DSAT(P)}$, until sufficient deposited charge has been removed so that the NMOS drain diode current is less than $I_{DSAT(P)}$. At this time the transient of the circuit output node at or near V_{SS} transitions back to V_{DD}. The resulting SET may cause erroneous circuit operation, particularly in dynamic circuits. SET durations of about 1 ns are typically reported [3,4], although durations as long as 3 ns have been experimentally measured [5]. SET duration is entirely a function of the LET and driving circuit. Therefore, upsizing the circuits can limit the SET duration, improving immunity.

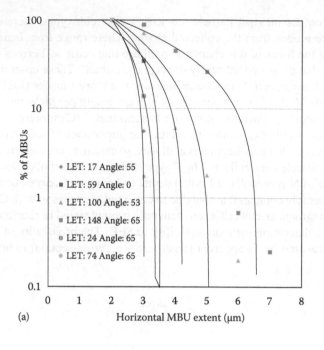

(a)   Horizontal MBU extent (μm)

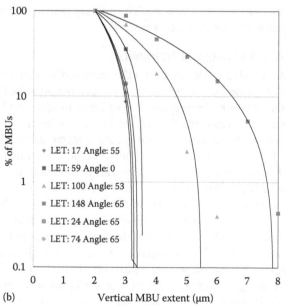

(b)   Vertical MBU extent (μm)

FIGURE 7.3 Measured multicell upsets in a 90-nm SRAM. (a) Parallel with the N-well direction. (b) Across the N-wells. At LET below 100, very few upsets extend beyond a few microns, regardless of the orientation.

7.1.2.2 Multiple-Node Charge Collection

In older processes charge was likely to be collected by a single N+ or P+ diffusion. However, in modern scaled processes, deposited charge is often collected by multiple circuit nodes [6–10]. Figure 7.3 shows multiple static random access memory (SRAM) cell upsets from the single-ion strikes. SRAM results can provide useful guidance for the hardening of other circuits. We have long used, as have others, SRAM multiple cell upsets as a metric for measuring multiple-node charge collection (MNCC) upset extent [6]. In particular, MNCC is easily analyzed using SRAMs.

The results shown in Figure 7.3 are obtained by subjecting a 90-nm SRAM test chip to ions at different LETs and orientations at the Texas A&M cyclotron. The FF NMOS feedback node transistor sizes are very close to those in the logic rule SRAM cell: SRAM NMOS transistor sizes are W/L = 190/100 nm versus 200/100 nm in the FFs. Additionally, the SRAM cell height at 1.64 μm is reasonably close to that of the 7-track library targeted by the FF designs presented here, at 1.96 μm. Last, the wells are horizontal in both this SRAM and the library. Referring to Figure 7.3, the horizontal (along the N-well) and vertical (across the N-well) simultaneous multiple-bit upset (MBU) extents due to MNCC are shown in Figure 7.3a and b, respectively.

A latch design that can tolerate only a single-node upset must ensure that the charge collecting diffusions are separated by more than this MNCC extent to assure hardness. Fortunately, for most applications, hardening to an LET of 50 MeV-cm²/mg (the highest that spacecraft normally encounter) is adequate so the required spacing is on the order of a few microns. In terrestrial applications, where alpha particle and neutron secondary ion LET is lower, even less spacing may be sufficient. On 90-nm processes, this is greater or equal to about two standard cell heights. This may be several standard cell heights in a modern sub-45-nm process.

7.1.3 HARDENING BY PROCESS

For many years hardened processes delayed the latch upset by adding one or more resistors to the feedback loop (see Figure 7.4). The RC time constant allows time t_{DELAY} for the driving circuit to remove the deposited charge before the feedback loop turned off

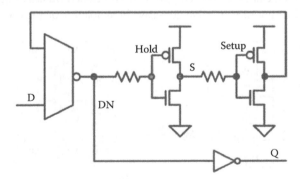

FIGURE 7.4 Classical hardened by process latch. The RC time constant provided by the explicit resistor and the capacitance of the driven circuit allows time for deposited charge to be removed.

the driver, completing the upset [11,12]. For instance, charge deposited at node S does not affect the setup node until about $t_{DELAY} = 0.69$ RC. Unfortunately, the resistance R required is on the order of hundreds of kΩ on 0.35- and 0.25-μm processes. Since C scales by 0.7 per generation, the R must scale up commensurately, which is increasingly difficult to achieve in a small area. Most rad-tolerant processes have used SOI substrates to attenuate the amount of collected charge, thereby reducing the required resistance.

A key aspect of SEE hardening digital circuits through latch hardening is the timing impact. Figure 7.4 makes it evident that the t_{SETUP} is increased by up to 2 × t_{DELAY} depending on the design. Conversely, the hold time is reduced by up to t_{DELAY} (i.e., the D input can go to the incorrect value on t_{DELAY} before the clock falling (latch closing) edge and the latch retains its state). This makes many hardened ICs very robust to poor design such as high clock skew, but potentially much slower than their unhardened counterparts.

Fabrication facility cost has made specialized processes relatively unaffordable, driving emphasis to radiation hardened by design (RHBD) approaches. RHBD relies solely on clever circuit design and layout to mitigate radiation effects while using a wholly conventional process [13,14]. Moreover, RHBD techniques have the advantage of being applicable to terrestrial circuits, where hardening at least key components is increasingly important. We focus first on hardened latch design, since as shown above FFs are comprised of two latches in series.

7.2 DESIGN-BASED MITIGATION APPROACHES FOR LATCH AND FF SOFT-ERROR MITIGATION

7.2.1 CIRCUIT REDUNDANCY

The simplest redundant latch uses three copies (i.e., triple mode redundancy [TMR]), as shown in Figure 7.5. The outputs can be combined by a voter (a simple majority gate common to adder circuits) to provide a single value that follows two of the latch outputs when one copy has been upset. Figure 7.5b shows the majority voter circuit—the output is the inversion of the two inputs that agree. Transistors MP3 and MP5, as well as MN3 and MN5, can be combined. On modern gridded transistor gate fabrication processes this does not save area, but does save input loading on the C input. Since this

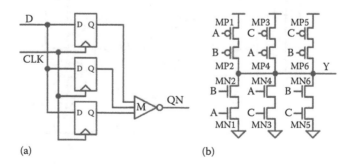

(a)   (b)

FIGURE 7.5 (a) TMR FFs to provide SEU immunity. The majority gate labeled M votes the outputs to provide the correct value after one FF has been upset. (b) Majority gate schematic.

circuit uses a common clock and D input, a SET on the D input during the rising clock edge is captured by all three FFs, resulting in a failure. Additional failure modes are produced by SETs on the clock input. Thus, this design is hard to SEU only. Moreover, the design is not self-correcting, so if not updated, errors may accumulate from multiple upsets. The majority gate delay adds slightly to t_{CLK2Q}, but otherwise timing is conventional. Entire digital blocks can be combined in a similar fashion. A primary difficulty with many TMR designs besides their larger power and area is the need to resynchronize the stored state after one module has been upset.

7.2.1.1　Dual Redundant Approaches

The built-in soft-error resilience (BISER) FF is a straightforward variation of the design in Section 7.2.1 requiring only two FFs [15]. The originally published circuit is illustrated in Figure 7.6a although numerous variations have been proposed. In this circuit, two dual-mode redundant (DMR) FFs drive a Muller C-element. The C-element (see Figure 7.6b) dates back to the earliest very-large-scale integration (VLSI) circuits as a synchronizer (see Mead and Conway [16]) where both the A and B inputs must match to change the output. Since the output is tristated after an upset, the FF output QN must be maintained by a jam latch (so called since there is no enable—the value is jammed in against the feedback, which as a consequence, must be very weak). The circuit in Figure 7.6a has an exposed storage node on the jam latch and clearly suffers from the output-charge-sharing noise issue discussed above. After a FF state upset, the latch is critical and if subject to noise it will consistently fail. A safer version buffers the output (Figure 7.6c) at the expense of a small increase in t_{CLK2Q}. Like the circuit in Section 7.1.4.1, this design is hard only to SEU and errors accumulate.

The area total savings over the TMR version is exactly one latch (1/6), but this circuit is slightly slower due to the series jam latch. Clock loading is reduced by one-third, but since these circuits are not SET-immune, they are most applicable to logic where they are infrequently clocked, such as configuration bits. A key to such an application is configuring SRAM redundancy. Since a change in the redundancy configuration due to a soft error will expose failing (and in fact unwritten) SRAM blocks, rows, or columns to the machine state, it is likely to be catastrophic.

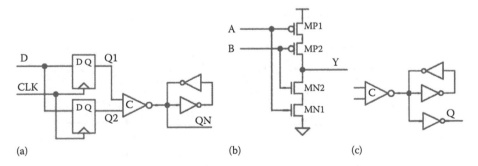

(a)　　　　　　　　　　　　　(b)　　　　　　　(c)

FIGURE 7.6　The BISER FF. Two redundant FFs are used (a) and if they mismatch, the Muller C-element output QN is tristated but retained by the jam latch. Details of the C-element are shown in (b). The output latch with good noise immunity (c) has precisely the same number of transistors and area as a latch.

The BISER FF is not self-correcting—an upset DMR FF stays upset until another clock edge, leaving the possibility of upset due to multiple events. Smaller and self-restoring, albeit much more complex, circuits intended for these types of configuration applications have been proposed [17].

7.2.1.2 Dual-Interlocked Cell

The most widely employed redundant latch design is the dual-interlocked cell (DICE) latch, shown in Figure 7.7 [18]. It has the advantage over the simple redundancy that it automatically corrects the upset nodes. The DICE latch is derived by connecting two latches (see Figure 7.7a) together by adding inverters between them. However, these added connections must be bidirectional, creating four interlocked latches, as shown in Figure 7.7b. Cleverly removing the redundant portions of the resulting eight inverters allows each node to be restored by one of two other nodes if upset. Referring to Figure 7.7c, the PMOS transistors from the odd-numbered inverters and the NMOS transistors from the even-numbered inverters are retained. The resulting circuit has logical 0 drive flowing in the clockwise direction and logical 1 drive flowing in the counterclockwise direction around the loop. When a node is upset (e.g., N1 storing a logic 1 is driven to logic 0 by collected charge), transistor MP11 restores it to a logic 1. During recovery, node N4 is briefly in contention as MN22 and MP21 are on simultaneously and node N2 is tristated since MN42 is cut off. However, the logic upset does not propagate around the feedback loop, which is now four instead of the usual two inversions. To write the DICE latch, at least two of the nodes are driven to the desired voltage. The DICE latch has found wide application despite being SEU but not SET hard. The timings are also similar to the conventional latch.

7.2.2 TEMPORAL REDUNDANCY

The latches thus far are hard to SEUs but not to SETs on the logic inputs whether for data or controlling (asynchronous) inputs such as reset or clock. The impact of SETs

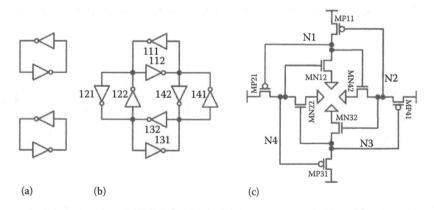

(a)   (b)   (c)

FIGURE 7.7 Derivation of the DICE latch. Beginning with dual redundant latches (a) the latches are connected so that each may repair the other (b). This results in four interconnected latches. Removing the redundant transistors results in the DICE circuit (c).

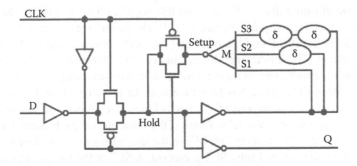

FIGURE 7.8 Temporal latch. The TMR feedback is separated in time so that an upset is propagated through it serially in time at nodes S1 through S3. The setup node is not disturbed. Consequently this design is the first proposed to be hard to both SEU and SET.

on the hardness of a digital design increases linearly with the clock frequency [19] and thus is becoming more important in modern high-performance designs. It is thus useful to explore latch designs that mitigate SETs as well as SEUs.

The design proposed by Mavis and Eaton [20] uses temporal redundancy (Figure 7.8). Here, delay elements spread the feedback into distinct delayed signals separated by zero, one, and two delay intervals. In the feedback (hold) mode, clock CLK is low and a loop is formed by the hold, S1, S2, and S3, and setup nodes. If node hold is upset, S1 immediately follows after one inverter delay. However, the delay elements, with input to output delays of t_δ, provide time for the majority gate to remove the charge. Throughout the transient, nodes S2 and S3 have the correct value so node hold is restored. Behavior is similar for upsets at the S1, S2, S3, or setup nodes. In the event of a SET on CLK, if D has a different value than S1, it will cause a glitch on hold that is similarly absorbed, as is a SET on D as the clock is closing, if its duration is less than t_δ. The penalty is that the setup time is given by

$$t_{SETUP} = t_{SETUP0} + t_\delta, \qquad (7.2)$$

where t_{SETUP0} is the nominal D latch setup time. It is further slightly increased by the slightly slower majority gate. Power dissipation is increased considerably depending on the delay element design. Delay elements turn out to be quite difficult to design with low power and size but high delay, as discussed in detail in Section 7.2.4.

7.2.3 COMBINED APPROACHES

At this point we have reviewed the basic latch and FF hardening approaches. Most other designs have used variations on or combinations of these approaches. While it is impractical to examine all of them and many are uninteresting due to limited improvement in hardness, we review some of the most interesting combinations next.

7.2.3.1 DF-DICE

DICE-based FFs are obviously susceptible to upset on the clock and preset/reset nodes. The delay-filtered DICE (DF-DICE) design addresses the DICE latch's shortcoming of immunity to SEU only by filtering the inputs to mitigate SETs. The DF-DICE schematic comprises Figure 7.9 [21]. The difference between the transistor count of the standard FF in Figure 7.2 and the DICE-based design in Figure 7.9 is readily apparent. The unhardened FF is comprised of 20 transistors. The DF-DICE design requires about three times the number of transistors compared to the unhardened FF if each delay filter has four inversions to produce the delay (and this is a very conservative undercount of the needed area but follows the assumptions in [21]). This design shows two delay filters (DFs) with one on each input. Clock, set, and reset nodes can be protected by DFs in the last stage of the clock or reset trees. However, the nodes are not fully SET-protected. A SET at the DF circuit output will propagate and cause an upset if it occurs in the sensitive timing window. Depending on the required hardness level, this may or may not be adequate—these nodes have similar target cross sections as the storage nodes. If the DFs drive multiple FFs the failure rate is higher. When a DF drives a single FF there is a 50% chance the FF will be driven to the same state, resulting in no failure. With N FFs, this chance of getting lucky is reduced to $1/2^N$. Consequently, this approach is less hard than the temporal latch-based FF, which with the correct layout mitigates all upsets.

This design increases the t_{CLK2Q} since both the clock and D input are delayed by the DF circuit. Thus, the internal clock occurs t_δ after it occurs at the pin, so Q appears delayed and the setup time appears unaffected (i.e., it must be slightly ahead of the clock edge). Note that the FF dead time, $t_{SETUP} + t_{CLK2Q}$, is increased by t_δ, as it is in the Mavis approach. This larger dead time is a consistent penalty with all approaches that mitigate SETs at the sequential element inputs.

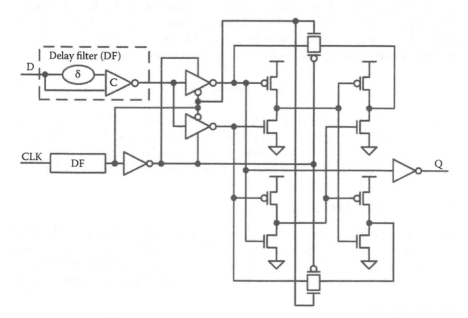

FIGURE 7.9 DF-DICE latch. Breaking the feedback to write greatly increases the transistor count and size but is generally required in modern processes.

7.2.3.2 Temporal DICE FF

While the DICE design is SEU but not SET hard, the temporal latch is fully hardened to both, but at the expense of six unit delay elements, where each delay is slightly longer than the worst-case SET duration to be hardened against. As shown in Section 7.2.4.1, delay elements are expensive in terms of both area and power dissipation. The latter is particularly problematic when applied to the clocks, since they have a high activity factor, as in the DF-DICE design. Using a temporal master with a DICE slave produces the temporal-DICE design in Figure 7.10. This design is hard against D input SETs but not to clock SETs when the clock is low [21]. A clock SET when the clock is high may upset the DICE slave latch as shown; however, a SET glitching the clock high when it is low will at worst propagate a glitch, which has the same effect as a SET at any downstream logic.

The DICE elements in this FF differ considerably from those in Figure 7.9. The DF-DICE completely turns off the feedback to write the two requisite latch nodes at significant circuit complexity. This design jams the values in, requiring the PMOS transistors to overcome the drive of the latch feedback devices. However, there are no transistor savings, since writing a zero through the feedback is difficult and thus all four DICE nodes must be written. The DF-DICE latch adds 12 transistors to the

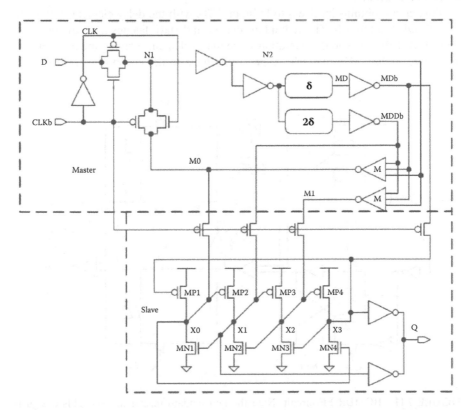

FIGURE 7.10 Temporal DICE FF combination. (After J. Knudsen and L. Clark, *IEEE Trans. Nucl. Sci.*, Vol. 53, No. 6, pp. 3392–3399, Dec. 2006.)

base DICE design of Figure 7.7c for the write function (two tristate inverters and two CMOS transmission gates). The approach used in the DF-DICE design is arguably better—on advanced processes gating the feedback for a write is more robust. Note that this cell does not buffer the input and is falling-edge-triggered.

7.2.3.3 Bistable Cross-Coupled DMR FF

The bistable cross-coupled DMR (BCDMR) FF is a modified BISER, as shown in Figure 7.11 [22]. This design is interesting in that it illustrates the elegance of the DICE approach by being nearly the opposite. Whereas the DICE removed unnecessary extra transistors so that there was redundancy but minimal overhead, the BDCMR design has six latches, four C-elements, and one delay circuit. Moreover, the designers opted to use nonstandard latches in one incarnation of the design. Referring to Figure 7.11, the master latches use a combination of N and P pull-up and pull-down transistors in the master latches, presumably since the PMOS pass gates have difficulty overcoming the feedback inverter NMOS transistors. The slave latches use a conventional NMOS pull down in the slave latches—here the NMOS pulls against a PMOS to jam the logic 0 into the side pulling low. These latches use more transistors than a conventional latch (compare with Figure 7.1b) but a clock inversion is saved.

This circuit obtains hardness to D input SETs with the delay element, following the DF-DICE approach. The redundant master and slave latches have the opposite polarities. This is key to the operation since the C-elements driving the third master and slave latches, KM and KS, respectively, have to drive opposite polarities. In

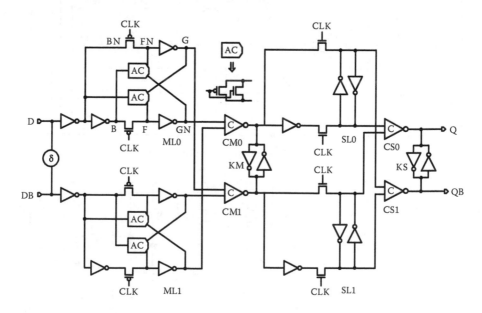

FIGURE 7.11 BCMDR FF circuit. Note the nonstandard master latches. This design is larger than the TMR FF. (After R. Yamamoto, C. Hamenaka, J. Furuta, K. Kobayashi, and H. Onodera, *IEEE Trans. Nucl. Sci.*, Vol. 58, No. 6, pp. 3053–3059, Dec. 2011.)

the event that one of the master latches ML0 and ML1 captures the correct value but the other does not, the value is not propagated and the third master latch that is driven by the C-elements contains the correct value. While latches ML0 and ML1 are transparent, their state is transferred to the third (KM) latch. If they subsequently have mismatching values, the C-elements are tristated. However, the correct value is contained in KM, which drives the slave latches. Like the temporal-DICE design, a positive clock SET will capture the master state into the slave erroneously.

A significant risk of such complicated designs is the large number of added nodes, which in RHBD may also collect charge and contribute to an upset. It also complicates analysis, particularly for susceptibility to MNCC-induced upsets. Like the BISER, this design should buffer the output jam latch, since both C-elements see the same inputs and will thus tristate the output jam latch simultaneously. This FF's t_{SETUP} is among the worst, since it not only adds a delay element but requires propagation through two latches in series. t_{CLK2Q} suffers from the series latches as well.

7.2.3.4 Filter Elements in the Latch Feedback Loop

Figure 7.12 shows a FF design which, like the temporal-latch, add delay elements within the feedback loop [23]. The FF in Figure 7.12 provides hardness against all SEU and SETs but reduces the FF delay element count from six to four, an area and power dissipation improvement over FFs using the temporal latch. This design, and one previous [23,24], introduced full layout interleaving to alleviate MNCC induced failures by interleaving the circuits of multiple FFs. Heavy-ion beam measurements proved the layout interleaving efficacy, which here interleaved portions of four FFs. Modern synthesis and automated placement and routing (APR) tools easily accommodate multiple-bit FF cells, which are also common in unhardened designs.

The design in Figure 7.12 works by adding DFs between the setup and hold nodes. The feed-forward DFs at first seem sufficient—there is only one node left susceptible to upset if the feedback DF is omitted from each latch. Referring to Figure 7.12, without the feedback delay if the output of the feed-forward C-element node MSetup is upset, the value propagates around the feedback loop and disables that C-element, tristating its output. When this happens, there is no circuit removing the charge from

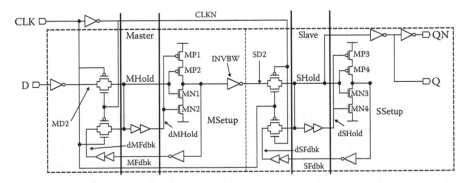

FIGURE 7.12 Temporal FF design with four delay elements that mitigates all SEU and SET. (After B. Matush, T. Mozdzen, L. Clark, and J. Knudsen, *IEEE Trans. Nucl. Sci.*, Vol. 57, No. 6, pp. 3180–3184, Dec. 2009.)

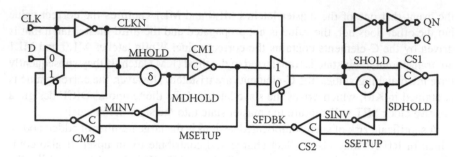

FIGURE 7.13 Temporal FF design using two delay elements that mitigate all SEU and SET. (After L. Clark and S. Shambhulingaiah, *Proc. ISVLSI*, July 2014.)

the C-element output MSetup, and a very long duration SET occurs that is bound to outlast the feed-forward delay element duration. When this happens, the latch is upset. In this design, back-writing from the slave to the master presents a failure mechanism that required adding one inverter between the master and slave, labeled INVBW in Figure 7.12. Collected charge at the MHold node tristates the following C-element CM1. If the clock goes high at this point, the slave latch writes the master feedback loop rather than writing the slave latch, propagating the slave value through the feedback delay filter and thereby writing the slave. Inverter INVBW eliminates this possibility.

We complete the review of derivative FF designs with a modification of this design that uses only two delay elements. This design, illustrated in Figure 7.13 closely follows the design in Figure 7.12, but replaces the feedback delay element with a single C-element [25]. The C-element is implemented in two poly pitches, while the delay element requires at least four, but more likely 12 to achieve a large delay. An extra inverter is required in the feedback path, producing one with four inversions and one from the delay with two inversions. This design appears to be the smallest, lowest power temporal based RHBD FF to date. With appropriate layout and circuit design, it is fully hardened against both SEU and SETs. On a 90-nm low-power (LP) process the removal of the second delay element from both the master and slave latches reduces the FF energy per cycle at full activity factor by 27%. The area is reduced by 19.5%. t_{SETUP} and the FF dead time are similar to the other temporal designs.

7.2.3.5 Layout and Circuit Design Interactions

In unhardened commercial cell libraries, the layout density is a key consideration. This can drive the choice of stack ordering as well as CMOS transmission gate versus tristate inverter, as variations allow greater layout density depending on the process design rules. FFs and latches are increasingly difficult to lay out with a single metal layer on advanced nodes and in high-density libraries, where the cell height can be as small as seven metal pitches tall. For hardened design, density is often sacrificed, as evident from redundant latches, which obviously increase area. However, the impacts are not always straightforward to estimate and the standard approaches to improving density can negatively impact the actual hardness.

We implemented the design in Figure 7.13 in a 90-nm low standby power (LP) process with cells compatible with a seven-track cell library. This cell height requires quite narrow transistors. Conventionally, FF circuit nodes are ordered to provide the smallest area by maximizing contiguous diffusions in the layout. The resulting dense circuit implementation of the FF master latch is shown in Figure 7.14a. Conventionally, the D input should control the outermost devices on the stack but results in a less dense layout. The C-element, multiplexer, and tristate inverter have contiguous diffusions. Unfortunately, the three-deep NMOS and

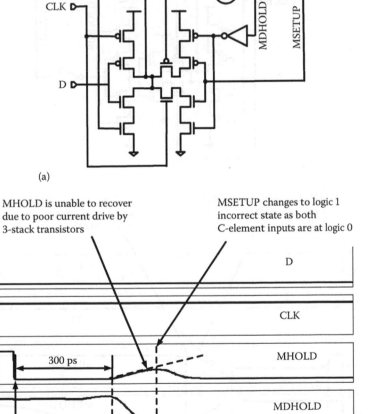

(a)

MHOLD is unable to recover due to poor current drive by 3-stack transistors

MSETUP changes to logic 1 incorrect state as both C-element inputs are at logic 0

300 ps

Low SET induced

(b)

FIGURE 7.14 Initial densest latch design has a three transistor deep PMOS stack (a) that does not provide timely recovery in the event of an SET in the feedback loop (b).

PMOS stacks feeding back to node MHOLD provide very poor drive current, making MHOLD vulnerable to upset. Figure 7.14b shows a low SET of duration 300 ps induced on MHOLD in the clock high phase with logic 0 on data input. At time t1, MHOLD node starts to recover to its original logic state (logic 1). However, due to the poor current drive provided by the three stacked transistors, the recovery time is excessive (see the dotted lines in Figure 7.14). At time t2, the logic state (logic 0) on MHOLD node fully propagates to MDHOLD node after passing through the

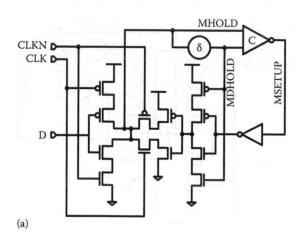

(a)

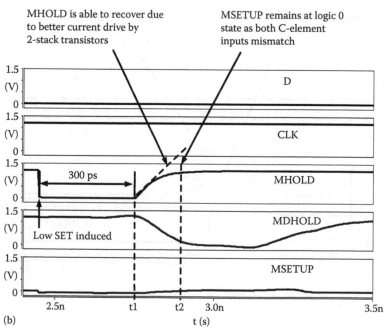

(b)

FIGURE 7.15 Hardness is improved by limiting the stack depth to two (a). The resulting circuit adds poly tracks but has a much better response to upsets in the feedback loop (b).

delay element. At this time, the MHOLD node voltage is still below the switching threshold of the C-element. Since both C-element inputs are at logic 0, its output (MSETUP) switches to the logic 1 state. Positive feedback drives node MHOLD back to logic 0, retaining the incorrect state.

A modified circuit with the same function but reduced stack depth is illustrated in Figure 7.15a. The three-transistor-deep stacks are removed by adding an inverter subsequent to each C-element, resulting in a stack depth of at most two. The feedback loops now have four inversions, but the overall setup time, dominated by the feed-forward delay elements as in all temporally protected designs, remains similar. Figure 7.15b shows the MHOLD node now properly recovering under the same simulation conditions. With the improved slew rate, the MHOLD node voltage is sufficient to cause inputs of the C-element to be in opposite logic states, causing MSETUP node to remain at logic 0 state.

7.2.4 DELAY ELEMENT CIRCUITS

The temporal approaches, which can protect against SET, require delay circuits. The ideal delay circuit has minimum area and power dissipation but large delay. Process scaling results in an approximately 0.7 delay and linear dimension reduction per generation. This in turn reduces the capacitance of the scaled digital circuit by a similar amount, providing power and delay reduction. The delay reduction has until recently provided significant performance improvements but makes producing a large delay increasingly difficult.

7.2.4.1 Producing Delays with Inversions

In hardened processes, the resistors used to harden latches use very little area and do not collect charge. However, delay circuits in RHBD designs are larger, as they are comprised of CMOS transistors, and each of their diffusion nodes can collect charge, so they contribute to the SER. The standard delay circuit comprised of series inverters is shown in Figure 7.16a. A CMOS gate or inverter delay can be approximated by

$$\tau = \frac{CV}{2I_{\mathrm{EFF}}}, \tag{7.3}$$

where τ is the gate delay, C is total node capacitance, and I_{EFF} is the effective current drive of the transistor. Since our goal is to increase the delay in a delay element, a larger capacitance increases the delay linearly. However, power dissipation is solely a function of the circuit capacitance, so increasing C is a clear power efficiency trade-off. Alternatively, a longer channel reduces I_{EFF} and increases the C presented by each stage. In sub-45-nm (and in some 45-nm) technologies, across chip line width (gate length) variation minimization mandates constant gate pitch and lengths, so using a longer channel is ruled out. In this case, at some area penalty, stacked gates (Figure 7.16b) will reduce the driving current. Since more than one inverter is generally required to achieve the needed delay, stacked or longer channel gates also generally increase the capacitive load presented to the preceding inverter as well.

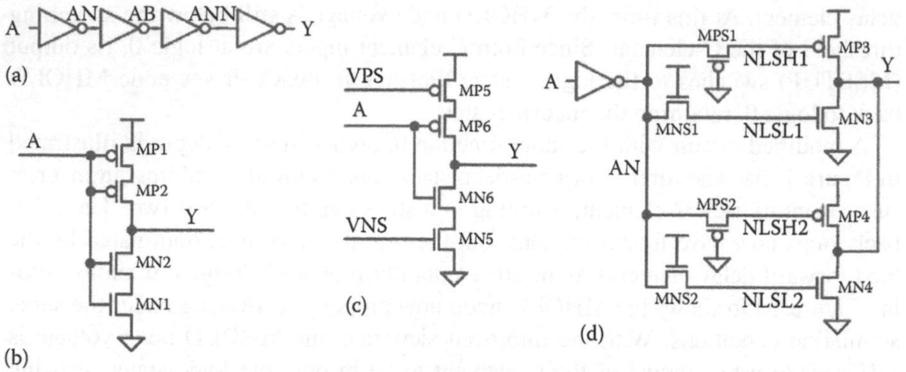

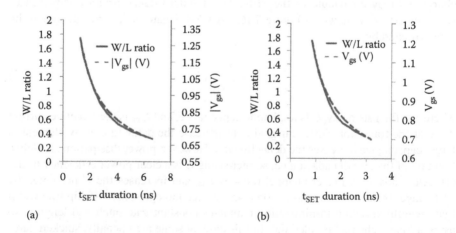

FIGURE 7.16 Delay element circuits: (a) Series inverters, (b) series transistors, (c) current-starved inverter, and (d) redundant low-swing delay element.

There is a limit to the practical channel length or stack depth used in the delay elements. Since I_{EFF} is also the current drive removing collected charge from the circuit, it also controls the duration of SETs that propagate forward when the delay nodes collect charge due to ionizing radiation. Figure 7.17 shows the relative SET duration produced by varying the W/L of the driving transistors. Clearly, t_{SET} increases as I_{EFF} is reduced. Consequently, to avoid their becoming the limiting circuits (i.e., those generating the worst-case t_{SET}), the τ from Equation 7.3 in the delay elements should not be greater than that for the lowest drive gates used in the design. The lowest drive library logic cells are usually four-input NAND (NAND4) and three-input NOR (NOR3) gates, but due to long channels used in hold buffers, they may be worse.

FIGURE 7.17 Comparing PMOS (a) and NMOS (b) SET duration (t_{SET}) versus W/L ratio and with reduced gate overdrive on a 130-nm process. (After S. Shambhulingaiah et al., *Proc. RADECS*, pp. 144–149, 2011.)

7.2.4.2 Current-Starved Delay Elements

The designs in [5] and [26] used current-starved delay elements (Figure 7.16c). They have the advantage of having low current drive in a minimal area but require two analog control voltages that limit current by reducing the gate overdrive V_{GS}–V_T, to produce a long delay. For long delays the biasing nodes (VPS and VNS) voltage magnitudes are reduced to near V_T, which means that small amounts of noise significantly affect the delay. We have abandoned this approach due to concerns that the analog voltages require shielding to be reliable and that this significantly complicates automated place and route (APR). Moreover, the longest bulk CMOS SET durations, exceeding 3 ns, have been measured while using current-starved delay elements in the measurements [5].

7.2.4.3 Low Gate Voltage Redundant Delay Element

Figure 7.16d shows a design that produces a long delay by reducing the gate overdrive on stacked transistors [27]. The goal of this design was to produce a long delay without producing the worst (i.e., limiting) t_{SET} on the design. The circuit is redundant, so we focus first on the top circuit comprised of nodes AN, NLSH1, and NLSL1. Transistor MPS1 produces a low swing at node NLSH1 of V_{DD} to V_{TP}. Similarly, transistor MNS1 produces a limited voltage swing on node NLSL1 of 0 V to V_{DD}–V_{TN}. These reduced swing nodes drive PMOS and NMOS transistors MP3 and MN3, respectively, providing a reduced gate overdrive V_{GS}–V_T, which provides the low drive current-starving function without the need to route an analog voltage. Node NLSH1 has only P-type diffusions, so it can only be driven high by charge collection at that node, cutting off transistor MP3. However, the redundant MP4 transistor then drives the output node Y. Similarly, redundant transistors MN3 and MN4 protect nodes NLSL1 and NLSL2 from SETs, and since these nodes are N-type diffusions, they can only collect electrons to drive them low, cutting off one of the NMOS devices. By spatially separating the low swing nodes, the worst-case delay is twice the nominal delay. The circuit provides about 25% power reduction over series inverters of the same delay. However, the area reduction is modest.

7.2.5 Taxonomy and Comparison

We can now summarize the relative area, timing, and soft-error resiliency of the designs. We estimate the delay elements as six inversions, thus comprising an area of six inverters. However, this is optimistic; test layouts on 90 nm through 32 nm have required up to 12 tracks when using CMOS gates for delays greater than 400 ps. Table 7.1 compares the different designs.

The metrics are open to some interpretation. The t_{SETUP} and t_{CLK2Q} follow from counting inversions. This "hand" analysis avoids differences in sizing and focuses on the relative merits of the topologies. The transit time through a CMOS pass gate is counted as one-half an inversion delay. The DICE FF uses the version with clocked feedback, which is the most commonly used in modern designs. The BISER and BCDMR designs assume a buffered output—making a latch storage node a pin is unreliable, as mentioned previously. The BCDMR gains in t_{CLK2Q} by not requiring a clock inversion to open the slave latch. The temporal DICE design benefits similarly.

TABLE 7.1

Comparison of Hardened FF Designs with an Unhardened FF Baseline

FF Type	Ref.	DICE	Temporal	Simple	t_{SETUP}	t_{CLK2Q}	SET	SEU	Transistors
		Basis					**Relative Hardness**		
Unhardened					2	2.5	0	0	20
DICE	[28]	X			2	3	0	1	48
Temporal	[25]		X		9.5	2.5	1	1	112
TMR FF				X	2	3	0	1	70
BISER	[29]			X	2	4	0	1	50
BCMDR	[30]		X	X	11	3.5	0.9	1	82
DF-DICE	[28]	X	X		2	9	0.9	1	72
Temp-DICE	[22]	X	X		10.5	2	0.8	1	68
4-DF FF	[27]		X		8	2.5	1	1	76
2-DF FF	[9]		X		8	2.5	1	1	64

Hardness is considered 0 for designs that do not mitigate an effect and 1 if all such effects are mitigated up to the t_{SET} designed for them. All of the designs evaluated are SEU hard with appropriate critical node spacing, which is assumed in the table. Many of the designs have been demonstrated to have failures, particularly DICE on modern processes [31,32], without careful layout to mitigate MNCC. The DICE FF has the least increase in transistor count over the baseline FF and approximately the same as the BISER, which is also only SEU hardened. The temporal latch-based FF has the most transistors but excellent hardness. Reaching the same hardness at low transistor count is problematic. The BCMDR, DF-DICE, and Temp-DICE are soft to specific but not all clock SETs. The four- and two-delay element delay filter-based FFs have the expected increase in t_{SETUP} and dead time but achieve hardness to all SEU and SET with considerable transistor count savings over the temporal latch-based FF.

7.3   DESIGN-LEVEL METHODS FOR HARDNESS ANALYSIS

Cell design is most efficient at the circuit design and simulation level. At the beginning of the design process, we have not yet chosen the circuit, let alone the specific layout, so device-level simulations require at that point in time undetermined layout-specific information. In this section, we describe basic circuit-simulation-based upset modeling and procedures for accurately verifying hardened latch and FF hardness. The analysis extends to MNCC analysis and hardening by methodically determining MNCC failure scenarios and automatically determining the requisite critical node spacing.

7.3.1   CIRCUIT SIMULATION-LEVEL MODELING

7.3.1.1   Upset Modeling

To simulate upsets, some sort of circuit-level upset model is required. For SOI ICs, a bipolar collection model has been demonstrated to be effective [28].

Nonetheless, the most commonly cited option for determining circuit response to collected charge is to model the charge with double exponential current source. This is simple and supported directly in commonly used SPICE-based circuit simulators. The response to such a current source is shown in Figure 7.18. Note that the response is nonphysical, greatly exceeding the power rail in amplitude, maybe alarmingly so (negative 3 V) depending on the current source parameters chosen. It is nonphysical because as the voltage substantially exceeds the supply rail, the charge-collecting diode becomes forward-biased, favoring injection into the substrate rather than collection from it. Moreover, a change in the driving circuit will not necessarily affect the response in a physical way, particularly since the driving transistor drain voltage may be grossly in error. The duration of the upset is entirely a function of the model, with the driving circuit having limited impact.

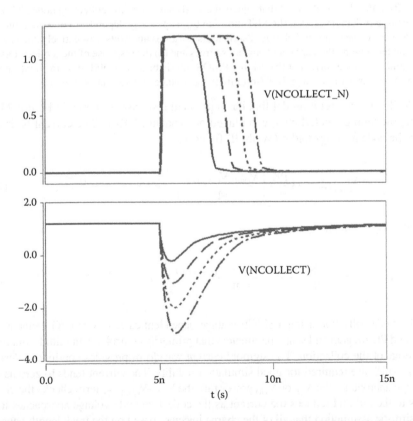

FIGURE 7.18 Voltage response of the collecting node NCOLLECT and the voltage response of the node after one inversion (node NCOLLECT_N). The former is grossly nonphysical, producing voltages well beyond the power rail depending on the choice of parameters.

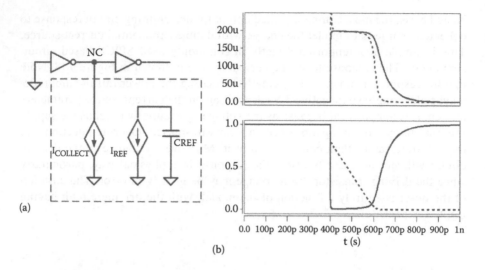

(a)

(b)

FIGURE 7.19 Physical model that injects the total charge into the collecting node NC. With this charge collection model, the SET duration t_{SET} is properly dependent on the driving circuit as well as the deposited charge. Y-axis units are amps and volts, respectively. (a) The circuit model showing the dependent current sources and (b) the response of the Icollect (dashed line) and total node current at the driving PMOS transistor drain (solid line) transients at top; (b) bottom shows the voltage SET (solid line) and voltage on Cref (dashed line).

A SPICE subcircuit model that has a physical response is shown in Figure 7.19a. Here, the total collected charge is placed on capacitor CREF. The current is driven into the node as a sigmoidal (Weibull) function

$$I_{COLLECT} = G\frac{PK}{PL}\left\{ \frac{V_{CREF}V_{COLLECT}}{PL} \right\}^{PK} e^{-(V_{CREF}V_{COLLECT}/PL)^{PK-1}}, \qquad (7.4)$$

for N+ node collection, and

$$I_{COLLECT} = G\frac{PK}{PL}\left\{ \frac{V_{CREF}(V_{DD} - V_{COLLECT})}{PL} \right\}^{PK} e^{-\{V_{CREF}(V_{DD}-V_{COLLECT})/PL\}^{PK-1}}, \quad (7.5)$$

for P+ node collection using a SPICE voltage-dependent current source (G-element). G, PK, and PL are gain and shape parameters that primarily control the amplitude and trailing edge of the collection. The sigmoid current waveform provides continuous derivatives, which are required for good simulation stability. The current tends to zero as the voltage approaches the V_{SS} or V_{DD} power rail; the V_{DD}–$V_{COLLECT}$ term checks the closeness to the rail and reduces the current as the collection node voltage approaches it. A pessimistic assumption that all of the charge injected, based on the track length beneath the target diffusion node and LET, is collected and can be used to determine the initial charge on capacitor CREF as $Q_{COLLECT}$. Thus, CREF begins with $Q_{COLLECT}$ as an initial condition by sizing it appropriately (for a 1 V initial condition, C and Q are the same).

A current-dependent current source removes charge from CREF as it is injected into the collection node, so that following Equations 7.4 and 7.5, injection ceases when the charge is exhausted. The simulated circuit response is shown in Figure 7.19b. The collection current is the top dashed waveform and the total current comprises the solid line. The latter extends the duration as the RC time constant of the driving transistor and the load capacitance (i.e., when collection ceases, the transistor drives the node to the original rail, but removing the charge stored on the node produces the standard RC response). The bottom voltage waveforms show the collecting node NC voltage (solid line) and the voltage on capacitor CREF (dashed line). This model always provides a physical result, and changing the load circuit correctly changes t_{SET}. The SET duration is properly a function of the driving circuit's ability to remove the charge [33], which can be approximated as

$$t_{SET} \propto \frac{Q_{COLLECT}}{I_{EFF}}, \tag{7.6}$$

where I_{EFF} in this case is that for the driving circuit. When the driving transistor saturates, the charge removal is limited. The transient continues as collected charge is removed from the node but fed in at the same rate, maintaining the upset. Consequently, the response is very physical, with SET duration as a function of the driving circuit and total charge collected.

For many simulations a simpler circuit, which drives the affected node to the power rail for a fixed period of time, is adequate. This simple model, comprised of switches with low or high impedance that simply force a node to the rail for the t_{SET} desired in the analysis, is used in much of the subsequent analysis.

7.3.2 MITIGATING MNCC

Broad-beam ion testing uses directional ionizing radiation particles, but ionizing particles causing upset in normal IC operation can impinge from any direction (i.e., are isotropic). Secondary particles from proton or neutron radiation are also isotropic. Consequently, ion testing needs to be performed from all possible directions to completely quantify hardness. In [34] the authors described FF design iteration using broad-beam measurements around the entire sphere of possible ion track directions around the target node to determine critical node spacing. This approach is impractical for most designs from both a time and budget perspective. Alternatively, comprehensive 3-D device simulations, which have shown rough equivalence, can be used [32]. Device simulation is still orders of magnitude slower than circuit simulation and requires a known cell layout. Ideally, for design hardening, the layout can be rapidly iterated or better, determined algorithmically. Numerous approaches to estimating SER have been proposed [35]. Most are based on the rectangular parallel-piped approximation of the sensitive collection volume [36]. However, different devices and circuit structures have different sensitive volumes. Fulkerson et al. showed a method to estimate the sensitive volume boxes for designs that required an ion to simultaneously strike two such critical nodes to create an upset [37]. They showed good agreement to experimental beam measurements on SOI. In this section, circuit-design-level methodologies to systematically determine latch and FF upset tolerance, as well as which nodes must be separated to avoid upset due to MNCC, are described. They follow the approach outlined for SOI in [37]. Additionally, algorithmic approaches to grouping circuit nodes and a simple circuit simulation model of upsets are shown.

7.3.2.1 Spatial Node Separation to Ensure MNCC Hardness

Calin et al. first described DICE failures due to MNCC [38] where certain layout topologies were found to collect charge simultaneously, resulting in a defeat of the DICE hardening. In this 1.2-μm process, the natural node separations were greater than 2 μm. Laser probing showed that the culprit was shared source nodes and a mitigation technique based on separation with an intervening well tap was suggested. Figure 7.20 shows some interleaving approaches that have been used. Experimental measurements of hardened FF designs that did not carefully control critical node spacing have shown similar failure rates to unhardened designs [31]. Knudsen et al. introduced systematic spatial separation of FF nodes in [26], as shown in Figure 7.20a, and broad-beam testing showed it was effective. The insertion of white space, which affords good separation but at large area expense (Figure 7.20b) has been used [32,34]. Interleaving the constituent FF circuits or by interleaving multiple FFs in a multibit FF macro as shown in Figure 7.20c avoids the wasted circuit area but still provides adequate critical node spacing. These approaches have been demonstrated to be effective [23] but were based on *ad hoc* node clustering. Interleaving one FF's constituent circuits was also presented for both vertical and horizontal interleaving in [27], where a systematic analysis approach was first suggested. Horizontal interleaving was recently shown to be effective for DICE latches on an SOI process [39]. DICE latches are, however, very difficult to spatially separate efficiently on bulk CMOS processes. Vertical interleaving provides intervening N-wells, which are very effective at collecting charge in the substrate, particularly at angles near parallel to the surface that are the greatest threat. This approach is shown in Figure 7.20d. Vertical interleaving, while most effective, can result in awkward cell shapes. This is not necessarily a problem. Figure 7.20e shows nonrectangular FFs with APR. The APR tool can place other cells into the notches with some minimal CAD flow changes.

While interleaving multiple FF circuits in a multibit design is effective, in some cases it is overkill (i.e., sufficient but not necessary). The next sections present a general design methodology for determining which nodes need to be separated and for estimating the relative cross sections of designs without resorting to device simulations or beam testing.

7.3.2.2 Multiple Node Upset and Spatial Separation

The cross section of an upset caused by charge collection at a single node is can be approximated by the node area (plus the deposited charge track diameter). Obviously, no hardened FF should be susceptible to single-node upsets, although many are. For instance, the DF-DICE design is susceptible to upset when the output of the DF circuits is struck. Nominally, for an MNCC upset both nodes must be near the passing ion track. Figure 7.21 shows the projection of the second node on the surface of a sphere with its center at the first collecting node. The silicon plane passes through the equator of the sphere. We assume the particle passes through node A and close enough below node B for simultaneous charge collection. The depth of collection defines the vertical

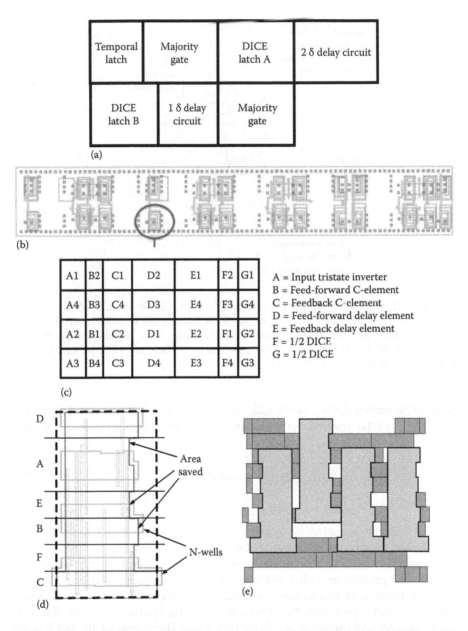

FIGURE 7.20 Layout topologies to separate FF and latch critical nodes. (a) Temporal-DICE circuit interleaving in 130-nm bulk CMOS. (b) DICE FF separation in 90-nm bulk CMOS. (c) Interleaved FFs in 90-nm and (d) vertically interleaved constituent components. (e) Nonrectangular FFs in an APR flow. (Adapted from and after J. Knudsen, M.S. Thesis, Arizona State University, Dec. 2006; J. Knudsen and L. Clark, *IEEE Trans. Nucl. Sci.*, Vol. 53, No. 6, pp. 3392–3399, Dec. 2006; B. Matush, T. Mozdzen, L. Clark, and J. Knudsen, *IEEE Trans. Nucl. Sci.*, Vol. 57, No. 6, pp. 3180–3184, Dec. 2009; S. Shambhulingaiah, S. Chellappa, S. Kumar, and L. Clark, *Proc. ISQED*, pp. 486–493, March 2014; K. Warren et al., *IEEE Trans. Nucl. Sci.*, Vol. 56, No. 6, pp. 3130–3137, Dec. 2009.)

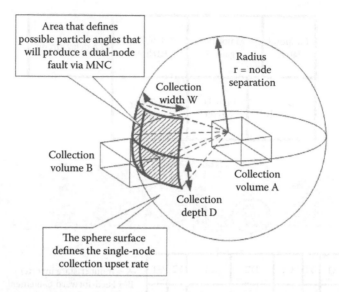

FIGURE 7.21 Multiple node charge collection from a single-ion track. A sphere centered at one of the collecting nodes has a radius that is the distance to the second node. The angle subtended by the arc matching the width (plus the charge track radius) provides the collection width W. The depth of collection provides the vertical angle. (After L. Clark and S. Shambhulingaiah, *Proc. ISVLSI*, July 2014.)

extent of the surface defined by the solid angle and the width of the second node, plus the charge track diameter, the horizontal extent. The area of the surface thus defined is

$$A = \int_{-\varnothing}^{\varnothing} \int_{-\theta}^{\theta} \sin\phi \, d\phi \, d\omega, \qquad (7.7)$$

where θ is the angle subtended by twice the vertical collection depth and ω is the angle ϕ subtended by the collection width.

We are interested in a design procedure, and the actual collection is sensitive to second-order parameters such as the size and voltage of adjacent diffusions that are not easily understood in a circuit-level analysis. Consequently, within the accuracy possible, we can approximate the area by assuming the sphere is relatively flat, estimating the sensitive surface as A = 2DW [25]. Twice D accounts for the fact that the collecting diffusion positions are interchangeable (i.e., the particle may pass through node B and under node A). The relative cross section of upsets caused by MNCC is then multiplied by the area of the rectangle in Figure 7.21 divided by the area of the sphere, which represents hitting node A from any angle. This produces an effective MNCC cross-section of

$$\sigma_{\text{NODE}_{\text{MNCC}}} = (\sigma_{\text{Node A}} WD)/(2\pi r^2), \qquad (7.8)$$

where $\sigma_{Node\ A}$ is the cross section of node A (herein approximated as its area). With Equation 7.8 we can estimate the cross section of MNCC-induced errors. The methodology below will catch any faults that can occur from single-node collection. These will dominate MNCC collection faults if adequate critical node spacing (r) is used.

7.3.2.3 Systematic Fault Analysis

In our own designs, we have experienced inadvertent hardness escapes when using *ad hoc* simulation-based verification. One example is the Temp-DICE design described in Section 7.2.3.2, where there was a significant failure cross section at high LET attributable to MNCC at two adjacent nodes. The FF design in Figure 7.12 was derived from the design in [23,24] that also had two delay elements per latch. Our goal was to reduce the required delay elements to one per latch, meeting the size of the DF-DICE design but with easier node interleaving. The original design, which lacked the feedback delay element (see [25]) had a single-node collection failure at the C-element output. This failure mode was removed by adding the feedback delay element, but at large area cost of two more delay circuits per FF. A systematic hardness validation approach that checks for all possible upsets avoids such design hardness escapes. The approach is related to dual-fault analysis long applied for reliable systems [30]. However, it differs in that the conventional dual-fault analysis is bottom up, whereas through circuit simulation, it is practical to simulate every possible dual-node MNCC case for a FF or latch. Analysis of failures then proceeds top down.

A rigorous simulation-based MNCC analysis of simultaneous node upsets can be performed by inducing temporary voltage reversals on all of the possible collecting node pairs [29]. In the simplest case, node voltage upsets are modeled using a SPICE voltage-controlled resistor (VCR) element, which is essentially an ideal switch. The nodes are held for a worst-case t_{SET} duration, regardless of the driving circuit strength. This simplification, while nonphysical, helps avoid minor circuit strength changes significantly affecting the analysis. This in turn, keeps the focus on the circuit topology rather than sizing. Figure 7.22 illustrates the methodology.

All transistor diffusions can collect charge except for those connected to power rails. As mentioned above, upsets at N+ diffusions drive nodes from V_{DD} to V_{SS} and upsets at P+ diffusions produce positive voltage upsets corresponding to electron and hole collection, respectively. Applying positive upsets to N+ and negative upsets to P+ can induce more errors (for instance, inadvertent writes of a logic one through NMOS transistors) but is not physical. A FF containing N nodes has $^{N}C^{2}$ total possible node pairs. In the analysis, upsets are injected on one pair of nodes at a time. To account for all possible FF states when a strike happens, the upsets are induced at times corresponding to the clock rising, clock high, clock falling, and clock low conditions as shown by windows A, B, C, and D, respectively, for both of the D input logic states (see Figure 7.22). Finally, the FF output depends on the master and slave hold node states, so the master and slave hold nodes are initialized to each of the four possible states as well. For each simulation run, the expected output for that case is determined by simulating with no upsets. Simulation runs that produce FF output or stored state mismatching the expected values indicates a hardening failure due to MNCC. There is a (we believe remote) possibility that a third node upset would

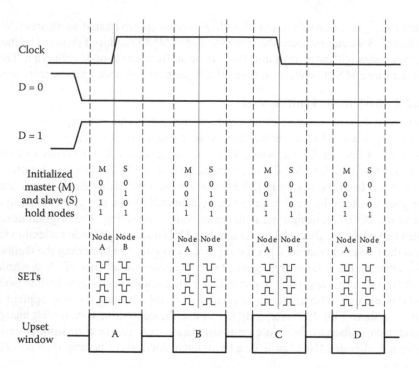

FIGURE 7.22 Simulation methodology. All FF conditions are simulated, master and slave latch opaque and transparent, D = 1 and 0, and clock rising and falling. All susceptible nodes have voltage transients applied for each condition. (After S. Shambhulingaiah, S. Chellappa, S. Kumar, and L. Clark, *Proc. ISQED*, pp. 486–493, March 2014.)

repair the incorrect logic state produced by two nodes collecting. Neglecting these cases makes the analysis conservative in that it predicts a slightly higher failure rate.

The use of this methodology is demonstrated through some examples. Figure 7.23a shows a DICE FF circuit and the resulting node groupings. The master and slave latches are identical. Since the DICE latch has four storage nodes, each has to be in a different group [29]. The resulting dual-fault result matrix comprises Figure 7.23b. Dual MNCC cases that pass (i.e., do not cause erroneous operation) are marked gray and failing node pairs (i.e., where the collected charge upset the FF state in the simulation) are black. The upper diagonal is symmetric, reducing the simulation effort by nearly one-half. Bold outlines represent the circuit groups (i.e., nodes that are laid out together). The groupings are determined postsimulation—algorithmic approaches for node grouping are discussed in Section 7.3.2.4. The matrix diagonal represents single node failures. Ideally there are none for a truly hard design, since single-node failures will dominate MNCC-induced failures. Since nodes in the same group presumably reside in close proximity (i.e., may be adjacent), groups on the diagonal should not have any failures. Nondiagonal groups can have failures but must not be spatially adjacent to mitigate MNCC redundancy defeats. It naturally follows that group pairs that do not have failures may be adjacent.

For the DICE design each latch is divided into six groups as shown. Since the DICE latch is not hard to clock SETs, upsets were induced only within the clock high and clock low phases (i.e., hold modes for the master and slave latches). Figure 7.23b shows that slave groups nodes G, H, I, J, K, and L have no failures with any of the master group nodes. Thus, a simple layout arrangement to harden the DICE against MNCC interleaves the master nodes with the slave nodes, resulting in an ordering A-G-B-H-C-I-D-J-E-K-F-L. Unfortunately, the DICE elements are quite small, with many occupying only two to four poly pitches. Moreover, proper ordering does not guarantee that sufficient separation is achieved. Referring to Figure 7.3, about 2 µm

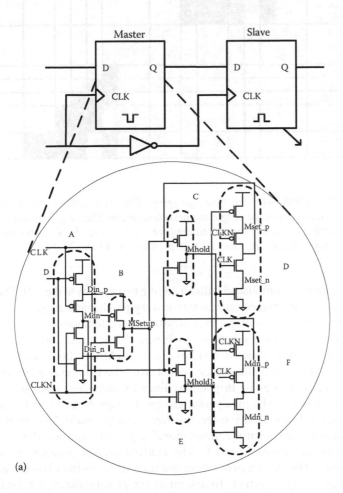

(a)

FIGURE 7.23 (a) DICE FF schematic. The slave follows the master with internal nodes having the same naming convention (and order). Circuit groups are outlined. Note the large number of circuit groups required for DICE.

(Continued)

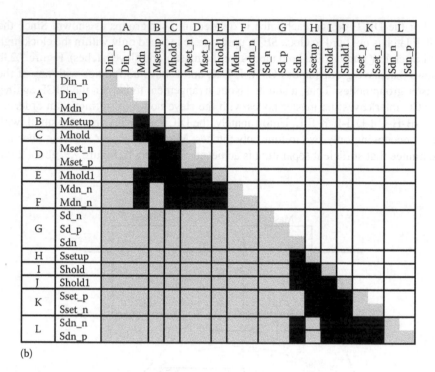

(b)

FIGURE 7.23 (CONTINUED) (b) Node grouping. The slave follows the master with internal nodes having the same naming convention (and order). Circuit groups are outlined. Note the large number of circuit groups required for DICE. (After S. Shambhulingaiah, S. Chellappa, S. Kumar, and L. Clark, *Proc. ISQED*, pp. 486–493, March 2014.)

is ideal, but cannot be accomplished without adding blank areas to the cell. The large number of circuit groups makes vertical interleaving problematic as well.

Application of the methodology to a majority voted temporal FF is shown in Figure 7.24. Referring to the schematic in Figure 7.24a, circuit groups B and E have two delay elements connected in series and nodes Md21mhold and Sd21shold are, respectively, the intermediate nodes connecting the two delay elements. The initial node grouping can be by inspection. Since the delay elements are hardening elements, they have to be in separate groups. Groups A, B, D, and E contain the delay elements. The majority voter, inverters, and the multiplexer in both the master and slave latches form a combinational logic path and are thus grouped together, comprising groups C and F. The analysis results produce the matrix shown in Figure 7.24b. As required, there are no failures in the diagonal groups, indicating the grouping is correct. To determine the group ordering for the layout, first the illegal adjacencies are determined. There are six groups to be interleaved and hence there are 15 possible group adjacencies. Referring to Figure 7.24b, the illegal adjacencies are found to be A-C, B-C, C-E, D-F, and E-F. One legal ordering for the full FF is A-D-E-B-F-C.

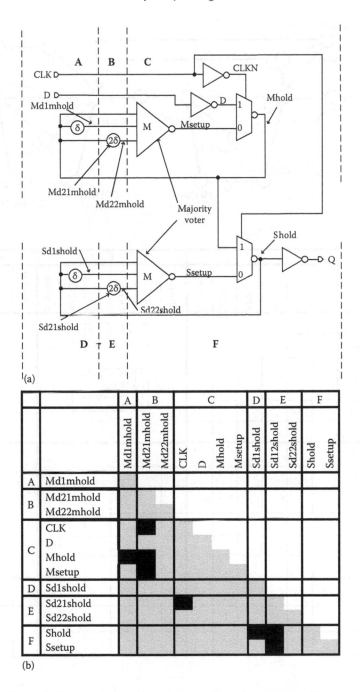

FIGURE 7.24 (a) Temporal FF with majority voting schematic and (b) the node matrix obtained after SET simulations. Note that the 2d delay element is produced with two series unit delays in the actual circuit, and hence has two collection nodes. (After S. Shambhulingaiah, S. Chellappa, S. Kumar, and L. Clark, *Proc. ISQED*, pp. 486–493, March 2014.)

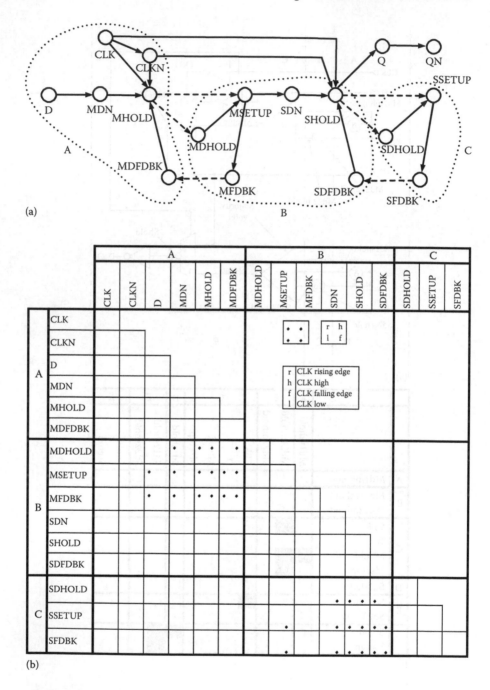

(a)

(b)

FIGURE 7.25 DAG showing the necessary and optional circuit groupings for the temporarily protected FF with delay elements in the feedback loop (a). Fault matrix with indications of the type of failure, which aids design analysis (b). (After L. Clark and S. Shambhulingaiah, *Proc. ISVLSI*, July 2014.)

7.3.2.4 Node Group Ordering

While node groups determined by inspection can be effective, an algorithmic approach to the node grouping based on heuristics comprises this section. Circuit node grouping is based on the following heuristics: First, all hardening element output nodes should be in different groups. Hardening elements are C-elements, majority gates, delay circuit output nodes, or DICE nodes. Second, nodes that comprise combinational logic paths should be in the same group. This follows intuitively—if one of the early stage nodes collects charge and transitions to a new voltage state, the SET propagates to affect other nodes in the logic paths with minimal delay. In the analysis, such combinational logic constituent nodes may be treated as one supernode.

To illustrate the methodology, we apply it to hardened FFs. Referring back to Figure 7.12, the temporal FF has six hardening elements: four delay elements and two C-elements. Determining the circuit groups is essentially a node clustering problem where representing the circuit as a directed graph (digraph) is helpful [25]. Figure 7.25a shows the digraph derived from the hardened FF circuit. Digraph edge arrows indicate the signal flow. Combinational paths connecting nodes D-MDN, MDN-MHOLD, CLK-MHOLD, CLK-CLKN, CLKN-MHOLD, and MHOLD-MDFDBK are connected with solid arrows, indicating they may reside in the same group. Hardening node connections MHOLD-MSETUP, MHOLD-MDHOLD, and MFDBK-MDFDBK are labeled with dashed arrows. Dashed edges indicate what must be cluster boundaries. All nodes within a delay element are considered as the same supernode, as they can affect the output similarly. Many nodes (e.g., CLK, CLKN, D, MDN, Q, and QN) may reside in any cluster. Since the design is intended to be fully hard to and input clock SETs, these could have originated externally, so an internal upset on these signals must be ok with any grouping. These nodes provide some measure of flexibility in balancing the circuit group sizes. The dashed lines indicate an initial clustering comprised of three groups, based on the digraph.

Figure 7.25b shows the dual-fault matrix with thick lines demarcating the three-node groups. This dual-fault matrix adds indications of when the failures occur. Dots in the upper left, upper right, lower left, and lower right indicate that the error occurred at the clock rising edge, during clock high, at the clock falling edge, and during clock low, respectively. We have found this very helpful in aiding debug and understanding of the fault analysis results.

The clusters can be analyzed as an adjacency matrix by putting a 0 in any group entry that may not be adjacent to groups that cause MNCC fails, with other entries indicating a legal adjacency as shown in Figure 7.26a. With these illegal adjacencies, there is no contiguous legal group ordering for the FF layout, as evident in the adjacency matrix with an all 0s column (dashes count as a 0, since a cell must be adjacent to itself). With these clusters the best area solution is to interleave the circuit groups of two FFs. Figure 7.26c shows the digraph derived from adding clusters to achieve an adjacency matrix that provides a legal ordering to allow a single, rather than spatially interleaved, multibit FF layout. The adjacency matrix in Figure 7.26b shows the results after the circuit has been repartitioned as indicated by the digraph. This adjacency matrix indicates that there are legal group orderings.

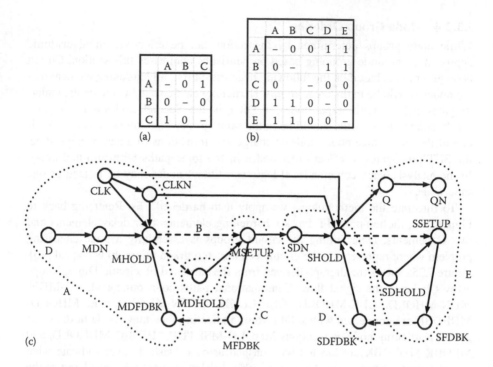

FIGURE 7.26 Adjacency matrix showing there is no single flip-flop circuit interleaving ordering using three groups (a). An adjacency matrix (b) for the five groups shown in (c). Locations of the clock and output inverters are arbitrary and should thus derive from the best layout area. (After L. Clark and S. Shambhulingaiah, *Proc. ISVLSI*, July 2014.)

Figure 7.27a shows the dual-fault matrix and digraph derived from the improved version of this FF (Figure 7.13) that uses only two delay elements. Again, dashed connectors indicate hardening circuit output nodes. The digraph is shown in Figure 7.27b. The illegal adjacencies are A-B, A-C, B-C, C-D, C-F, and E-F, resulting in a usable adjacency matrix. Group C has only one legal adjacency, E, so C must be at the end, followed by E. One legal grouping is C-E-B-F-A-D.

7.3.2.5 Estimating MNCC Cross-Section Improvement

The analysis can again be extended to allow estimation of the overall FF cross sections. Errors on the diagonal of the dual-fault matrix indicate that a failure will occur if that node is struck. The timing indicated by the dot notation in Figures 7.25b and 7.26a aids in that we assume the clock is high half the time and low half the time but that susceptibility to capturing an SET at a clock edge has a small timing window. The dual-fault matrices generated by the analysis are inputs to estimate the relative FF cross-sections σ_{FF} by summing the individual node cross sections; that is

$$\sigma_{FF} = \sum_{i=0}^{N} \sigma_{NODE_i}. \tag{7.9}$$

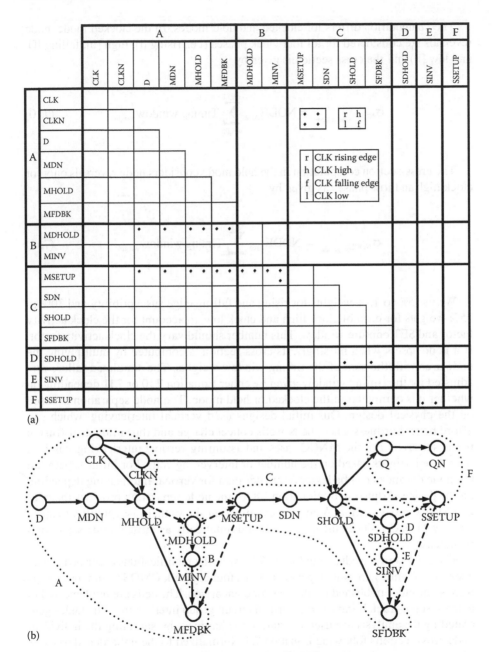

(a)

(b)

FIGURE 7.27 (a) Dual fault matrix for the two delay circuit temporarily hardened FF. (b) DAG showing the circuit groupings to avoid MNCC induced upset. (After L. Clark and S. Shambhulingaiah, *Proc. ISVLSI*, July 2014.)

The computation differs for clocked and hold modes. In the clocked mode, node reversals are considered for all four clock phases (i.e., rising [r], high [h], falling [f], and low [l]) and the cross section is given by

$$\sigma_{\text{NODE}_{\text{Single-clocked}}} = \text{NODE}_{\text{Area}} \sum_{\text{weight}}^{r,h,f,l} \text{Timing window}_{\text{weight}}. \qquad (7.10)$$

The cross-section computation in the hold mode considers node reversals only for clock high and low phases as given by

$$\sigma_{\text{NODE}_{\text{Single-hold}}} = \text{NODE}_{\text{Area}} \sum_{\text{weight}}^{h,l} \text{Timing window}_{\text{weight}}. \qquad (7.11)$$

We use 5% to 15% weights for rising and falling edge susceptibility and 30% to 45% weights for each of clock high and clock low, to account for the clock activity factor and SET capture window. This number should vary the clock activity factor. If a node upsets when hit singly, its cross section is computed by multiplying the node area with the sum of the timing window weights. The simulated failure rate adjusted for the timing window given by either Equation 7.10 or 7.11 depending on whether σ is computed in the clocked or hold mode. The node separation is based on the physical design. Our initial designs used vertical interleaving, which can afford better hardness since the N-wells collect charge and thus mitigate diffusion in the substrate. For the MNCC cases and assuming vertical interleaving, r is 1, 2, 3, or 4 cell heights based on the number of intervening cells. For inline separation the node distances can be lesser or greater than for vertical interleaving depending on the design. Adjusting by Equation 7.8 for the node separation provides the relative MNCC cross section. A cursory examination shows that susceptibility to upset only by MNCC reduces the cross section dramatically as critical node separation is increased.

The FF design introduced in Figure 7.13, with the modifications described subsequently in Figure 7.15, was implemented in a foundry bulk CMOS 90-nm process as were some other published FF designs for comparison. The cells are implemented in a dense seven-track library using vertical circuit group interleaving. The tables generated by the analysis are used to analyze the hardness by summing the individual node cross sections following Equation 7.9. Normalized to the unhardened baseline FF cross-section, the Figure 7.13 FF when vertically interleaved has a cross-section that is reduced by over 98% [25]. The relative cross section of the design with four delay elements per latch (Figure 7.12) is reduced compared to the baseline unhardened FF by 97%. The larger overall area (more MNCC susceptible nodes) accounts for the larger cross section in the four-delay element design. Different SET vulnerability timing assumptions will affect the numbers, but the method is effective for comparing design hardness without resorting to 3-D device simulation.

7.4 CONCLUSIONS

There are numerous hardened latch and FF designs in the literature. This chapter has reviewed the basic approaches as well as some designs using combinations. Careful analysis of many proposed FFs shows their basis in the original redundant, DICE, or temporal designs. We have found that carefully redrawing the schematics is often necessary to make this apparent. In general, an SEU hardened design, with reasonable critical node physical separation, can achieve about 90% cross-section reduction at high LET assuming anisotropic ions, depending on the operating frequency. Hardening for SETs on the clock and D inputs provides about an order of magnitude further reduction depending on the assumed timing windows and data activity factors. A significant drawback of SET hardening is the impact of the delay elements in the setup or t_{CLK2Q} timing on the overall maximum circuit frequency. Minor mistakes and poor electrical design can significantly affect the hardness of any approach. Consequently, careful validation is required for high-quality hardened FF design.

We have reviewed simple simulation-based approaches to hardness verification and for providing correct critical-node separation in hardened cell layouts. Broadbeam testing will always be required to verify hardness. However, it may not be possible within reasonable budgets and schedules to fully analyze a FF design over the entire 360° arc of potential upsets. By assuming that the worst-case angle can occur, circuit-simulation-based analysis can ensure that there are no inadvertent hardness oversights in a particular circuit's topology and layout. Moreover, simulation is ideal to allow understanding of the upsets that are uncovered, whereas root causes can only be inferred when proton, neutron, or ion testing produces an unexpected failure. It is also sufficient for reasonably accurate circuit hardness comparisons.

REFERENCES

1. P. Hazucha and C. Svensson, Impact of CMOS technology scaling on the atmospheric neutron soft error rate, *IEEE Trans. Nucl. Sci.*, Vol. 47, No. 6, pp. 2586–2594, Dec. 2000.
2. P. Dodd and F. Sexton, Critical charge concepts for CMOS SRAMs, *IEEE Trans. Nucl. Sci.*, Vol. 42, pp. 1764–1771, 1995.
3. T. Makino et al., Soft-error rate in a logic LSI estimated from SET pulse-width measurements, *IEEE Trans. Nucl. Sci.*, Vol. 53, Vol. 6, pp. 3575–3579, Dec. 2006.
4. B. Narasimham et al., Characterization of digital single event transient pulse-widths in 130 nm and 90 nm CMOS technologies, *IEEE Trans. Nucl. Sci.*, Vol. 54, No. 6, pp. 2506–2511, Dec. 2007.
5. J. Benedetto, P. Eaton, D. Mavis, M. Gadlage, and T. Turflinfer, Digital single event transient trends with technology node scaling, *IEEE Trans. Nucl. Sci.*, Vol. 53, No. 6, pp. 3462–3465, Dec. 2006.
6. J. Black et al., Characterizing SRAM single event upset in terms of single and multiple node charge collection, *IEEE Trans. Nucl. Sci.*, Vol. 55, No. 6, pp. 2943–2947, Dec. 2008.
7. E. Cannon, Soft errors from neutron and proton induced multiple-node events, *Proc. IRPS*, pp. SE2.1–SE2.7, 2010.
8. G. Gasiot, D. Giot, and P. Roche, Multiple cell upsets as the key contribution to the total SER of 65 nm CMOS SRAMs and its dependence on well engineering, *IEEE Trans. Nucl. Sci.*, Vol. 54, No. 6, Dec. 2007.

9. D. Giot, P. Roche, G. Gasiot, and R. Sorenson, Multiple-bit upset analysis in 90 nm SRAMs: Heavy ions testing and 3D simulations, *IEEE Trans. Nucl. Sci.*, Vol. 54, No. 4, pp. 904–911, Aug. 2007.

10. D. Heidel et al., Single-event upsets and multiple-bit upsets on a 45 nm SOI SRAM, *IEEE Trans. Nucl. Sci.*, Vol. 56, No. 6, pp. 3499–3504, Dec. 2009.

11. S. Deihl, J. Vinson, D. Shafer, and T. Mnich, Considerations for single event immune VLSI logic, *IEEE Trans. Nucl. Sci.*, Vol. 30, No. 6, pp. 4501–4507, Dec. 1983.

12. T. Hoang et al., A radiation hardened 16-Mb SRAM for space applications, *Proc. IEEE Aerospace Conf.*, pp. 1–6, 2006.

13. G. Anelli et al., Radiation tolerant VLSI circuits in standard deep submicron CMOS technologies for the LHC experiments: Practical design aspects, *IEEE Trans. Nucl. Sci.*, Vol. 46, No. 6, pp. 1690–1696, Dec. 1999.

14. R. Lacoe, J. Osborn, R. Koga, S. Brown, and D. Mayer, Application of hardness-by-design methodology to radiation-tolerant ASIC technologies, *IEEE Trans. Nucl. Sci.*, Vol. 47, No. 6, pp. 2334–2341, Dec. 2000.

15. M. Zhang et al., Sequential element design with built-in soft error resilience, *IEEE Trans. VLSI*, Vol. 14, No. 12, pp. 1368–1378, Dec. 2006.

16. C. Mead and L. Conway, *Introduction to VLSI Systems*, Addison-Wesley, Reading, MA, 1980.

17. A. Drake, A. Klein Osowski, and A. Martin, A self-correcting soft error tolerant flip-flop, *Proc. 12th NASA Symp. VLSI Design*, pp. 1–5, Oct. 2005.

18. T. Calin, M. Nicolaidis, and R. Velazco, Upset hardened memory design for submicron CMOS technology, *IEEE Trans. Nucl. Sci.*, Vol. 43, No. 6, pp. 2874–2878, Dec. 1996.

19. R. Reed et al., Single event cross sections at various data rates, *IEEE Trans. Nucl. Sci.*, Vol. 43, No. 6, pp. 2862–2867, 1996.

20. D. Mavis and P. Eaton, Soft error rate mitigation techniques for modern microcircuits, *Proc. IEEE IRPS*, pp. 216–225, 2002.

21. R. Naseer and J. Draper, DF-DICE: A scalable solution for soft error tolerant circuit design, *Proc. IEEE Int. Symp. on Circuits and Systems*, pp. 3890–3893, May 2006.

22. R. Yamamoto, C. Hamenaka, J. Furuta, K. Kobayashi, and H. Onodera, An area efficient 65-nm radiation-hard dual-modular flip-flop to avoid multiple cell upsets, *IEEE Trans. Nucl. Sci.*, Vol. 58, No. 6, pp. 3053–3059, Dec. 2011.

23. B. Matush, T. Mozdzen, L. Clark, and J. Knudsen, Area-efficient temporally hardened by design flip-flop circuits, *IEEE Trans. Nucl. Sci.*, Vol. 57, No. 6, pp. 3180–3184, Dec. 2009.

24. J. Knudsen, Radiation hardened by design D flip-flops, M.S. Thesis, Arizona State University, Dec. 2006.

25. L. Clark and S. Shambhulingaiah, Methodical design approaches to radiation effects analysis and mitigation in flip-flop circuits, *Proc. ISVLSI*, July 2014.

26. J. Knudsen and L. Clark, An area and power efficient radiation hardened by design flip-flop, *IEEE Trans. Nucl. Sci.*, Vol. 53, No. 6, pp. 3392–3399, Dec. 2006.

27. S. Shambhulingaiah et al., Temporal sequential logic hardening by design with a low power delay element, *Proc. RADECS*, pp. 144–149, 2011.

28. D. Fulkerson and E. Vogt, Prediction of SOI single-event effects using a simple physics-based SPICE model, *IEEE Trans. Nucl. Sci.*, Vol. 52, No. 6, pp. 2168-2174, 2006.

29. S. Shambhulingaiah, S. Chellappa, S. Kumar, and L. Clark, Methodology to optimize critical node separation in hardened flip-flops, *Proc. ISQED*, pp. 486–493, March 2014.

30. W. Vesely, F. Goldberg, N. Roberts, and D. Haasl, *Fault Tree Handbook*, U.S. Nuclear Regulatory Commision, Pub. NUREG-0492, Jan. 1981.

31. N. Gaspard et al., Technology scaling comparison of flip-flop heavy-ion single-event upset cross sections, *IEEE Trans. Nucl. Sci.*, Vol. 60, No. 6, pp. 4368–4373, Dec. 2013.

32. K. Warren et al., Heavy ion testing and single event upset rate prediction considerations for a DICE flip-flop, *IEEE Trans. Nucl. Sci.*, Vol. 56, No. 6, pp. 3130–3137, Dec. 2009.

33. D. Kobayashi, T. Makino, and K. Hirose, Analytical expression for temporal width characterization of radiation-induced pulse noises in SOI CMOS logic gates, *Proc. IRPS*, pp. 165–169, 2009.
34. M. Baze et al., Angular dependence of single event sensitivity in hardened flip/flop designs, *IEEE Trans. Nucl. Sci.*, Vol. 55, No. 6, pp. 3295–3301, Dec. 2008.
35. E. Peterson, V. Pouget, L. Massingill, S. Buchner, and D. McMorrow, Rate predictions for single-event effects—Critique II, *IEEE Trans. Nucl. Sci.*, Vol. 52, No. 6, pp. 2158–2167, Dec. 2005.
36. J. Pickel and J. Blandford, Cosmic-ray-induced errors in MOS devices, *IEEE Trans. Nucl. Sci.*, Vol. 27, No. 2, pp. 1006–1012, 1980.
37. D. Fulkerson, D. Nelson, and R. Carlson, Boxes: An engineering methodology for calculating soft error rates in SOI integrated circuits, *IEEE Trans. Nucl. Sci.*, Vol. 53, No. 6, pp. 3329–3335, Dec. 2006.
38. T. Calin et al., Toplogy-related upset mechanisms in design hardened storage cells, *Proc. RADECS*, pp. 484–488, 1997.
39. M Cabanas-Holmen et al., Robust SEU mitigation of 32 nm dual redundant flip-flops through interleaving and sensitive node-pair spacing, *IEEE Trans. Nucl. Sci.*, Vol. 60, No. 6, pp. 4374–4380, Dec. 2013.

8 Assuring Robust Triple-Modular Redundancy Protected Circuits in SRAM-Based FPGAs

Heather M. Quinn, Keith S. Morgan,
Paul S. Graham, James B. Krone,
Michael P. Caffrey, Kevin Lundgreen, Brian Pratt,
David Lee, Gary M. Swift, and Michael J. Wirthlin

CONTENTS

8.1   INTRODUCTION

FPGAs with volatile programming memory, such as the Xilinx Virtex families, have made inroads into space-based processing tasks [1–3]. These components are attractive for a number of reasons. Static random access memory (SRAM)-based field-programmable gate arrays (FPGAs) can provide custom hardware implementations of applications that are often faster than traditional microprocessor implementations without the cost of manufacturing application-specific integrated circuits (ASICs). Furthermore, using commercial-off-the-shelf (COTS) components with available, mature design tools should reduce the cost of designing space-based systems. Finally, reprogrammability also allows designers to reconfigure the component while deployed with either new applications or new implementations of existing applications, which should increase the usable lifetime of the entire system.

In this chapter, we will focus on only the Xilinx Virtex reconfigurable SRAM-based FPGAs. Unlike most SRAM-based FPGAs, Xilinx has published several reports verifying latchup immunity [4,5], which have made them the preferred choice for space usage. This family of components implements logic in lookup tables (LUTs), where logic is reduced from gates to a four-input and one-output equation that is stored in configuration memory. Furthermore, the wiring is programmable so that design flow tools can determine how best to optimize routing signals from one LUT to another LUT through routing switches. Therefore, unlike traditional SRAM components that store data, in a SRAM-based FPGA much of the stored data defines the user circuit, including whether particular routes or LUTs are used. In the more modern components, embedded cores, such as multipliers and microprocessors, have become more common. On-chip SRAM, called BlockRAM, for storing intermediate processing values is increasing in size for each generation of component.

Single-event upsets (SEUs) caused by ionizing particles are a problem for these components, as SEUs change values stored in SRAM. For a SRAM-based FPGA, SEUs could cause changes in the programmable logic and routing, which could potentially cause the user's circuit to malfunction. To this end, most FPGA-based systems attempt to mask SEUs by protecting the user's circuit with triple-modular redundancy (TMR) [6–8]. The current suggestion for space-based FPGA designs is to triplicate all logic (modules and voters) and all signals (inputs, outputs, clock, and reset). The viability of TMR-protected circuits, particularly on a single chip, remains an open question. While our past research [8] on the Virtex-I has demonstrated through fault injection and accelerated testing that logic-level TMR with programming data scrubbing effectively mitigates single-bit upsets (SBUs), other researchers have shown analytically that SBUs can defeat TMR on the Virtex-I [9]. Our later research on the Virtex-II has shown that when the above guidelines are followed it is possible to completely remove all unprotected cross section [10] and the design will be susceptible only to multiple-cell upsets (MCUs), multiple independent upsets, and single-event functional interrupts. As discussed later in Chapter 8, MCUs can be problematic for mitigated circuits. As the occurrence of single-bit SEUs dominate events on these components, we believe that most designs that use these TMR criterion should be adequately protected on orbit if implemented properly. While Chapter 8 focuses primarily on the earlier generation Xilinx FPGAs, we found these results to be consistent in the Virtex-4 and Virtex-5. Right now, very little is known about the efficacy of mitigation on the latest Xilinx FPGAs, the series-7 components. As discussed later in Chapter 8, these FPGAs have some architecture differences that might change how well current mitigation techniques work.

Unfortunately, applying TMR techniques to user's circuits can be error-prone, leading to unprotected cross section in the protected circuits. Designers are not necessarily at fault in these scenarios. In particular, a number of problems can be tied to the design flow tools, which can only be circumvented entirely by avoiding many of the design automation tools—a choice most designers will not make. In other cases, designers are forced to apply TMR only partially to a design to meet component or resource constraints. In those scenarios, the unprotected cross section will only be partially removed.

In all of these scenarios, hardness assurance issues with TMR-protected circuits can be very difficult to ascertain, especially in complex systems. In the past, we have used fault injection [8] to estimate hardness assurance issues that might exist in FPGA designs. Unfortunately, designers cannot always perform fault injection effectively on their designs due to flight system limitations or the limitations of hardware prototypes amenable to fault injection. In these cases, a non-hardware method for estimating the unprotected cross section and for finding design flaws is necessary.

In this chapter, we will discuss the ability to use TMR to protect radiation-induced faults and the efficacy of modeling tools to determine design-level problems with the application of TMR. In Section 8.3 we will provide an overview of the sensitivity of Xilinx SRAM-based FPGAs to radiation-induced upsets. In Sections 8.3 and 8.4 we will discuss the use of TMR to protect FPGA user circuits. Finally, in Section 8.5 we will introduce modeling tools that might be helpful in determining problems with the application of TMR.

8.2 OVERVIEW OF SEU AND MCU DATA FOR FPGAs

Before continuing our discussion, we would like to discuss static test results that we have collected that show the sensitivity of the components to ionizing radiation. We have performed accelerator tests on five generations of Xilinx Virtex components. We have used a similar test fixtures, as shown in Figure 8.1, for static testing of the Virtex-II, Virtex-4, and Virtex-5 components. The fixture consists of both hardware and software components. The hardware test fixture provides support for reading (readback) and writing (programming) the configuration data in the SRAM-based FPGA. The software test fixture controls the programming and reading back the FPGA. Similar test fixtures were used for the Virtex-I and Virtex-7 components.

The hardware test fixture the authors used for the Virtex-5 results is shown in Figure 8.1. It uses two Xilinx AFX series development boards (one Virtex-II and one Virtex-5) biased nominally. A third, smaller board contains a universal serial bus (USB) connection to a host computer that allows the computer to control the operation of the test fixture. The hardware test fixture uses custom software that performs programming, differencing, and readback, as well as keeping the graphical user interface (GUI) updated with minimal statistics to help the testers determine whether the test fixture remains operational. The FPGA is completely reprogrammed and error locations, called *the differential bitstream*, are saved to the host computer's hard drive every second in this scheme, which allows us to test continuously at high fluences without accumulating too many upsets per readback. With this scheme we can collect approximately 3600 differential readbacks per hour. Custom software is

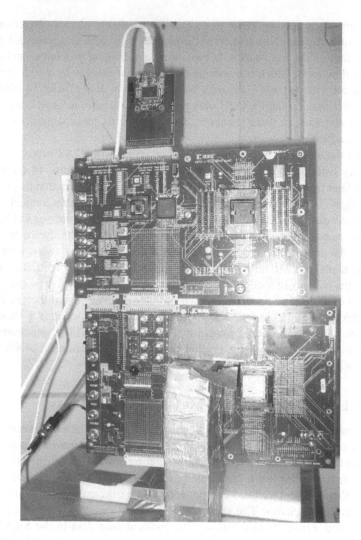

FIGURE 8.1 Hardware test fixture for the Xilinx Virtex-5 component.

used to analyze the differential readbacks for the component's sensitivity to errors and to categorize the errors by size and location.

From testing the Xilinx Virtex components, we have been able to observe several trends [11]. The components' bit cross sections (Figure 8.2 and Table 8.1)* have been within an order of magnitude over 10 years [12,13], but the percentage of MCUs for the entire component (Figure 8.3) has rapidly increased. Table 8.2 lists the frequency of MCUs in protons and Figure 8.3 shows the frequency of MCUs for heavy ions for Virtex family components. Both of the proton and heavy-ion data sets have shown that MCUs have become more frequent in the newer components. Figure

---

* The Virtex-I components are a combination of normal incidence and angular data, but the rest of the curves are solely from normal incidence data.

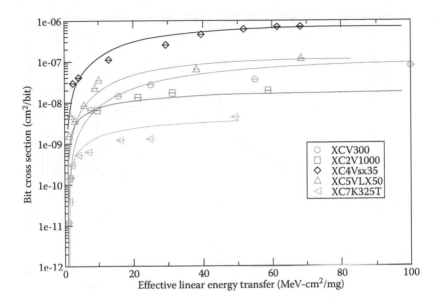

FIGURE 8.2 Heavy-ion bit cross sections for Virtex family components. (From H. Quinn, P. Graham, J. Krone, M. Caffrey, and S. Rezgui, *IEEE Transactions on Nuclear Science*, Vol. 52, No. 6, pp. 2455–2461, December 2005; D. Lee, M. Wirthlin, G. Swift, and A. Le, *to be published in the IEEE Radiation Effects Data Workshop [REDW]*, Dec. 2014; A. Le, Boeing Corporation, Tech. Rep., 2013.)

TABLE 8.1
Bit Cross Section for SEUs for Protons for Several Xilinx FPGAs

Component	Energy (MeV)	σ_{bit} (cm²/bit)
XCV1000	63.3	$1.32 \times 10^{-14} \pm 2.69 \times 10^{-17}$
XC2V1000	63.3	$2.10 \times 10^{-14} \pm 4.64 \times 10^{-17}$
XC4VLX25	63.3	$1.08 \times 10^{-14} \pm 2.71 \times 10^{-17}$
XC5VLX50	65.0	$7.57 \times 10^{-14} \pm 1.35 \times 10^{-15}$
XC5VLX50	200.0	$1.07 \times 10^{-13} \pm 5.37 \times 10^{-16}$

Source: H. Quinn, P. Graham, J. Krone, M. Caffrey, and S. Rezgui, *IEEE Transactions on Nuclear Science,* Vol. 52, No. 6, pp. 2455–2461, December 2005; H. Quinn, K. Morgan, P. Graham, J. Krone, and M. Caffrey, in *IEEE Radiation Effects Data Workshop (REDW),* July 2007.

8.3 also shows how MCU frequency increases with energy. At the highest tested linear energy transfer (LET) there are 21% MCUs on the Virtex-II (XC2V1000) at 58.7 MeV-cm²/mg and there are 59% MCUs on the Virtex-5 (XC5VLX50) at 68.3 MeV-cm²/mg. While most of the events on both the Virtex-II and the Virtex-5 involve four or fewer bits, the distribution of event sizes changed. As shown in Figure 8.4, the dominant SEU sizes for the 150-nm Virtex-II are 1-bit and 2-bit events (Figure 8.4a), whereas 3- and 4-bit events total 25% of all events (Figure 8.4b) for the Virtex-5.

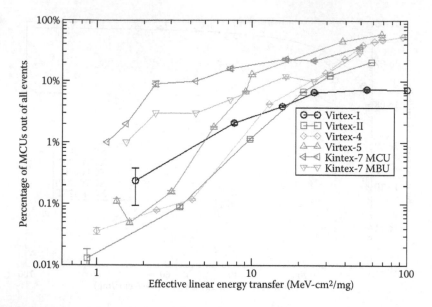

FIGURE 8.3 Percent of MBU and MCU events out of all events induced by heavy-ion radiation for five Xilinx FPGAs.

TABLE 8.2
Frequency of Upset Events and Percent of Total Events Induced by Proton Radiation (63.3 and 65 MeV) for Five Xilinx FPGAs

Family	Total Events	1-Bit Events	2-Bit Events	3-Bit Events	4-Bit Events
Virtex-I	241,166	241,070	96	0	0
		(99.96%)	(0.04%)	(0%)	(0%)
Virtex-II	541,823	523,280	6293	56	3
		(98.42%)	(1.16%)	(0.01%)	(0.001%)
Virtex-II Pro	10,430	10,292	136	2	0
		(98.68%)	(1.30%)	(0.02%)	(0%)
Virtex-4	152,577	147,902	4567	78	8
		(96.44%)	(2.99%)	(0.05%)	(0.005%)
Virtex-5 (65 MeV)	2963	2792	161	9	1
		(94.23%)	(5.43%)	(0.30%)	(0.03%)
Virtex-5 (200 MeV)	35,324	31,741	3105	325	110
		(89.86%)	(8.79%)	(0.92%)	(0.43%)

This phenomena has caused the average SEU event size to increase from 1.3 bits in the Virtex-II at 58.7 MeV-cm^2/mg to 2.6 bits in the Virtex-5 at 68.3 MeV-cm^2/mg at normal incidence.

We have also been watching the trend for MCU shapes because it indicates the amount of spacing that would be needed to correct TMR defeats. With the diversity of

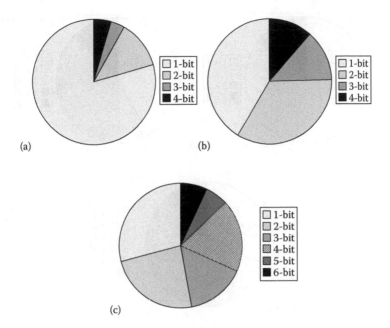

FIGURE 8.4   Distribution of event sizes: (a) 100%, (b) 99%, and (c) 99%.

possible MCU shapes, we report shapes as bounding boxes.* As shown in Figure 8.5, most of the events on the Virtex-II can be confined within two rows and two columns (Figure 8.5a), whereas most of the Virtex-5 events are confined by three rows and two columns (Figure 8.5b).

As shown in Figures 8.4c and 8.5c, both MCUs and bounding boxes worsen for nonnormal incidence radiation strikes. At a LET of 72.8 MeV-cm²/mg, striking the component with Kr at a 60-degree angle has a 72% probability of an MCU, and the average SEU size is 4.2. These figures show an increase in the percentage of larger MCUs, including a 13% probability of 5- and 6-bit events, and that only 94% of MCUs are confined between four rows and two columns.

While much of this data might appear dire, SRAM-based FPGAs are less likely to experience MCUs than traditional SRAM components. Gasiot [14] shows that MCUs could comprise 23%–81% of all events that occur on the component in neutron radiation, depending on the well design. Furthermore, Tosaka [15] reports that 2-bit upsets occur at approximately 10% of the frequency of SBUs in neutron radiation. Tosaka also noted that MCUs occurred more frequently in smaller feature-sized components than larger feature-sized components. As neutron and proton radiation cause similar reactions in CMOS components, these two articles indicate that MCUs are 3–27 times more likely in traditional SRAM components than SRAM-based FPGAs. As the structure of SRAM-based FPGAs is more heterogeneous in layout and the memory structures are not optimized for area like traditional SRAM components, these components are less likely to have MCUs than traditional SRAM components.

---

* A bounding box is the number of rows and columns that completely cover an MCU. A discussion of bounding boxes can be found in [16].

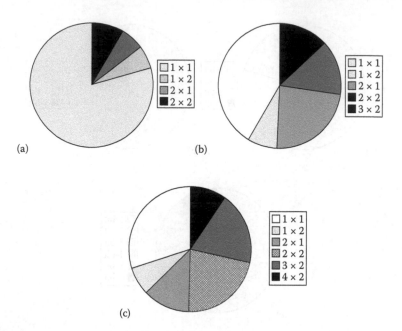

FIGURE 8.5 Worst-case distribution of heavy-ion-bounding boxes. Distribution of bounding boxes: (a) 100%, (b) 99%, and (c) 94%.

It also appears that the MCUs in the newest generation of FPGA, the Xilinx 7-series FPGAs, are different than the earlier generation of FPGAs. As seen in Figure 8.3, the 28-nm Kintex-7 FPGA is more sensitive to MCUs at low LETs than older components [17]. As the energy increases, the MCU rate tracks the older Virtex-II series FPGA, suggesting that the manufacturer invested additional effort to protect the configuration cells from MCUs in this smaller-process geometry. Furthermore, manufacturers have begun to interleave the bits in the FPGA so that a 32-bit error correcting code can be used by internal scan circuitry to automatically fix all SBUs. In this case, some of the MCUs are multiple-bit upsets (MBUs), which are the MCUs overlaid onto the FPGA frame structure. Figure 8.3 also represents 28-nm Kintex-7 MBUs and highlights the advantage of configuration memory interleaving. By interleaving the configuration memory, the effective MBU rate is lower than all but the Virtex-I FPGA family.

8.3 TMR PROTECTION OF FPGA CIRCUITS

While the component is inherently radiation-tolerant and therefore SEU-sensitive while on-orbit, using TMR to protect the circuit should mask the effects of many SEUs as long as there is only one error in the system at a time. Even still, there are a number of ways in which either the design could be flawed or the design implementation toolset could render the final implementation of the design flawed. Furthermore, there might be design constraints placed on the circuit—such as not enough input/output pins for full triplication—that affect the reliability of the design. The potential reliability issues for TMR-protected designs for these components are threefold:

problems with the circuit design, design constraints, and architectural influences on the circuit design. These issues are presented next.

8.3.1 CIRCUIT DESIGN PROBLEMS

The first issue is the design of the TMR-protected circuit. Many FPGA circuit designers use a hardware description language (HDL), such as VHDL or Verilog, to describe the FPGA circuit. The circuit description is then optimized for area and translated to an industry-standard circuit representation called Electronic Design Interchange Format (EDIF), using circuit synthesis tools such as Synplify or Synopsys. Even the most careful descriptions of TMR-protected circuits are often undermined by the synthesis tools. As FPGA synthesis and implementation tools are designed to remove redundant logic to optimize the circuit for area and speed, these tools usually recognize and remove the functional redundancy intended to improve reliability. More subtly, though, sometimes the redundant modules remain but are no longer functionally equivalent or independent. In this case, part of the redundant logic is reduced to a single implementation in one module that is shared by all three modules. This problem is shown in Figure 8.6. In this situation, the inverter that is used for the least significant bit in the counter has been removed from all three modules and the inverted data is shared by all three counter modules. While the circuit is still functionally equivalent to a correctly TMR-protected design, untriplicated logic now exists in the circuit. In large-circuit designs, detecting this issue is difficult.

Figure 8.6 also highlights a common problem in TMR-protected circuits with feedback loops. Feedback loops in TMR-protected systems are also sensitive to *persistent errors* [18] and need to use triplication and voters to break the feedback loops. The counter in Figure 8.6 shows a feedback loop that has not been cut properly and the counters will not be able to autonomously resynchronize after the SEU is removed. In this scenario, while the first SEU in one feedback loop will be masked

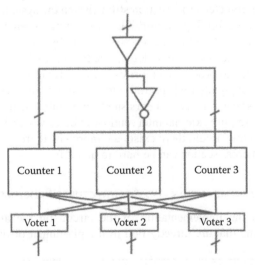

FIGURE 8.6   Example of a TMR-protected counter design with a number of design flaws.

by the voters, a second SEU in another feedback loop is not guaranteed to be masked. To fix the counter design, the output of the voters will need to be fed back to the input of counters to remove the persistent cross section.

To circumvent issues with the synthesis tools, the recommended approach for applying TMR is to apply TMR to the EDIF circuit descriptions. While this can be done in a text editor for small designs, the authors suggest using one of the automated tools (BLTmr [18], the Xilinx TMRTool [19], the Synopsis tool [20], or the Mentor Graphics tool [21]). These tools work with the postsynthesis circuit representation and therefore the synthesis tools are able to optimize the basic circuit without affecting the application of TMR. The optimization of the circuit after synthesis is usually limited to removing signals that do not route to output pins. Therefore, optimization of the redundant modules is unlikely. Also, these tools have been built with an understanding of persistence issues so that feedback loops are properly protected by TMR.

8.3.2 Component Constraint Problems

The second issue is about design constraints. Because these components can be pin- and area-constrained, designers are sometimes unable to implement a fully triplicated design. In particular, not being able to triplicate input, output, clock, or reset signals is common, and SEUs in the input/output blocks, routing, global clock network, and flip-flops could cause errors to manifest across all three logic modules. The counter in Figure 8.6 shows the three counters are sharing the same inputs. While this design is not uncommon in cases where the data stream originates from a single sensor, unprotected cross section exists between the input pins and the inputs of the counters. Furthermore, we have found that, when not using automated tools to apply TMR to a design, the optimization by the synthesis tools of the TMR-protected circuit with shared inputs is more likely to remove most of the reliability-based redundancy. While it is possible to triplicate some of these signals internally on the component,* an unprotected cross section still exists in the system between the input pins and the triplicated flip-flops responsible for splitting the signal.

Designers might also find themselves constrained by the component's size, and are unable to fully triplicate the circuit logic. The BLTmr tool addresses this problem by balancing the need to protect the most essential parts of the design and meeting area constraints by applying TMR partially to the circuit. BLTmr gives highest priority to sub-circuits that may reach a persistent error state due to feedback, because error recovery may require external intervention. In cases where TMR has only been partially applied to the circuit, there exists an unprotected cross section. The effect of this unprotected cross section can be hard to quantify.

8.3.3 Circuit Implementation and Architectural Problems

The third issue is the implementation of the circuit on the architecture. There are several problems that are directly tied to the placement of the circuit onto the

---

* Clocks should only be triplicated using the global clock buffers, and skew should be carefully monitored.

component, such as domain crossing errors and logical constants. These components are very complex and have a number of architectural features, such as the resources for fast carry-chains, shift registers, and embedded arithmetic functions, to improve the speed, power, and silicon utilization of user circuits. As an artifact of translating a design to the specific resources available on the FPGA, sometimes the inputs to carry-chains and multipliers need to be tied to a ground, such as when the multiplication is using fewer inputs than the embedded multipliers have. These grounds are tied to a logical constant on the power network, called the *global logic network*. The power network for the Virtex-I and Virtex-II is a virtual network of grounds and VCCs that use constant LUTs. Because the power network is load-balanced by the design flow tools, redundant logic could share the same power network, introducing potential single points of failure into the design. Further complicating the issue, the power network is implemented in SEU-sensitive logic, which could translate to unprotected cross section in the design. Because the load-balancing affects the number of constant LUTs that are used, the exact quantity of single points of failure caused by them cannot be determined until after the design is placed.

Both BLTmr and TMRTool tools address this issue by extracting the half-latches and the constant LUTs to input/output pins to provide these constant logic values in a TMR domain-aware manner. Because this solution elevates logical constants to a global signal, like the clock tree, the input/output pins used for the logical constants will need to be triplicated. Los Alamos National Laboratory also developed a tool, called *RadDRC*, for providing half-latch mitigation [22,23]. The mitigation process involves finding the half latches in the XDL representation of the user circuit and replacing the half latches with constant LUTs. Both BLTmr and RadDRC have been tested using fault injection and radiation testing.

The final reliability problem involves the placement of the design on the component. Because many of the tools involved in converting a designer's circuit description to a bitstream are attempting to minimize the implemented circuit's area and maximize the clock speed, redundant logic can be placed in close proximity. We have shown in the Virtex-II that when area and timing constraints cause the component to be highly utilized, there is a chance an MCU can defeat TMR by introducing errors into multiple redundant modules, a situation referred to as a domain crossing error (DCE) [10]. Given the complexity of DCEs, we will discuss them in greater detail in the following section.

8.4 DCEs

A DCE occurs when two or more redundant copies (domains) of the TMR circuit are corrupted such that the voter selects the wrong value. As shown in Figure 8.7c, the ionizing particle would need to change at least two TMR domains to the same wrong value to cause a DCE. Because two domains have matching answers, the system does not detect the incorrect operation. Therefore, unless erroneous output data can be detected or locations on the component that have known DCE issues are accounted for, these errors could remain undetected. As MCUs can manifest in the system as independent errors, it is more likely that an MCU could trigger this condition than a single-bit SEU.

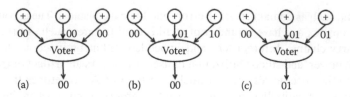

FIGURE 8.7 Example of a DCE in a 2-bit adder with TMR and bitwise voting: (a) correct operation, (b) masking vote, and (c) DCE.

There are many factors that can make systems both more or less vulnerable to DCEs, such as the robustness of the design, voter design, component utilization, and sensitivity of the system to errors. The most robust implementation of TMR has triplicated voters, data signals, and control signals because untriplicated they would become single points of failure. When properly triplicated, identical failures in two domains are needed to propagate the error. Likewise, bitwise voting can mask many potential TMR vulnerabilities, as shown in Figure 8.7b, because failures affecting different bits would vote out. Designs that use most of the component could potentially heighten the risk of a DCE because there appears to be a correlation between high component utilization and DCEs. Finally, the sensitivity of the circuit to errors can play a factor in whether errors can propagate in a system. For example, logic masking lowers the probability that a DCE manifests as an observable output error. All of these situations will be discussed in greater detail in the results section.

The rest of this section focuses on our test methodology, the test results, analysis of results, and a simple probability model for determining the likelihood of occurrence.

8.4.1 Test Methodology and Setup

In this section, we present our test methodology for both our fault injection test fixture and our accelerator test fixture as well as an overview of the test circuits used for this study. While the hardware aspect of the test fixtures is the same, the experimental approach and the software test fixtures are different.

1. *Test circuits*: The test circuits implemented are listed in Table 8.3. All of these circuits were designed for the Virtex-II XC2V1000 component. These designs intentionally represent the worst-case scenarios for TMR limitations. While synthetic in nature, these circuits are representative of corner cases for circuits that can be used as part of a larger design. As noted in Table 8.3, there is a mixture of feed-forward and feedback circuits within the complete set of circuits.

   Each circuit has two TMR implementations, except for the linear feedback shift register (LFSR). One TMR version has triplicated voters interspersed frequently in the design under test (DUT) circuit, and the other only votes once off-chip. The LFSR test circuit was made from an intellectual property module made available by Xilinx, and the TMR implementation only votes off-chip. The TMR implementations were designed with the

TABLE 8.3
Circuit Resource Utilization

Circuit	Type	Voting	Flip-Flop%	LUT%	Slice%
Shift register	Feed-forward	Frequent	96%	97%	97%
		Off-chip	96%	0%	96%
Adder tree	Feed-forward	Frequent	44%	48%	71%
		Off-chip	44%	22%	46%
Divider tree	Feedback	Frequent	81%	33%	98%
		Off-chip	81%	27%	97%
AND tree	Feed-forward	Frequent	45%	90%	100%
		Off-chip	45%	45%	45%
OR tree	Feed-forward	Frequent	45%	90%	100%
		Off-chip	45%	45%	45%
LFSR	Feedback	Off-chip	89%	2%	100%
Pseudo-LFSR	Feedback	Frequent	50%	99%	99%
		Off-chip	50%	49%	50%

accepted best practices for creating FPGA TMR circuits with triplicated data, control signals, and voters.

Each circuit was designed so that most of the component is utilized. In the case of the off-chip voting circuits, the design is functionally the same as the frequent voting circuits so the component utilization is lower for these designs. We also explicitly created designs that used the special features of the Virtex-II component, such as the fast carry chains and the embedded multipliers. The OR tree and AND tree circuits exclusively use LUTs to implement logic. While the Virtex-II component has BlockRAM resources for on-component temporary data storage, the cross section and the mitigation methods for the BlockRAM are substantially different than with the reconfigurable fabric. Given space limitations, circuits using BlockRAM resources are not highlighted in this study.

2. *Fault-inject test methodology*: The test circuits were fault-injected using a Virtex-II SEU emulator we have used for previous studies [24]. Figure 8.8 shows a picture of the hardware test fixture. The SEU emulator operates two Xilinx Virtex-II AFX demonstration boards in lockstep with a USB interface to a host computer. One AFX board has the golden component and the other has the DUT. While both components have the same circuit to test, the golden component has additional computation. This extra computation supplies the input vectors to both test circuits, receives the output vectors from both test circuits, and determines if there is mismatch between both sets of output vectors. Any mismatches are relayed to the host PC for logging.

The SEU emulator software test fixture is designed to inject faults across the entire component with user-specified patterns (1-bit upsets, 2-bit vertical upsets, etc.). In this manner, it is easier with fault injection to gain complete coverage of the entire component than with accelerator testing.

FIGURE 8.8   Fault injection and accelerator hardware test fixture for the Virtex-II.

Even still, because logic masking can play a strong role in whether errors propagate to outputs, it is necessary to run the SEU emulator multiple times for each test circuit and each test pattern to get a representative set of DCEs.

The software aspect of the SEU emulator injects faults in the following manner. First, a fault (or faults in the case of an MCU) is injected into the DUT at a specified location through programming data (i.e., bitstream) manipulation and partial reconfiguration, which simulates the most common SEU on FPGAs. Once a fault is injected, the two boards are reset to synchronize the design and to clear the state of the component so that each fault-injection trial is independent from previous trials. The designs operate in lockstep for many cycles to allow errors to propagate to the outputs. During this time period approximately 250,000 randomly generated test vectors are sent through both boards. Next, the software test fixture checks the golden board to determine if a miscompare has occurred and records the result. The software test fixture then removes the fault from the programming data through partial reconfiguration, and the boards are resynchronized to make certain the DUT returns to normal operation. Once this process is completed, the next fault can be injected.

In order to constrain the fault-injection tests we injected the patterns that occurred most frequently in accelerator testing. Our Virtex-II heavy-ion accelerator data indicates that 99% of SEUs at 58.7 MeV-cm^2/mg can be classified as follows: 1-bit upsets (79%), 2-bit vertical events (6%), 2-bit horizontal events (6%), 3-bit corner events (4%), and 4-bit squares (5%). As this data point is the highest tested heavy-ion LET, these percentages indicate a worst-case scenario for MCUs. While lower LET heavy ions or protons have fewer MCUs in frequency, the MCUs can be still be classified as one of the shapes from the previous list. Because these shapes represent most of the events that will occur on the Virtex-II component, simulating these patterns across the entire component will provide good coverage of Virtex-II DCEs.

3. *Accelerator test methodology*: The same hardware test fixture is used at the accelerator to validate the fault-injection results; however, the software aspect of the test fixture is different. At the accelerator, the software test fixture performs a readback of the component's bitstream, compares the readback to a reference bitstream to determine the upset locations, records the upset locations, records the result of polling the golden component for miscompares, performs a partial reconfiguration of the component to remove the faults in the upset locations, and resynchronizes the two boards through a design reset. All of these actions are performed while the part is being irradiated to simulate what would be done on orbit. Flux is deliberately kept low to minimize the number of DCEs due to uncorrelated upsets. We were able to conduct accelerator testing at Indiana University Cyclotron Facility in 2007. We tested for a total fluence of 6.6 × 10^{11} protons/cm^2 in a little over 2 hours with two XC2V1000 parts. We also rotated the test fixture to a 45° angle to increase the MCU cross section. With this setup we averaged 1–3 upsets/readback cycle.

8.4.2 Fault-Injection and Accelerator Test Results

The fault-injection results can be found in Table 8.4. The number of DCEs are listed in two forms: the raw number from fault injection and the analyzed version that represents only DCEs created by that shape. The analyzed data is in parentheses. The voting circuit column indicates whether the design votes frequently or once off-chip.

We were able to gather some preliminary accelerator data on DCEs using one design (adder tree, frequent voter) at Indiana University Cyclotion Facility (IUCF) using 200-MeV protons and the component angled at 45°. The intent of this test was to prove that DCEs would occur with both radiation-induced and fault-injection methods. During a 2-hour test we were able to observe 31 DCEs for a cross section of 6.6 × 10^{-11} ± 3.8 × 10^{-13} cm^2. We have been able to correlate 42% of these DCEs to known fault-injection DCEs. In the future, we hope to correlate more of the DCEs to the fault-injection data.

In comparison, during the same test we also observed 19 single-event functional interrupts (SEFIs) for a cross section of 4.1 × 10^{-11} ± 4.9 × 10^{-13} cm^2. While the MCU cross section is several orders of magnitude larger than the SEFI cross section in the Virtex-II, the DCE cross section may be on the same order of magnitude of the SEFI cross section for this design due to the fraction of MCUs that cause DCEs. In fact, 1%

TABLE 8.4

Fault Injection Results

Circuit	Voting	All Pairs (w/o overlap)	13-Bit Corner (w/o overlap)	4-Bit 2 × 2 Square (w/o overlap)
Shift register	Freq	6355	4545	9186
		(6355)	(539)	(489)
	OC	2185	1364	2352
		(2185)	(253)	(489)
Adder tree	Freq	18,733	11,264	19,464
		(16,843)	(1116)	(1783)
	OC	1166	715	1310
		(1104)	(101)	(213)
Divider tree	Freq	1556	1056	1966
		(1556)	(259)	(0)
	OC	1276	767	1335
		(1226)	(169)	(274)
AND tree	Freq	0	2	0
		(0)	(2)	(0)
	OC	0	0	0
		(0)	(0)	(0)
OR tree	Freq	1784	1645	3333
		(1784)	(202)	(814)
	OC	5	2	5
		(5)	(0)	(1)
LFSR	Freq	4966	2711	4709
		(4966)	(297)	(803)
Pseudo-LFSR	Freq	26,105	18,023	31,606
		(26,105)	(2194)	(4092)
	OC	44	28	58
		(44)	(6)	(7)

Note: Freq = frequently, OC = once off-chip.

of the component is affected by DCEs and the DCE cross section is approximately 15 times smaller than the MCU cross section. Finally, the analogy toward SEFIs is a useful one in how to approach DCEs. SEFIs, while possible, are not the first-order effect for these components and can be approached as a manageable problem.

8.4.3 DISCUSSION OF THE RESULTS

While we have both fault-injector and accelerator results, complete test coverage is easier with fault injection. Therefore, much of our focus in this section will be on the fault-injection results. Our most important result from our testing is that we were able to observe DCEs in both fault-injection and accelerator testing. The discussion in this section will cover circuit design and architectural characteristics that might be causing DCEs to manifest in the Virtex-II designs we tested.

1. *DCE characteristics*: While Sterpone [9] analytically showed that SBU DCEs existed, no SBU DCEs occurred in our Virtex-I fault-injection and accelerator testing. However, in the Virtex-II SBU DCEs manifested only for the adder tree and multiplier tree circuits. All of the SBU DCEs were tied the global logic network that was providing constant zeros to the designs. Most of these SBU DCEs had similar characteristics, where a multiplexor that routes data internally in a slice from a LUT to the slice's output signal was altered by the SEU. In these cases, further analysis showed that the slice's output was receiving signals from the wrong multiplexor input. However, there was one case, where the multiplexor that is attached to the slice's output signal was corrupted by a SEU. In this singular case, analysis indicated that the SEU caused the output multiplexor to use a different slice's output signal. This one case indicates a possibility that a single-bit SEU could cause similar DCE corruption, but so far it appears to be rare.

   Most of our test circuits were vulnerable to a number of MCU-induced DCEs, except in the case of the two AND tree implementations. We believe, as discussed later in this section, that the AND tree is particularly insensitive to errors that caused the minimal response from the circuit implementations. As shown in Figure 8.9, of the circuits that exhibited DCEs between 0.0001%–2.6% of the component is affected by DCEs. On average, only 0.9% of the bitstream was involved in MCU-induced DCEs for the frequent voting circuits, as compared to only 0.1% for the off-chip voting circuits.

2. *Architectural concerns*: Approximately 99% of the MCU DCEs happened in the configurable logic blocks (CLBs). The remaining DCEs occurred in the input/output blocks, the global clock, and the BlockRAM interconnect (used for routing). On average, 75% of CLB DCEs occurred in the CLB routing network, 22% spanned the routing network and the LUT region,

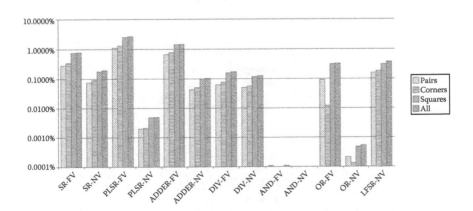

FIGURE 8.9   Percentage of the entire component affected by DCEs.

and 2% occurred in the LUT region. Of the ones that occurred solely in the LUT region, 80% span multiple frames, CLBs, or slices.

The CLB routing network is a concern because it is the single largest resource type for the entire component, with 53% of the configuration data in the XC2V1000. In accelerator testing, 95% of CLB SEUs and 48% of all SEUs involve the routing network. A schematic of one routing switch with its attached CLB for the Virtex-II is shown in Figure 8.10. Each CLB in the Virtex-II consists of four slices. Each slice has two LUTs, two flip-flops, and a number of bits that define the mode of the slice. Every routing switch has two main functions: routing data and control signals to the attached CLB, and routing data and control signals to other routing switches. Even with multiple options for switch-to-switch communication, with a four-to-one ratio of slices to routing switches, routing can become congested. Furthermore, each of the four slices attached to a single switch can have four separate sets of data and control signals. Therefore, in the case of TMR circuits, it is possible that all three domains are placed in one CLB and routed through one routing switch matrix. In this scheme, routing switches become single points of failure.

In further analysis, we looked at what was being corrupted by the MCUs that caused DCEs. Many of the DCEs were caused by changes in global signals. We found many instances where the clocks for two different domains were switched, which could introduce subtle timing problems in the affected domains. We also found many instances where one domain's clock signal and another domain's reset signal would be switched, causing one domain's flip-flops to not be clocked and the other domain's flip-flops to be reset every clock cycle.

To reduce the impact of MCUs on the circuit, the placement and routing tools need to place the three separate domains far enough apart to not be

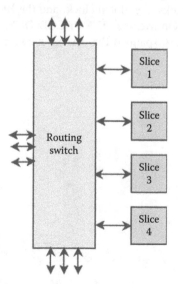

FIGURE 8.10 Routing switch and attached CLB.

affected by MCUs. Even simple changes, such as not allowing domains to share a CLB, could decrease the number of DCEs, but might lead to poor component utilization.

3. *Voting and component utilization*: Our first attempt to analyze the fault injection results was to correlate the data to the component utilization statistics for each circuit; however, this analysis was fruitless. The worst design for DCEs, pseudo-LFSR, used approximately the same number of LUTs, slices, and PIPs as the shift register design that had 73% fewer DCEs. While the design with the smallest component utilization has one of the lowest numbers of DCEs, the design with no DCEs, the AND tree, uses nearly all of the slices and LUTs. Furthermore, AND tree's twin, OR tree, has all of the same utilization statistics with different LUT functionality and has DCEs. Therefore, the obvious component utilization statistics does not play a clear role in the number of DCEs.

We found that increasing voting increased the number of DCEs, but a simple correlation cannot be made. Many circuits saw a large drop in DCEs in the off-chip voter implementations, such as the pseudo-LFSR circuit where the off-chip voting implementation had 1% as many DCEs as the frequent voter implementation. On the other hand, there were a few circuits that only saw a limited improvement in DCEs by voting off-chip, such as divider tree where the off-chip voting version of the circuit had 68% of the DCEs as the frequent voter implementation. Apparently, there are many factors that influence the role of voters in these results.

The most obvious cause would be spacing, because MCUs are only problematic to closely spaced logic. The off-chip implementations all have lower component utilizations, which would allow the design tools to place the domains further apart or keep large blocks of the domain together while still meeting timing requirements. Voting also causes the three domains to converge at LUTs to be voted, forcing the synthesis tool to place the domains closer together to meet timing requirements. Therefore, the voters force the domains to be proximately located. Because single-bit voters can be implemented in one LUT and each slice has two LUTs, it is possible that up to eight different voters are attached to one single routing switch. Therefore, voting frequently not only uses more resources, but also causes the placement of the circuit on the component to become congested and entangled.

4. *Design sensitivity*: The OR tree and AND tree circuits were designed to closely mimic each other in terms of layout and component utilization. The only difference is the logic realized in the LUTs. While they are essentially the same circuit, their DCE characteristics are different. In fact, the OR tree circuit differed from most of the other designs, as only 44% of DCEs are solely in the routing network and 53% are spanning LUTs and routing switches. Therefore, while the routing network is more suspect in most circuits, the OR tree circuit is more vulnerable to SEUs changing the routing switch and the LUT simultaneously. Likely, these errors are manifesting as multiple independent errors in the system. There are also many potential scenarios, such as the inputs to two LUTs getting stuck at zero, which would cause the OR

tree to have many observable errors while the AND tree could logically mask most of the same errors. Therefore, the OR tree is possibly more sensitive to how the LUT/routing network MCUs manifest, but the AND tree is not.

8.4.4 PROBABILITY OF DCEs

We found that the number of DCEs increase with SEU size. The DCE space for each SEU size is comprised of two parts: DCEs caused by smaller events and DCEs unique to the event shape. For example, in Figure 8.11, the 2-bit vertical DCE is over-lapped with a 3-bit vertical MCU. Therefore, the event space for 3-bit vertical MCU DCEs can be partitioned into DCEs caused by overlapping the 2-bit vertical DCE locations and DCEs uniquely triggered by the 3-bit vertical. Furthermore, the 2-bit vertical DCE event space occurs twice in the 3-bit vertical event space because each 2-bit vertical DCE is triggered by two 3-bit vertical MCUs. Therefore, the number of DCEs for a given SEU size is larger than or equal to the number of DCEs for all of the smaller SEU sizes that it overlaps. Table 8.4 indicates the number of unique DCEs for each shape in parenthesis. For 3-bit and larger DCEs the event space is dominated by the smaller-sized DCEs.

As the SEU size increases, the probability that a DCE is triggered goes to one. Fortunately, while the probability of a DCE approaches one, the probability that the event occurs goes to zero. Because the probability of a 5-bit or larger MCU for the Virtex-II component is small, the probability of a DCE is dominated by the prob-ability of 1- to 4-bit MCUs.

We have created a simple model for estimating the probability of a DCE occur-ring. The model is based on this reasoning:

$$P(\text{DCE}) = \sum_{i=1}^{\max} P(\text{upset}_i) P(\text{DCE} \mid \text{upset}_i)$$

$$= \sum_{i=1}^{\max} P(\text{upset}) \frac{N(\text{DCE}_i)}{C_i}$$

(8.1)

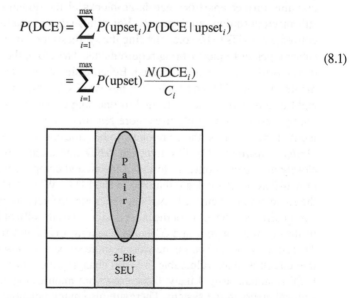

FIGURE 8.11   3-Bit MCU that overlaps with a 2-bit DCE.

where $P(\text{upset}_i)$ is the probability that an upset of i bits occurs based on accelerator data, $N(\text{DCE}_i)$ is the number of DCEs triggered by an SEU of size i, and C_i is the number of combinations for an SEU of size i. We used this model to determine the probability of a Virtex-II DCE from our results by using our normal incident heavy-ion static test data collected at Lawrence Berkeley National Laboratory for the $P(\text{upset}_i)$ values and the fault injection data for the $N(\text{DCE}_i)$ values. As shown in Figure 8.12a, the probability of a DCE in one of the Virtex-II test designs is still fairly low with a worst-case probability of 0.36%. We then extended our model to include the Virtex-5 using the Virtex-II fault-injection DCE data for the $N(\text{DCE}_i)$ data and the Virtex-5 accelerator data collected at Lawrence Berkeley National

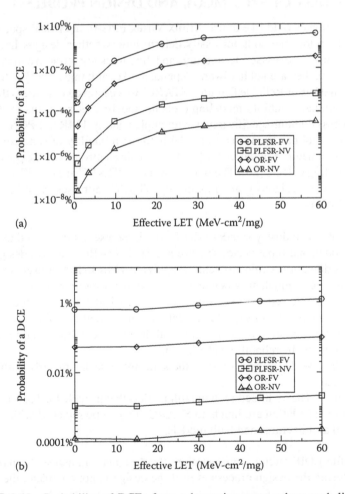

(a)

(b)

FIGURE 8.12 Probability of DCEs from a heavy-ion event, where each line represents a tested design (FV = frequent voter and NV = no voter). (a) 2V1000 fault injection and (b) 5VLX50 for a range of incident angles (projected based on data using Kr from LBNL's 10-MeV/nucleon cocktail).

Laboratory for Kr at five different angles. These projections, shown in Figure 8.12b, are for a range of angles based on data taken using Kr from LBNL's 10-MeV/nucleon cocktail with a LET of 36.4 MeV-cm^2/mg at normal incidence. This model predicts that DCEs could be up to 1.2% of all 1- to 4-bit events. Because we cannot account for the larger MCUs that occur on the Virtex-5 component, the real probability of a DCE is likely higher. Given these probabilities we determined the space rate for TMR defeats for a GPS orbit of 20,200 km and 55 degrees. In the best-case scenario (solar maximum) only 0.6 upsets will happen per-component/day and in the worst-case scenario (peak) 3700 upsets will occur per-component/day. For the Virtex-II this translates to between 0.003–19 DCEs per-component/day.

8.5 DETECTION OF SBUs, MCUs, AND DESIGN PROBLEMS

Many organizations would like to use Xilinx Virtex FPGAs in critical space applications, and therefore methods for easily determining whether designs have been mitigated properly is necessary. Reliability modeling tools are attractive under these scenarios because they are not hardware-dependent like fault injection. The Scalable Tool for the Analysis of Reliable Circuits (STARC) was designed to address the limitations of traditional reliability modeling tools in modeling user circuits for FPGAs as well as address domain-specific issues with implementing TMR in FPGA circuits. In the past, this tool has been used to model both the reliability of supercomputers in the presence of neutron radiation [25] and nanoscale electronics in the presence of permanent yielding defects [26]. The main drivers for STARC are usability, computational complexity, scalability, and modularity. STARC addresses these limitations with these solutions:

- *Usability*: The industry-standard EDIF circuit representation is used for the input model, and input vector sets are not used. STARC was also designed to assess domain-specific problems of applying TMR to FPGA user circuits and can detect imbalances between the modules, find untriplicated logic, estimate unprotected cross section, and detect logical constant usage.
- *Computational complexity*: Memoization* of reliability values reduces recomputation of similar components and the use of combinatorial reasonings simplifies the reliability calculation.
- *Scalability*: Without input vectors, the state space scales linearly with the circuit size.
- *Modularity*: The architectural and fault models that provide the basis of the reliability calculation are inputs to STARC and can be replaced with user-specified architectures and fault models.

By using the EDIF circuit representation, the designer can assess the reliability of a circuit during the design process even if the design is not complete, the design

---

* Memoization is a combinatorial optimization method from computer science that breaks problems into subproblems. The technique solves the subproblems and then substitutes these solutions, where possible, into the problem.

does not work, or the hardware is not available. Without the use of input vector sets reliability is determined through the probability of component or input failure and is not dependent on specific input data sets. Without input data sets, the reliability of components is determined by type, such as a two-bit adder and can be memoized for reuse. In this manner, large-scale circuits are analyzed in a fraction of the time and memory required by traditional approaches, making design exploration more worthwhile.

Just as STARC can estimate the hardness assurance of FPGA user circuits within minutes, STARC can also be used for designers facing area and resource constraints. Under these circumstances, it is possible to generate a range of designs in BLTmr with different balances of unprotected cross sections and resource utilization. In this manner, STARC can help designers choose among a range of possible design choices by quantifying the remaining unprotected cross section for each.

There are a few disadvantages to this approach. First, because EDIF does not contain information about the routing, information regarding placement and routing is absent from the calculation. Since routing can have a large impact on the protected and unprotected cross sections, the routing cross section is estimated statistically based on an analysis we did of several designs using JBits [27]. The point of the statistical model is to provide a good estimate of the single-bit cross section as the only way to fix unprotected configuration bits in the routing is to mitigate the unprotected logic. Furthermore, currently there is no way to assess placement-related issues, such as MCU-induced TMR defeats. We are currently working on a solution for this limitation for designs that have completed the design flow. Second, without input vector sets, logical masking cannot be taken into account, and STARC estimates the worst-case failure rate. While this value may be lower than the value determined by other tools [28], STARC provides a useful lower bound on the circuit's reliability.

8.5.1 RELATED WORK

Traditionally, circuit reliability has been determined using purely analytical approaches [29,30] or techniques that model Boolean networks as probabilistic systems [31–38]. These modeling techniques represent circuits as probabilistic transfer matrices, stochastic Petri-nets, Markov chains, or Bayesian networks. The combinatorics-based analytical approaches have been found to be error-prone and computationally complex for the analysis of large designs. Similarly, a number of limitations have been identified for many modeling-based approaches. First, model creation and input data sets greatly increase the time commitment of using these tools. Transforming circuits into intermediate probabilistic system models is an additional, computationally complex task. Within an analytical tool a state space is generated from the input model and input data vector set. The state space encodes all of the possible failure states in the circuit and grows exponentially with circuit size. The exception to these problems is the SETRA tool [39] that directly addresses the state space issues as well as automated model generation. Attempts at reducing computational complexity through circuit partitioning and hierarchical modeling of large circuits requires additional modeling effort. These limitations lead to the

STARC tool, which uses EDIF circuit representations, no input data vectors, and simpler combinatorial reasoning to decrease the time commitment for the designer and reduce computational complexity in the tool.

Besides these differences between STARC and traditional reliability analysis tools, the tool methodologies differ greatly. There are two distinct methods [35,37,38,40] that can be used to analyze the reliability of circuits: generalized or instance-based. The *generalized* approach entails the combinatorial modeling of circuits without considering specific failure distributions of the inputs, gates, and interconnects. A circuit's output's probability distribution is computed through combinatorics under the assumption that each gate can fail independently. Thus, the reliability is evaluated in stages using conditional probabilities. Generalized techniques to compute the reliability of large circuits require complex combinatorial reasonings. Reusing subcircuit analysis to reduce the combinatorial complexity in the analysis of a larger circuit is difficult. Because specific input probability distributions are not considered during analysis, the generalized approach determines either the circuit's lower or upper bound on reliability.

Several *instance-based* methodologies have been proposed recently [28,33,34,36]. Instance-based reliability circuit analysis uses probability distributions on the primary inputs as well as gate and interconnect failure probabilities to develop an instance of the circuit. Each instance is then transformed into probabilistic circuit models. This method computes the exact reliability of the circuit for the input distribution. The main drawback of these tools is that several instances of the circuit needs to be analyzed to predict performance trends, which can be computationally expensive. Therefore, the input vector set needs to be limited to bound the computational cost yet provide enough intuition on the circuit's reliability.

The STARC methodology is a hybrid of the two approaches. STARC, as with other generalized approaches, is independent of specific input vectors and their probability distributions, yet uses specific gate distribution instances. Hence, this approach avoids the complex combinatorial reasonings that cause bottlenecks in generalized approaches and also bounds the computational complexity that affects instance-based methods. STARC computes a lower bound on reliability. When we have compared STARC with a purely instance-based approach based on PRISM [34,41], the results of our comparison of STARC and PRISM were favorable. We tested four different designs with two probability-of-failure models based on estimated yield defects on a Dell Linux machine with 4 GB of RAM and dual 3.4 GHz Xeon microprocessors. We then compared the calculated reliability values and execution times. The ratio of the two calculated reliability values indicated that STARC was within three to seven digits of significance to PRISM. STARC also executes faster that PRISM and for several designs was more than nine times faster.

It should be noted that a reliability analysis tool called SEUper fast [42], designed by Boeing in the 1990s, uses many of the same reasonings as probability transfer matrix tools. This tool approached the problem far more generally than STARC and was hindered by solving a much more complex reliability equation than STARC uses. While currently not as generally applicable as SEUper-fast, we believe we will be able to generalize this technique to different problems without having to employ the more complex reliability analysis technique.

8.5.2 STARC Overview

In this section, we will provide an overview of STARC. The reliability of the circuit is determined from dependency graphs of the circuit that are created during a hierarchical exploration of the circuit. By using the EDIF circuit representation, the hierarchy in the circuit should be preserved. Because designers tend to create complex circuits by creating less complex components or subcircuits, maintaining this structure can be very useful in calculating the reliability. In particular, STARC can determine the reliability of a circuit hierarchically. STARC navigates through the layers of the circuit hierarchy to determine the smallest circuit component that needs to have its reliability calculated. Once an entire layer of the circuit hierarchy is completed, these values can be used to determine the reliability of the next-higher layer. This hierarchical nature allows circuits to be examined at the highest level of abstraction or the most minute level of detail. STARC automatically determines the appropriate level of the hierarchy that needs to be explored.

Because input vectors are not used in the reliability calculation, the reliability is determined by component type. For example, one component type might be a two-bit adder. The first time a two-bit adder is found during hierarchical exploration, these three steps are executed:

1. A dependency graph is determined.
2. The reliability of the dependency graph is calculated.
3. The reliability value of the dependency graph is memoized.

The next time another two-bit adder is found in a design, the memoized value is used and the first two steps of the process are eliminated. It is in this way the state space of the circuit grows with the circuit size, as the state space is limited to the unique number of components in the circuit. Even if a circuit has very little component reuse, the state space will never grow larger than the number of components in the circuit. Because the size of the state space has a first-order effect on the speed of computation, STARC is able to analyze the reliability of a circuit in polynomial time, instead of the exponential time necessary for most traditional reliability tools. Therefore, STARC should be able to compute the reliability of circuits with thousands of components in the design in a matter of minutes.

As stated above, during hierarchical exploration dependency graphs are determined for each unique component. For maximum reuse, dependency graphs for each primary output at each level of the hierarchy is determined. These dependency graphs indicate all of the components that exist in the path between a particular output and the reachable inputs. Because not all logic or inputs are reachable from every output, this technique removes unrelated logic from dependency graph, and hence, the reliability calculation.

Once the dependency graph for an output is determined, the reliability can be calculated. In unmitigated designs, the cross section is the total area of the dependency graph:

$$A(O) = \sum_{i=0}^{m} A(C_i) \qquad (8.2)$$

where $A(X)$ is the sensitive area of X (where X is either a wire or a component) and $C = \{C^0,..., Cm\}$ is the set of components that can be reached from output wire O. STARC also applies a modular approach to the fault model and the architectural model. Because reliability is determined hierarchically in STARC, the only components that need to be precalculated are the primitives for the given architecture. Figure 8.13 shows our methodology for library characterization. The primitives for a hardware platform are defined in an architectural model. Fault models for transient and permanent defects are combined with the architectural models to create the characterized primitive library. Traditional probability of failure equations are also available to calculate the reliability of defect-based architecture models. Our automation framework is designed so that users can define primitive libraries for their own architectural models or use our models for basic logic and the Xilinx architecture. To be used in our methodology, user-defined libraries have to be characterized for specific fault models to define their reliability.

In this manner, STARC was designed to be architecturally independent. While this chapter focuses on reliability as it relates to Xilinx FPGAs, STARC is modular in nature and the Xilinx cross-section model is an input to this system. The tool has also been used for probability of failure calculations for nanoscale electronics based on yield estimates. In the future, we would like to expand into models for probability of failure and cross-section models for structured ASICs, as these components are frequently being used in space-based systems as well.

Finally, STARC was also designed to help designers find problems in the application of TMR. For mitigated circuits, the sensitive area is confined to the part of the

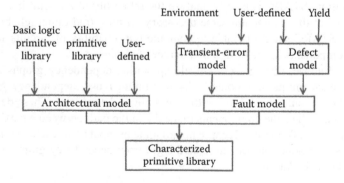

FIGURE 8.13   Library characterization.

design that is not triplicated, as triplication will mask errors as long as there is one voter for each redundant module. STARC also checks to make certain the modules have equivalent components. Any logical elements that might be shared by two or more TMR domains are considered unprotected cross section even if the elements reside within one of the modules. STARC also checks to make sure the feedback loops are properly triplicated and cut. If persistent cross section is found, a warning is displayed to inform the designer that a particular component has not had TMR applied correctly.

Recently, we have been adding support for placement-related information in STARC to provide DCE predictions. For designers that are further along in the design process, it is possible to get placement-related information from the Xilinx Design Language (XDL) representation of the circuit. Like EDIF, XDL provides a human-readable circuit representation. Unlike EDIF, the component names from the circuit are slightly obscured. In our initial attempts, however, we have been able to map the XDL circuit representation onto the EDIF circuit representation. Because our analysis showed that the most common problems with DCEs were caused by domains sharing the same CLB, our initial attempt also includes the ability to gauge how many CLBs are populated with more than one domain.

In all of these cases, STARC provides warnings and information about the design to the designer. The output of the tool provides the designer a list of subcircuits that are untriplicated, a quantity for the unprotected cross section, and warnings about potential single points of failures from functionally nonequivalent modules and logical constants. Because EDIF is tightly coupled to the circuit design, the designer should be able to directly use STARC's output to find and fix the design flaws in the user circuit.

8.5.3   CASE STUDY: TRADESPACE OF RELIABILITY ISSUES UNDER AREA CONSTRAINTS

In this section we present a case study of two image-processing algorithms that use STARC to explore the tradespace of reliability issues under an area-constrained design process. The two-image processing algorithms we examined are an edge-detection algorithm and a noise-filtering algorithm. The edge-detection algorithm uses the Sobel convolution masks [43] as the computational basis. These convolution masks are well matched to FPGA implementation, because the multiplication can be reduced to shifts. The noise-filtering algorithm breaks the image into a series of small windows. The pixel in the center of the window is replaced with the minimum pixel value in the window. Both of these circuits are feed-forward and therefore do not have error persistence issues. Because both algorithms use nine 8-bit pixels as input, the algorithms both use the same data input circuit.

Several implementations of these circuits were developed: without TMR, with full TMR, and two partial TMR approaches. To avoid design issues with applying TMR, BLTmr was used. It should be noted that STARC has been modified to automatically recognize designs that have been mitigated through BLTmr and the Xilinx TMRTool. Logical constants were also extracted to input pins. For the partial TMR

TABLE 8.5

STARC Results for Two Image-Processing Algorithms

Design	Implementation	Total Unprotected Cross Section (bits)	Unprotected Logic (bits)	Unprotected Routing (bits)	Number of Components	Time to Calculate (sec)
Edge detection	No TMR	15,418	3641	11,777	1356	56
	Partial TMR (1)	21,800	19	21,781	3787	426
	Partial TMR (2)	24	16	8	3793	401
	Full TMR	0	0	0	3799	230
Noise filter	No TMR	14,914	4522	10,392	1603	95
	Partial TMR (1)	14,332	19	14,313	4273	785
	Partial TMR (2)	24	16	8	4279	565
	Full TMR	0	0	0	4285	309

approaches, we had BLTmr triplicate the logic in both implementations for both algorithms and varied how the input and output signals were handled. In the partial TMR 1 implementations we had BLTmr not triplicate any input or output signals, and in the partial TMR 2 implementations we had BLTmr triplicate only the reset, logical constant, and clock input signals.

STARC was used to determine the unprotected cross section of all of the implementations, as shown in Table 8.5.

The first thing to note from these values is that applying TMR to just the design's logic (partial TMR 1) provided little improvement for the noise filter and actually increased the cross section for the edge-detection algorithm. When we looked through the STARC results we found the large unprotected cross section in the partial TMR 1 versions were due to the unmitigated signals. As Table 8.5 shows, all of the unprotected cross section for these implementations are in the routing network, indicating that the logic was properly triplicated. Because the triplicated logic has three times as many flip-flops, the untriplicated clock, reset, and logical constant trees now have to route to three times as many locations. In a heavily pipelined design, like the edge-detection algorithm, this decision was disastrous. When we went back to BLTmr and chose to triplicate the logic and the global signals, the unprotected cross section for both designs was 99.8% smaller than the unprotected cross section in the unmitigated design. When full TMR is applied to both algorithms, there was no unprotected cross section.

Finally, STARC was able to find the hardness assurance issues that existed in the implementations without TMR and with partial TMR. In both algorithms the implementations without TMR used the component-provided logical zeros and STARC correctly identified this as a potential problem. Also, the implementation of the two algorithms with partial TMR had input signals, a voter, and input/output registers that were not triplicated. STARC was able to find these untriplicated signals and logic, report them, and properly calculate the cross section for them.

We have recently begun validation of the STARC tool. Table 8.6 shows some results from fault injection of the unmitigated implementations of the two image-processing algorithms. While the edge detection algorithm is within 93.8% of the STARC-predicted cross section, the noise-filtering algorithm is not as close at 63.8%. When looking at the numbers closer, for both designs the routing estimates look reasonable, but the logic is overestimated in both cases. We believe that the reason why there is such a gap in the logic values is due to logical masking on the fault-injection

---

TABLE 8.6

STARC Validation Results for the Unmitigated Implementation of Two Image-Processing Algorithms

Design	Total Unprotected Cross Section (bits)	Unprotected Logic (bits)	Unprotected Routing (bits)
Edge detection	14,461	2291	12,170
Noise filter	9507	1462	8045

hardware. In particular, we found that the outputs of the edge detection algorithm are much more sensitive to data changes than the noise filter. In examining the execution times we found that the tool was able to compete on average 12 components/second. Note that the execution time tripled from the unmitigated implementations to the mitigated implementations. As BLTmr flattens the circuit hierarchy while applying TMR, the entire circuit's state space must be analyzed to determine the reliability of the circuit.

8.6 CONCLUSIONS

In this chapter, we have provided an overview of a number of topics regarding assuring the robustness of TMR-protected user circuits in Xilinx FPGAs. We have presented a number of hardness assurances, including redundant modules that share logic, the inability to fully triplicate designs, component-provided logical constants, and domain crossing errors. Our studies into domain crossing errors have shown that the CLB routing network has proven to be fragile in TMR applications with highly utilized and congested routing scenarios. We have also introduced a tool, STARC, that automates the process for identifying hardness assurance issues with TMR-protected circuits for Xilinx FPGAs as well as estimating their unprotected SEU cross sections. As an illustration, we used STARC to analyze four implementations of two different image processing algorithms with different approaches to TMR. These results showed that full TMR provided a 100% reduction in cross section, and that triplicating just the logic, clock, and reset could reduce the unprotected cross section by 99.8%.

REFERENCES

1. E. Fuller, M. Caffrey, P. Blain, C. Carmichael, N. Khalsa, and A. Salazar, Radiation test results of the Virtex FPGA and ZBT SRAM for space based reconfigurable computing, in *Proceeding of the Military and Aerospace Programmable Logic Devices International Conference (MAPLD)*, Laurel, MD, September 1999.
2. H. Quinn, P. Graham, K. Morgan, Z. Baker, M. Caffrey, D. Smith, M. Wirthlin, and R. Bell, Flight experience of the Xilinx Virtex-4, *IEEE Transactions on Nuclear Science*, Vol. 60, No. 4, pp. 2682–2690, August 2013.
3. H. Quinn, D. Roussel-Dupre, M. Caffre et al., The Cibola Flight Experiment, *to be published in the ACM Transactions on Reconfigurable Technology and Systems (TRETS)*, Vol. 8, No. 1, pp. 3:1–3:22, Mar. 2015.
4. G. M. Swift, Virtex-II static SEU characterization, Xilinx Radiation Test Consortium, Tech. Rep. 1, 2004.
5. G. Allen, G. Swift, and C. Carmichael, Virtex-4VQ static SEU characterization summary, Xilinx Radiation Test Consortium, Tech. Rep. 1, 2008.
6. C. Carmichael, Triple module redundancy design techniques for Virtex FPGAs, Xilinx Corporation, Tech. Rep., November 1, 2001, XAPP197 (v1.0).
7. F. Lima, C. Carmichael, J. Fabula, R. Padovani, and R. Reis, A fault injection analysis of Virtex FPGA TMR design methodology, in *Proceedings of the 6th European Conference on Radiation and its Effects on Components and Systems (RADECS 2001)*, 2001.

8. N. Rollins, M. Wirthlin, M. Caffrey, and P. Graham, Evaluating TMR techniques in the presence of single event upsets, in *Proceedings fo the 6th Annual International Conference on Military and Aerospace Programmable Logic Devices (MAPLD)*. Washington, DC: NASA Office of Logic Design, AIAA, September 2003, p. P63.

9. L. Sterpone and M. Violante, A new analytical approach to estimate the effects of SEUs in TMR architectures implemented through SRAM-based FPGAs, *IEEE Transactions on Nuclear Science*, Vol. 52, No. 6, pp. 2217–2223, 2005.

10. H. Quinn, K. Morgan, P. Graham, J. Krone, M. Caffrey, and K. Lundgreen, Domain crossing errors: Limitations on single device triple-modular redundancy circuits in Xilinx FPGAs, *IEEE Transactions on Nuclear Science*, Vol. 54, No. 6, pp. 2037–2043, 2007.

11. H. Quinn, P. Graham, J. Krone, M. Caffrey, and S. Rezgui, Radiation-induced multi-bit upsets in SRAM-based FPGAs, *IEEE Transactions on Nuclear Science*, Vol. 52, No. 6, pp. 2455–2461, December 2005.

12. D. Lee, M. Wirthlin, G. Swift, and A. Le, Single-event characterization of the 28 nm Xilinx Kintex-7 field-programmable gate array under heavy-ion irradiation, *to be published in the IEEE Radiation Effects Data Workshop (REDW)*, Dec. 2014. doi: 10.1109 /REDN.2014.70054595.

13. A. Le, Single event effects (SEE) test report for XC7K325T-2FFG900C (Kintex-7) Field Programmable Gate Array, Boeing Corporation, Tech. Rep., 2013.

14. G. Gasiot, D. Giot, and P. Roche, Multiple cell upsets as the key contribution to the total SER of 65 nm CMOS SRAMs and its dependence on well engineering, *Nuclear Science*, Vol. 54, No. 6, pp. 2468–2473, 2007. Available: http://dx.doi.org/10.1109 /TNS.2007.908147.

15. Y. Tosaka, H. Ehara, M. Igeta, T. Uemura, H. Oka, N. Matsuoka, and K. Hatanaka, Comprehensive study of soft errors in advanced CMOS circuits with 90/130 nm technology, in *IEEE International Electron Devices Meeting Technical Digest*, 2004, pp. 941–944. Available: http://dx.doi.org/10.1109/IEDM.2004.1419339.

16. H. Quinn, K. Morgan, P. Graham, J. Krone, and M. Caffrey, Static proton and heavy ion testing of the Xilinx Virtex-5 device, in *IEEE Radiation Effects Data Workshop (REDW)*, July 2007.

17. M. Wirthlin, D. Lee, G. Swift, and H. Quinn, A method and case study on identifying physically adjacent multiple-cell upsets using 28-nm, interleaved and SECDED-protected arrays, *submitted to the IEEE Transactions on Nuclear Science*, Vol. 61, No. 6, pp. 3080–3087, Dec. 2014.

18. K. Morgan, M. Caffrey, P. Graham, E. Johnson, B. Pratt, and M. Wirthlin, SEU-induced persistent error propagation in FPGAs, *IEEE Transactions on Nuclear Science*, Vol. 52, No. 6, pp. 2438–2445, 2005.

19. Xilinx TMRTool user guide, http://www.xilinx.com/products/milaero/ug156.pdf.

20. A. Sutton, Creating highly reliable FPGA designs, 2013. Available: http://www.synop sys.com/Company/Publications/SynopsysInsight/Pages/Art5-fpga-designs-IssQ1-13 .aspx,last accessed 8/15/2013.

21. R. Do, The details of triple modular redundancy: An automated mitigation method of I/O signals, in *Proceedings of the Military and Aerospace Programmable Logic Devices*, 2011, https://nepp.nasa.gov/respace mapld11/talks/thu/MAPLD C/0800-Do .pdf, last accessed 8/15/2013.

22. P. Graham, M. Caffrey, M. Wirthlin, D. E. Johnson, and N. Rollins, SEU mitigation for half-latches in Xilinx Virtex FPGAs, *IEEE Transactions on Nuclear Science*, Vol. 50, No. 6, pp. 2139–2146, December 2003.

23. H. Quinn, G. R. Allen, G. M. Swift, C. W. Tseng, P. S. Graham, K. S. Morgan, and P. Ostler, SEU-susceptibility of logical constants in Xilinx FPGA designs, *IEEE Transactions on Nuclear Science*, Vol. 56, No. 6, pp. 3527–3533, December 2009.

24. M. French, M. Wirthlin, and P. Graham, Reducing power consumption of radiation mitigated designs for FPGAs, in *Proceedings of the 9th Annual International Conference on Military and Aerospace Programmable Logic Devices (MAPLD)*, September 2006.
25. H. Quinn, D. Bhaduri, C. Teuscher, P. Graham, and M. Gohkale, The STAR systems toolset for analyzing reconfigurable system cross-section, in *Military and Aerospace Programmable Logic Devices*, 2006, p. 162.
26. H. Quinn, D. Bhaduri, C. Teuscher, P. Graham, and M. Gohkale, The STARC truth: Analyzing reconfigurable supercomputing reliability, in *Field-Programmable Custom Computing Machines*, 2005.
27. JBits: A Java-based interface to FPGA hardware. Available: http://www.io.com/guccione/Papers/Papers.html.
28. D. Bhaduri and S. Shukla, NANOLAB—A tool for evaluating reliability of defect-tolerant nanoarchitectures, *IEEE Transactions on Nanotechnology*, Vol. 4, No. 4, pp. 381–394, 2005.
29. J. A. Abraham, A combinatorial solution to the reliability of interwoven redundant logic networks, *IEEE Transactions on Computers*, Vol. 24, No. 6, pp. 578–584, May 1975.
30. J. A. Abraham and D. P. Siewiorek, An algorithm for the accurate reliability evaluation of triple modular redundancy networks, *IEEE Transactions on Computers*, Vol. 23, No. 7, pp. 682–692, July 1974.
31. C. Hirel, R. Sahner, X. Zang, and K. Trivedi, Reliability and performability using SHARPE 2000, in *11th International Conference on Computer Performance Evaluation: Modeling Techniques and Tools*, Vol. 1786, 2000, pp. 345–349.
32. F. V. Jensen, *Bayesian Networks and Decision Graphs*. New York: Springer-Verlag, 2001.
33. S. Krishnaswamy, G. F. Viamontes, I. L. Markov, and J. P. Hayes, Accurate reliability evaluation and enhancement via probabilistic transfer matrices, in *Design, Automation and Test in Europe (DATE'05)*, Vol. 1. New York: ACM Press, 2005, pp. 282–287.
34. G. Norman, D. Parker, M. Kwiatkowska, and S. Shukla, Evaluating the reliability of nand multiplexing with prism, *IEEE Transactions on CAD*, Vol. 24, No. 10, pp. 1629–1637, 2005.
35. A. Zimmermann, Modeling and evaluation of stochastic petri nets with TimeNET 4.1, in *2012 6th International ICST Conference on Performance Evaluation Methodologies and Tools (VALUETOOLS)*, 2012, pp. 54–63.
36. S. J. S. Mahdavi and K. Mohammadi, SCRAP: Sequential circuits reliability analysis program, *Microelectronics Reliability*, Vol. 49, No. 8, pp. 924–933, August 2009.
37. E. Maricau and G. Gielen, Stochastic circuit reliability analysis, in *2011 Design, Automation Test in Europe Conference Exhibition (DATE)*, March 2011, pp. 1–6.
38. N. Miskov-Zivanov and D. Marculescu, Circuit reliability analysis using symbolic techniques, *IEEE Transactions on Computer-Aided Design of Integrated Circuits and Systems*, Vol. 25, No. 12, pp. 2638–2649, December 2006.
39. D. Bhaduri, S. K. Shukla, P. S. Graham, and M. B. Gokhale, Reliability analysis of large circuits using scalable techniques and tools, *IEEE Transactions on Circuits and Systems–I: Fundamental Theory and Applications*, Vol. 54, No. 11, pp. 2447–2460, November 2007.
40. D. Bhaduri, S. K. Shukla, P. Graham, and M. Gokhale, Comparing reliability-redundancy trade-offs for two von Neumann multiplexing architectures, *IEEE Transactions on Nanotechnology*, Vol. 6, No. 3, pp. 265–279, May 2007.

41. D. Bhaduri and S. Shukla, Nanoprism: A tool for evaluating granularity vs. reliability trade-offs in nano-architectures, in *14th GLSVLSI*. Boston: ACM, April 2004, pp. 109–112.

42. M. Baze, S. Buchner, W. Bartholet, and T. Dao, An SEU analysis approach for error propagation in digital VLSI CMOS ASICs, *IEEE Transactions on Nuclear Science*, Vol. 42, No. 6, pp. 1863–1869, December 1995.

43. A. K. Jain, *Fundamentals of Digital Image Processing*. Upper Saddle River, NJ: Prentice Hall Information and System Sciences, Series 1989.

9 Single-Event Mitigation Techniques for Analog and Mixed-Signal Circuits

Thomas Daniel Loveless and
William Timothy Holman

CONTENTS

9.1   INTRODUCTION

The interaction of ionizing radiation with a circuit's semiconductor components can manifest as spurious transient signals within the circuit. A single-event transient (SET) may disrupt the valid electric signals or compromise the functionality of the circuit. SETs may result in single-event upsets (SEUs) in digital circuits if the transients alter the state of memory (e.g., a memory cell or data register can be changed from a logic 0 state to a logic 1 state or vice versa). SEUs can lead to circuit errors if corrupted data propagates throughout the circuit and is observable at the output. These upsets are often termed soft errors, as they do not result in permanent damage within the circuit, although they can result in application or mission failure if the soft error disrupts a critical function.

While the meaning of a soft error in a digital circuit (i.e., a corrupt memory or logic state) is fairly straightforward (this in no way implies that the effects are trivial), there exists no standard metric for soft errors in analog and mixed-signal (hybrid analog and digital) systems, as the effect of a single event (SE) is dependent on the circuit topology, type of circuit (function), and the operating mode. For example, Figure 9.1 illustrates the voltage amplitude versus time-width (full-width at half-maximum of amplitude) of SETs observed at the output of an LM124 operational amplifier under heavy-ion broad-beam exposure (specifically 100 MeV Br, 150 MeV Mg, and 210 MeV Cl ions) [1]. A variety of positive- and negative-going transients were observed in the experiment with both long and short durations. The type of transient observed at the circuit output can be attributed to the location of the initial ion strike within the circuit.

As with digital components, the hardening of analog components can generally be accomplished through a brute force approach; that is, area, power, and/or bandwidth are sacrificed through the increase of capacitance, device size, and/or current drive in order to increase the critical charge required to generate analog SETs (ASETs) [2–4]. The challenge of radiation-hardened analog and mixed-signal circuit design is to develop techniques for mitigating ASETs while minimizing these design penalties.

SET mitigation, whether achieved through brute force or some other sophisticated design approach, can be implemented at various levels of abstraction: the technology process level (through physical process changes), layout level (through device and/or structural changes), circuit level (through topological changes), and/or system level (through architectural changes). SET mitigation generally involves one or both of the following, irrespective of the technology and abstraction level:

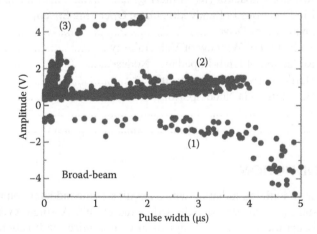

FIGURE 9.1 Amplitude versus time-width (measured full-width at half-maximum of amplitude) of SETs at the output of an LM124 operational amplifier under heavy-ion exposure (100 MeV Br, 150 MeV Mg, and 210 MeV Cl ions). (From Y. Boulghassoul, L. W. Massengill, A. L. Sternberg, R. L. Pease, S. Buchner, J. W. Howard, D. McMorrow, M. W. Savage, and C. Poivey, *IEEE Trans. Nucl. Sci.*, Vol. 49, No. 6, pp. 3090–3096, Dec. 2002.)

1. The reduction in the amount of collected charge (Q_{coll}) at a metallurgical junction [2]
2. The increase in the critical charge (Q_{crit}) required to generate an ASET [2]

The historical assumption is that (SEs) affect a single sensitive volume in the semiconductor material. Reports in the literature generally indicate that the single sensitive volume approximation breaks down somewhere between the 130- and 90-nm planar technology nodes (i.e., the effects of charge sharing between multiple sensitive volumes is not a dominant factor for 130-nm feature sizes and greater) [5–7]. The impacts of charge-sharing phenomena and novel device structures (such as nonplanar field-effect transistors [FETs]) in the generation and propagation of ASETs must be considered in more advanced integrated circuit processes.

This chapter overviews basic and state-of-the-art mitigation approaches, provides some examples of hardened analog and mixed-signal circuits, and attempts to classify the techniques based on the fundamental mechanisms of hardening. The primary focus of the chapter will be on layout and circuit-level approaches to single-event hardening.

9.2 REDUCING THE COLLECTED CHARGE

9.2.1 SUBSTRATE ENGINEERING

The reduction of Q_{coll} at a critical node of an integrated circuit requires a fundamental alteration of the technology process or the device implementation. One technology-level method for performing this reduction is substrate engineering. For example, the use of charge blocking layers in the substrate—shown in [8] for a silicon germanium (SiGe) heterojunction bipolar transistor (HBT) technology—can be effective for controlling delayed charge collection from events that occur outside of the deep trench in deep-trench isolation (DTI) technologies. Figure 9.2 illustrates a cross section of the buried layer concept in IBM's 5HP technology. The buried p-type layer is located

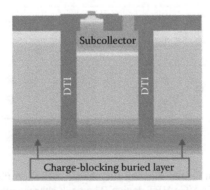

FIGURE 9.2 Cross section of the buried layer concept for the IBM 5HP technology. The p-type buried layer is located at the level of the bottom of the deep trench isolation, is 2-μm thick, and has a peak boron concentration of 1×10^{17} cm^{-3}. (From J. A. Pellish, R. A. Reed, R. D. Schrimpf et al., *IEEE Trans. Nucl. Sci.*, Vol. 53, No. 6, pp. 3298–3305, Dec. 2006.)

at the bottom of the deep trench isolation [8]. Figure 9.3 shows simulation results of the charge collection following the interaction with 36 MeV ^{16}O ions in the SiGe HBT device in Figure 9.2 [8]. The peak charge collection occurs for strikes within the DTI and the tails represent charge collection from events outside the DTI. Three versions of the p-type charge-blocking layer were utilized: 10^{16}, 10^{17}, and 10^{18} cm^{-3}. It is clear that increasing the doping of the blocking layer limits the charge collection from outside the DTI [8].

An additional example of substrate engineering is the use of a very thin silicon layer to limit the collection volume (e.g., silicon-on-insulator [SOI]) [9]. The dielectric isolation of transistors for prevention of single-event latchup has allowed SOI technologies to be prevalent in space and military applications for over 50 years [9]. In recent years with the shrinking of gate oxide thickness due to technology scaling, modern sub-100-nm SOI devices have seen a reduction in total ionizing dose (TID) effects, prompting additional interest in use of the technology for space applications. Some challenges associated with design in SOI technologies include additional parasitic elements not present in bulk technologies [9], floating-body and history effects [10], and higher likelihoods of wave-shaping effects (i.e., pulse propagation, attenuation, and broadening) [11]. These circuit-level challenges with SOI generally complicate the analysis of the single-event sensitivities of the technology. However, the smaller (and inherently more isolated) sensitive volumes when compared to like feature-size bulk devices (i.e., Q_{coll} is generally less for SOI than for bulk in similar feature size technologies) generally means shorter SET durations and smaller SE cross sections [12–15].

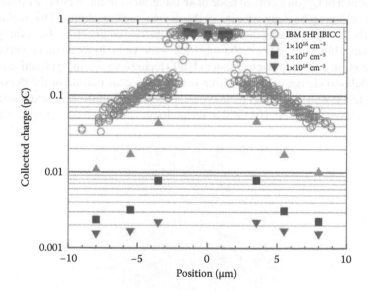

FIGURE 9.3 Simulation results showing the charge collection (following the interaction with 36-MeV ^{16}O ions) of a SiGe HBT device. The peak charge collection occurs for strikes within the DTI and the tails represent charge collection from events outside the DTI. Three versions of the p-type charge blocking layer were utilized: 10^{16}, 10^{17}, and 10^{18} cm^{-3}. (From J. A. Pellish, R. A. Reed, R. D. Schrimpf et al., *IEEE Trans. Nucl. Sci.*, Vol. 53, No. 6, pp. 3298–3305, Dec. 2006.)

9.2.2   LAYOUT-LEVEL MITIGATION

A reduction in Q_{coll} may also be obtained through layout-level mitigation techniques. Layout-level mitigation generally involves transistor- or circuit-level modification of layout cell arrangements for reducing the amount of collected charge at critical device junctions. Examples include guard rings [16–19], guard drains [19], and diodes [6] around bulk MOS devices. Guard rings and drains (sometimes called guard contacts) allow for quicker recovery of the well potentials (in bulk complementary metal-oxide semiconductor [CMOS]). An additional benefit is gained through the isolation created between neighboring sensitive volumes. The benefit is especially pronounced in p-type metal-oxide semiconductor (PMOS) devices that exhibit a propensity for parasitic bipolar enhancement. Figure 9.4 illustrates the layout of a typical structure that includes guard contacts between each PMOS and n-type metal-oxide semiconductor (NMOS) device (adapted from [16]). The guard contacts are formed from the same diffusion as the N-well and P-substrate contacts and have been shown to limit the charge collection at the circuit diffusions [16,18,19].

Similarly, the use of n-rings [20], substrate-tap rings [21], and nested minority-carrier guard rings [22] may be utilized in bipolar structures (illustrated for a SiGe HBT technology in [23]). Additional techniques for HBT devices (analogous to the aforementioned techniques for CMOS) include the addition of dummy collectors for charge collection in HBT devices [24] and the increase in substrate and well contacts (for reduction in substrate and well impedances) [17,25–27].

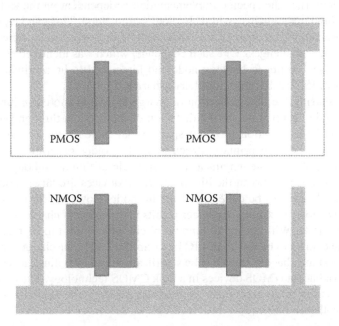

FIGURE 9.4   Generic layout that includes guard contacts between each PMOS and NMOS device. (Adapted from J. D. Black, A. L. Sternberg, M. L. Alles, A. F. Witulski, B. L. Bhuva, L. W. Massengill, J. M. Benedetto, M. P. Baze, J. L. Wert, and M. G. Hubert, *IEEE Trans. Nucl. Sci.*, Vol. 52, No. 6, pp. 2536–2541, Dec. 2005.)

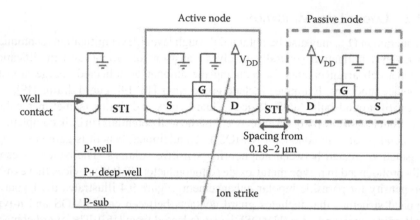

FIGURE 9.5 Cross section of two adjacent NMOS devices in a bulk CMOS technology. The active node is referred to as the original hit node whereas the passive node or device refers to any adjacent device that collects charge. (From O. A. Amusan, A. F. Witulski, L. W. Massengill, B. L. Bhuva, P. R. Fleming, M. L. Alles, A. L. Sternberg, J. D. Black, and R. D. Schrimpf, *IEEE Trans. Nucl. Sci.*, Vol. 53, No. 6, pp. 3253–3258, Dec. 2006.)

While the aforementioned layout-level approaches have generally been shown to reduce Q_{coll} of microelectronic devices, the approaches are generally dictated by the technology (i.e., the specific implementation is dependent on the technology or requires a fundamental change in technology parameters). This can be problematic in a rapidly changing technological environment (with newly developed technologies also comes new challenges). One such challenge, which was identified as a potential problem at the 130-nm technology node and forecasted to be a primary issue for smaller nodes [5], has been termed charge sharing.*

Charge sharing (i.e., the collection of charge by two or more *p-n* junctions following a single event) is due to the diffusion of the carriers through semiconductor material. The reduction in nodal spacing can increase the charge collection at nodes other than the primary struck node. For older generation technologies (greater than 130-nm gate lengths as a general rule of thumb, although not a hard limit), the distances between the hit and adjacent devices are large enough so that most of the charge can be collected at the hit node. However, for advanced technologies, the close proximity of devices results in diffusion of charge to nodes other than the hit node. With the small amount of charge required to represent a logic-HIGH state (shown to be less than 1 fC in 45-nm SOI [31]), the charge collected due to diffusion at an adjacent node may be significant. Figure 9.5 illustrates a cross section of two adjacent NMOS devices in a bulk CMOS technology. The active node is referred to as the original hit node whereas the passive node refers to any adjacent node that collects charge [5].

---

* It should be noted that at the present time the 10- to 16-nm technology nodes developed in FDSOI and bulk and SOI FinFET technologies are in commercial development and manufacturing phases. Many of the works are evaluating the impact of this paradigm shift on the radiation vulnerability of systems [28–30].

9.2.2.1   Nodal Separation and Interleaved Layout

One solution for mitigating the amount of charge shared between adjacent nodes is nodal separation [5,25]. Figure 9.6 illustrates the charge collected on the passive device versus the LET of the incident ion on the active device as a function of nodal separation. Both PMOS-to-PMOS and NMOS-to-NMOS charge sharing are illustrated and show a decrease in charge collection with increase in distance between devices. Although effective, nodal separation is not a practical solution when considering the demands for higher packing densities and increased speeds.

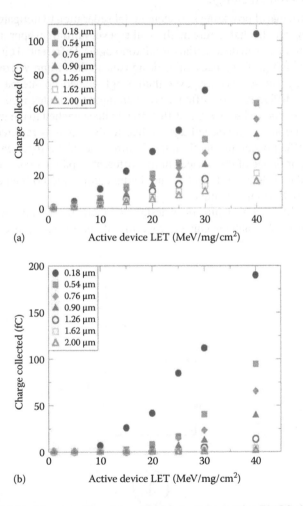

(a)

(b)

FIGURE 9.6   (a) Nodal separation of two PMOS devices: passive PMOS device shows a decrease in charge collection with increase in distance. (b) Nodal separation of two NMOS devices: passive NMOS device shows a decrease in charge collection with increase in distance. (From O. A. Amusan, A. F. Witulski, L. W. Massengill, B. L. Bhuva, P. R. Fleming, M. L. Alles, A. L. Sternberg, J. D. Black, and R. D. Schrimpf, *IEEE Trans. Nucl. Sci.*, Vol. 53, No. 6, pp. 3253–3258, Dec. 2006.)

Interdigitation, or interleaved layout, is a technique that takes advantage of the benefits of nodal separation while maintaining device density requirements. Provided that the designer has knowledge of the circuit nodes (or combinations of nodes) sensitive to SETs as well as those that pose less of a threat, the less sensitive transistors can be placed between pairs of sensitive devices and the nodal spacing between critical devices can be increased while maximizing density [25,32]. This method has been successfully implemented in dual-interlocked cell (DICE) [33] and other redundancy-based designs [34–39].

9.2.2.2 Differential Design

Given the challenge of providing sufficient nodal separation to mitigate the effects of charge sharing, another hardening method takes a very different approach by encouraging the phenomenon in differential circuit topologies. Differential circuits are commonly used in high-performance analog design due to their improved dynamic output range and better noise rejection over their single-ended counterparts. Figure 9.7 depicts a basic differential pair often used as an input to an integrated amplifier. Two transistors are connected so that any differential voltage applied to the inputs ($V_p - V_n$) is amplified and any common voltage applied to the inputs is rejected. However, a single-event effect occurring in circuitry feeding one of the input gates of the differential pair, or in one of the devices in the differential pair, can perturb the voltage at the input. This voltage perturbation, not being common to both inputs, will result in a transient in the output voltage.

While the differential nature of a design is inherently a circuit-level approach, various layout-level strategies can improve the SE performance over single-ended

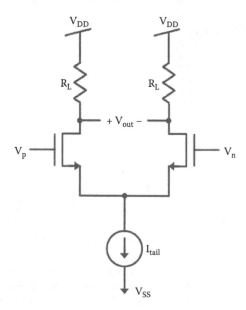

FIGURE 9.7 Basic schematic of a differential pair.

designs. Hypothesized in [23] and shown for the first time through simulations in [40] and experiments in [41], the layout of matched transistors in a differential data path can be modified in order to exploit the charge-sharing phenomenon, thereby causing any single-event transient to be rejected as a common-mode perturbation. The layout technique, termed differential charge cancellation (DCC), minimizes the distance between the drains of matched devices in the differential pair and maximizes the likelihood of an ion strike affecting both sides of the differential pair through common-centroid layout. It should be noted that the common centroid technique is typical for differential design to improve transistor matching but does not require that the distance between the drains in the differential pair be minimized as in DCC layout.

Figure 9.8 illustrates layout variations of the differential pair, including devices A and B before and after DCC layout for maximizing charge sharing. Each transistor in the DCC configuration is split into two devices and placed diagonally. The device pairs should be arranged in a common well with drains located as close as possible to promote common-mode charge rejection. Figure 9.9 shows surface plots of experimentally measured charge collected at points in the die scan for transistor A of the differential pair (device dimensions illustrated in Figure 9.8). Charge was injected using a laser two-photon absorption technique. Single-transistor charge collection is shown in the top row for the two-device configuration (left) and common-centroid layout (right). Differential charge is shown in the bottom row. The DCC layout configuration was projected to decrease the sensitive area by over 90% [41].

Since a DCC layout is essentially a variation of a common-centroid layout used to improve matching in analog circuits, its use in differential input stages results in minimal layout penalty for the designer. In many cases, a DCC layout requires only minor changes in one or two metal layers of a standard analog layout. Furthermore, the DCC layout technique can be extended anywhere along a symmetric differential signal path, as shown in Figure 9.10. Transistors M6 and M7 (or M8/M9, MC4/MC5, etc.) can also utilize DCC layouts. In this case, the layout penalty is slightly greater since those pairs of transistors would not normally require a common-centroid layout for matching purposes.

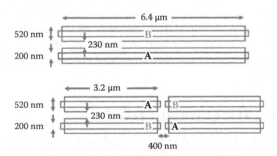

FIGURE 9.8   Differential pair including devices A and B before and after DCC layout for maximizing charge sharing. (From S. E. Armstrong, B. D. Olson, W. T. Holman, J. Warner, D. McMorrow, and L. W. Massengill, *IEEE Trans. Nucl. Sci.*, Vol. 57, No. 6, pp. 3615–3619, Dec. 2010.)

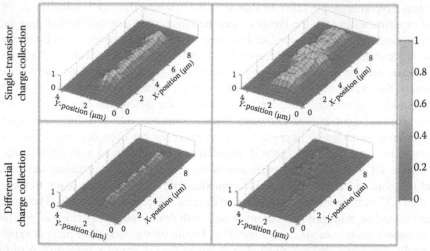

All Z: normalized charge

FIGURE 9.9 Surface plots of experimentally measured charge collected at points in the die scan for transistor A of the differential pair. Charge was injected using a laser two-photon absorption technique. Single-transistor charge collection is shown in the top row for the two-device configuration (left) and common-centroid layout (right). Differential charge is shown in the bottom row. (From S. E. Armstrong, B. D. Olson, W. T. Holman, J. Warner, D. McMorrow, and L. W. Massengill, *IEEE Trans. Nucl. Sci.*, Vol. 57, No. 6, pp. 3615–3619, Dec. 2010.)

As multiple node charge collection, or charge sharing, is becoming more commonplace, methods for utilizing charge sharing for improved SET performance have become promising in digital electronics as well. For technologies where the time constant for device-to-device charge transport is on the order of the gate-to-gate electrical propagation, the layout orientation, device spacing, and electrical signal propagation may be designed to interact in order to truncate a propagated voltage transient (pulse quenching). Pulse quenching has been identified as a factor in the analysis and measurement of digital SETs and may be a reasonable technique to harness for improved radiation performance [42]. Since the discovery of pulse quenching, many digital design techniques have employed the concept for improving the SE performance [34,43].

9.3   REDUCING THE CRITICAL CHARGE

Methods for reducing the critical charge of a device or circuit may seem at first glance (at least from a circuit designer's point of view) to be more straightforward. In other words, a designer may operate under the assumption that the physical process (thus the amount of collected charge) cannot change, thus requiring creative solutions to limit the impact on the circuit. Conventional, perhaps brute force methodology, designed to counteract superfluous charge, include [1] increasing the transistor

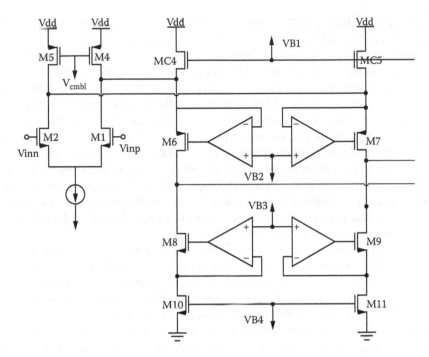

FIGURE 9.10 Schematic for a typical operational transconductance amplifier (OTA), with internal symmetry that can be exploited for DCC radiation-hardened layouts. (From A. T. Kelly, P. R. Fleming, W. T. Holman, A. F. Witulski, B. L. Bhuva, and L. W. Massengill, *IEEE Trans. Nucl. Sci.*, Vol. 54, No. 6, pp. 2053–2059, Dec. 2007.)

sizes (buffering) [3,4], increasing the drive currents [1], increasing the supply voltage [1], increasing capacitor sizes [1], and implementing redundancy.

Additionally, a variety of mitigation techniques for analog and mixed-signal (AMS) circuits can be found in the literature covering a wide range of topologies. Some common AMS circuits discussed in this chapter include operational amplifiers (OAs), low-noise amplifiers (LNAs), bandgap voltage references, voltage-controlled (VCOs) and injection-locked oscillators (ILOs), phase-locked loops (PLLs), serializer/deserializers (SerDes), comparators, and analog-to-digital converters (ADCs). Although many of the circuit-level mitigation techniques seem to be topology-specific, the fundamental hardening mechanisms have much in common. The remainder of the chapter presents an overview of the current state of the art (some that may qualify as brute-force approaches and some that may not) in single-event effect (SEE) mitigation in AMS circuits, and organizes the techniques based on the root cause mechanisms.

9.3.1 REDUNDANCY

While more common in digital circuits, triple modular redundancy (TMR) has been successfully used in mixed-signal circuits with digital output signatures, such as the voltage comparator. This approach was adopted in [44] where a single comparator was replaced by three parallel comparators driving a CMOS majority-voting block. The voting circuit was hardened by oversizing the transistors [1,44].

Olson et al. [45] goes further by evaluating trade-offs of comparator redundancy when implemented in a pipelined ADC. While TMR is effective at mitigating transients in the comparators, the single-event improvement reaches a point of diminishing returns when comparator TMR is applied to the first half of the pipeline (i.e., improvement is maximized when redundancy is implemented on the most significant bit [MSB]). Olson used the signal-to-noise ratio (SNR) metric to compare the single-event hardness of different mixed-signal circuit designs. By randomly injecting upsets into the circuit (in the design phase), and analyzing the response in the frequency domain, the SNR indicates the impact of the SEs on the overall response of the circuit. It is important to note that while this technique assumes a simulated SE error rate that is much too high to ever occur in an actual circuit, it does allow for hardness comparisons between different designs.

Figure 9.11 shows the SNR for increasing use of comparator TMR in a 10-bit pipelined ADC. Results shown are for a model with an individual comparator upset probability of 0.1% and 100%. The upset probability refers to probability of a SE strike during each data cycle. Figure 9.11 indicates that the application of comparator TMR to the first half of the 10-bit pipelined ADC produces the best trade-off in decreasing single-event vulnerability versus increasing area and power. Note that even assuming extremely high comparator upset rates, comparator TMR is most effective when applied to the first 50% to 70% of the total number of stages. Olson et al. show similar results regardless of ADC resolution [45].

Additionally, Loveless et al. [46] show a VCO topology hardened to single events using an approach based on TMR. Rather than running three stand-alone VCOs in

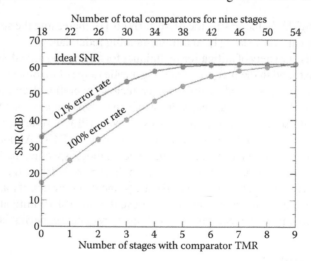

FIGURE 9.11 Signal-to-noise ratio improvement for increasing use of comparator TMR in a 10-bit pipelined ADC. The results shown are for a model with an individual comparator upset probability of 0.1% and 100% and are compared to the ideal SNR. (From B. D. Olson, W. T. Holman, L. W. Massengill and B. L. Bhuva, *IEEE Trans. Nucl. Sci.*, Vol. 55, No. 6, pp. 2957–2961, Dec. 2008.)

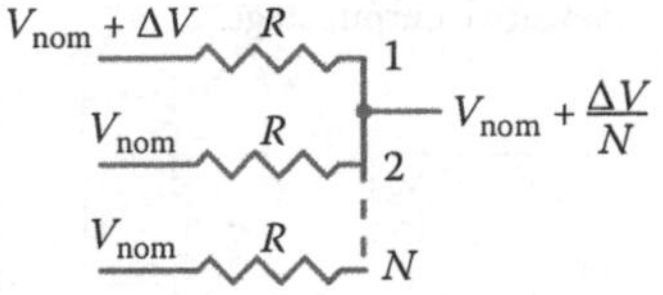

FIGURE 9.12   Analog averaging through the use of N identical resistors, R. A perturbation (ΔV) due to a SE strike on any one copy of the circuit is reduced to $\Delta V/N$.

parallel, three voltage-controlled-delay-lines (VCDLs), each with independent bias stages, are implemented in parallel with a single feedback path for jitter reduction. The design is shown to reduce the output phase displacement following ion strikes to below the normal operating noise floor [46].

9.3.2   AVERAGING (ANALOG REDUNDANCY)

Analog averaging is a form of hardware redundancy for the reduction of spurious transients. The averaging of an analog voltage can be accomplished by replicating and parallelizing a circuit N times and connecting the replicated nodes together through parallel resistors to a common node, as seen in Figure 9.12. A perturbation (ΔV) due to an SE on any one copy of the circuit is reduced to $\Delta V/N$. This technique has been offered as a solution to the observed vulnerability of a charge pump for PLLs [32] and implemented in the bias circuitry of VCO [47]. Kumar et al. propose a similar approach to harden the charge pump and VCO blocks of a PLL by including two independent charge pump/low-pass filter (LPF) blocks controlling two cross-coupled VCO circuits [48].

9.3.3   RESISTIVE DECOUPLING

Resistive decoupling was first published in 1982 as a technique for hardening memory cells by introducing series resistors in the cross-coupling lines of the inverter pairs [49,50]. The resistors effectively increase the time constant seen by the two storage nodes and limit the maximum change in voltage during a single event, thus increasing the minimum charge required to change the state of the memory. This technique is also used in AMS circuits for hardening digital latches, such as those present at the output of voltage comparators in an ADC [51]. A similar technique may be used to filter high-frequency transients by decoupling nodes sensitive to ASETs and introducing a time constant through a series resistor or low-pass filter. This was shown in [52] and [53] where the high-impedance output of a charge pump circuit was decoupled from the capacitive input to a VCO. As seen in Figure 9.13, the number of sensitive nodes present in the output stage of the charge pump (Figure 9.13a) may be reduced and subsequently decoupled from the VCO control voltage (Figure 9.13b). The improvement can be seen in Figure 9.14, where (a) the laser cross section (sensitive area determined through a two-dimensional [2-D] raster scan) of PLL subcircuits is plotted versus the square of laser energy (proportional to charge [54]) and (b) the number of erroneous clock pulses is plotted versus operating frequency at an incident laser energy of 30 nJ [53]. It should be noted that the improvement in

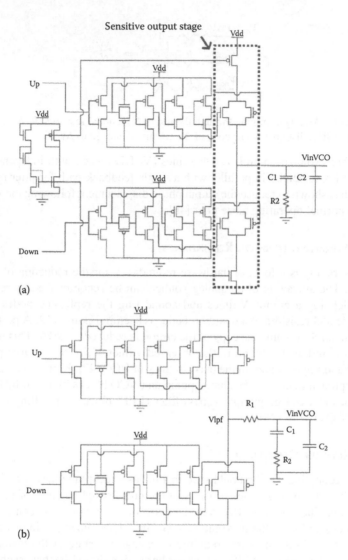

FIGURE 9.13 (a) A standard current-based charge pump configuration for PLL circuits. The sensitive output stage (high-impedance tristate configuration) is directly coupled to the control voltage of the VCO. (b) Single-event hardened voltage-based charge pump configuration. The number of sensitive nodes is reduced and subsequently decoupled from the VCO control voltage, resulting in approximately 2 orders of magnitude improvement in the magnitude of output transients resulting from single events. (From T. D. Loveless, L. W. Massengill, B. L. Bhuva, W. T. Holman, A. F. Witulski, and Y. Boulghassoul, *IEEE Trans. Nucl. Sci.*, Vol. 53, No. 6, pp. 3432–3438, Dec. 2006.)

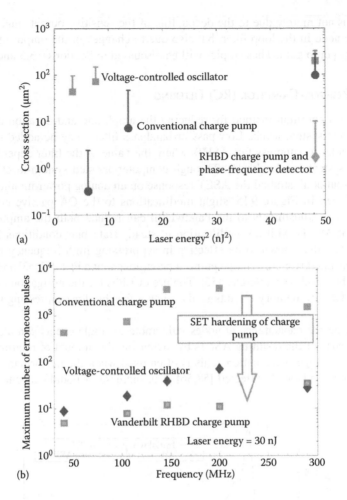

FIGURE 9.14 (a) The laser cross section (sensitive area determined through a 2-D raster scan) of PLL subcircuits is plotted versus the square of laser energy (proportional to charge [54]) and (b) the number of erroneous clock pulses is plotted versus operating frequency at an incident laser energy of 30 nJ. Approximately 2 orders of magnitude improvement in both the laser cross section and number of erroneous pulses is obtained through the use of the hardened charge pump circuit. (From T. D. Loveless, L. W. Massengill, B. L. Bhuva, W. T. Holman, R. A. Reed, D. McMorrow, J. S. Melinger, and P. Jenkins, *IEEE Trans. Nucl. Sci.*, Vol. 54, No. 6, pp. 2012–2020, Dec. 2007.)

this case is not merely due to the decoupling of the sensitive output stage with the capacitive node in the loop filter, but also due to changes in the output impedance and charge pump gain. These topics will be discussed in Sections 9.3.5 and 9.3.7.

9.3.4 RESISTOR-CAPACITOR (RC) FILTERING

Filtering is a common method for reducing the amplitude and duration of ASETs at circuit and system levels. Low-pass or bandpass filters may be added to critical nodes in order to suppress fast ASETs where the value of the filter depends on the circuit or system bandwidth [1]. Through computer-assisted system-level analysis, Boulghassoul et al. studied the ASET response on an analog power-monitoring network. As seen in Figure 9.15, slight modifications to the OA passive component networks (i.e., adjustments to the bandwidth) can reduce both the amplitude and duration of ASETs with no modification to steady-state bias conditions [55]. The approach has also shown to be effective in suppressing high-frequency noise and ASETs generated from the charge pump subcomponent of a PLL [52,53] and in hardening the bias nodes of a SerDes [56]. The use of LPFs for the mitigation of SETs in advanced CMOS memory circuits is also shown feasible for suppressing transients ≤50 ps [57].

Increasing the capacitance at nodes vulnerable to single events can also reduce the amplitudes of the resulting ASETs by increasing the amount of required charge to induce a voltage perturbation. This is often used when the performance specifications are not adversely affected [58,56]. The increase of nodal capacitance often

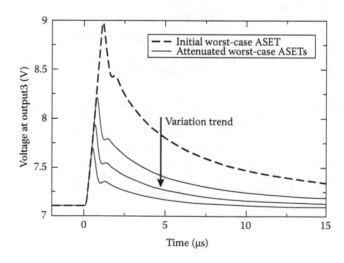

FIGURE 9.15 Simulated output voltage transients in a power-monitoring network for various resistor values in the OA feedback networks. Lowering of the resistance has the effect of reducing the amplitude and duration of transients. (From Y. Boulghassoul, P. C. Adell, J. D. Rowe, L. W. Massengill, R. D. Schrimpf, and A. L. Sternberg, *IEEE Trans. Nucl. Sci.*, Vol. 51, No. 5, pp. 2787–2793, Oct. 2004.)

alters characteristic parameters such as gain and bandwidth. The subsequent section will discuss mitigation techniques when such characteristics are paramount.

9.3.5 MODIFICATIONS IN BANDWIDTH, GAIN, OPERATING SPEED, AND CURRENT DRIVE

One effective way to reduce the circuit's sensitivity to ASETs is to reduce the part's bandwidth, thereby suppressing all transients outside of the frequency band. This rule of thumb can be seen as generally applicable to analog topologies that can be expressed as closed-loop amplifier structures. It has been shown to be applicable in various studies on OAs [1,55,58] and PLLs [59,60], both of which can be represented as a closed-loop amplifier. However, Boulghassoul et al. [55] and Sternberg et al. [58] also discuss the importance of examining the severity of an ASET as defined by the application for which it is a part. For example, the threshold for an application is typically defined by both ASET amplitude and duration. Sternberg et al. have pointed out that depending on the origin of the ASET, the duration of the pulse may increase as modifications are placed to decrease the amplitude. Therefore, specific consequences regarding the size of the resistors, compensation capacitors, and stage gains may occur and require special attention. In general, as seen by Loveless et al. regarding PLLs [60] and Sternberg et al. regarding OAs [58], it appears that maximizing speed and minimizing the open- and closed-loop gains may improve the ASET response.

Operating speed plays a curious role in determining the SET response of analog circuits. As previously mentioned, analog circuits have been shown to exhibit reduced ASET vulnerability for increased operating frequency [58,60]. This is contrary to that typically observed in digital systems, where increasing error cross sections as a result of SETs induced in combinational logic have been observed for increasing operating frequency [61]. In digital circuits, a SET can result in an SEU and lead to a circuit error if the corrupted data propagates throughout the circuit and is observable at the output. The ability of the SEU to reach the circuit output depends on the logical and electrical masking as well as the window of vulnerability (latch window masking). The result of latch window masking is that for equivalent SET pulse widths, faster circuits have a higher probability of being latched into memory. In analog electronics, however, increased speed is often accompanied by increased drive current and an improved ability to dissipate the deposited energy, making the circuit less vulnerable. It is thus important to attribute the improvement to either speed or drive strength, as increased bias current is a well-known technique and is often used in AMS circuits for improved SET performance [56]. The improved performance may or may not be as a result of increased speed but rather subtle changes in the individual device operating conditions such as bias, current drive, and load.

Loveless et al. discuss a more complex example of the importance of device conditions (not just speed) in regard to SET mitigation of mixed-signal PLL circuits [60]. For a particular oscillator design, for example, it is shown that the operating frequency should be maximized within the designed bandwidth (consistent with that shown in [58] for OAs). On the other hand, the natural frequency of the

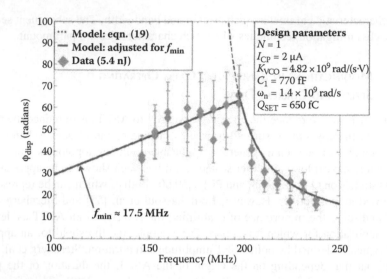

FIGURE 9.16 Average phase displacement versus lock frequency within the PLL's operating region for incident laser energy of 5.4 nJ plotted against the model. (From T. D. Loveless, L. W. Massengill, W. T. Holman, B. L. Bhuva, D. McMorrow, and J. H. Warner, *IEEE Trans. Nucl. Sci.*, Vol. 57, No. 5, pp. 2933–2947, Oct. 2010.)

PLL (analogous to the response time of the closed-loop PLL and not to be confused with the output frequency) is found to amplify transients in the PLL resulting from ionizing radiation and thus should be reduced to improve the SET response of the PLL. The authors go on to provide an analytical expression for determining an upper bound for reasonable radiation performance. Figure 9.16 illustrates this effect by showing the output phase displacement of a PLL versus the operating frequency. Data was obtained through a two-photon laser absorption technique and plotted versus the model presented in [60]. The phase displacement is observed to decrease with increasing frequency due to a complex interaction of the drive strength, natural frequency, and loop gain. The error (phase displacement) decreases with decreasing frequency in the low frequency regime (<200 MHz in this case) because the VCO following the strike reaches the designed minimum frequency of operation [60].

A similar effect can be observed for a VCO as a stand-alone subcircuit, albeit for different reasons. Figure 9.17 illustrates the simulated maximum number of errors generated by single-ion strikes (total injected charges of between 100 fC to 1 pC) as a function of the VCO input voltage (which is proportional to drive strength and frequency) [62]. The VCO was designed in a 130-nm CMOS process and has a center frequency of operation of 1 GHz. For injected charges less than or equal to 200 fC, the number of errors increases then decreases with increasing input voltage. The eventual decrease (an in some cases, elimination) of the number of errors is due to the increasing drive currents with increasing input voltage. Stronger drive currents will restore the nominal bias conditions faster, thus compensating for the current perturbation. For injected charges greater than or equal to 500 fC, the maximum number of errors increases with increasing input voltage. This is due to the increase

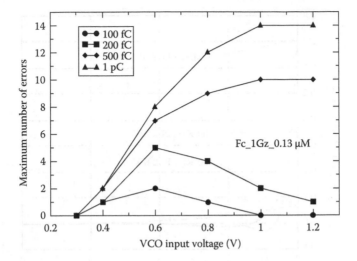

FIGURE 9.17 The simulated maximum number of errors generated by single-ion strikes (total injected charges of between 100 fC to 1 pC) as a function of the VCO input voltage. The VCO was designed in a 130-nm CMOS process and has a center frequency of operation of 1 GHz. (From Y. Boulghassoul et al., *IEEE Trans. Nucl. Sci.*, Vol. 52, No. 6, pp. 3466–3471, Dec. 2005.)

in oscillating frequency, resulting in a greater number of cycles with the characteristic time of the charge collection [62].

Moreover, Chung et al. show that the magnitude of the error response to transient perturbations in the PLL increases for increasing bandwidth, further indicating the importance of bandwidth in determining the SET response of the topology [59]. Figure 9.18 illustrates the simulated error response (in units of radians) of the PLL versus time for various PLL bandwidths. Increasing the PLL bandwidth is often accompanied by decreases in lock time (improved speed) and increased jitter (can be considered as noise for practical purposes). Trade-offs in operating speed, jitter, settling time, bandwidth, and SET performance should be carefully considered.

Through the efforts of Boulghassoul et al. in understanding the effects of scaling on the SET sensitivity of high-speed RF circuits, it is shown that the SET performance is not merely set by the bandwidth but by the gain-bandwidth product. For a given bandwidth, large gains result in degraded SET performance. Additionally, for the VCO circuits described, the optimum operating ranges are technology specific; the topologies discussed perform worse than a circuit in the same technology but with a smaller gain-bandwidth product, or worse than a circuit in an older technology at comparable speeds. More importantly, derating the frequency in a state-of-the-art technology node does not compensate for the increases in radiation vulnerabilities at that node [62].

9.3.6 REDUCTION OF WINDOW OF VULNERABILITY

The window of vulnerability is a well-known concept in the digital design community and describes the amount of time during a clock cycle that a circuit is vulnerable

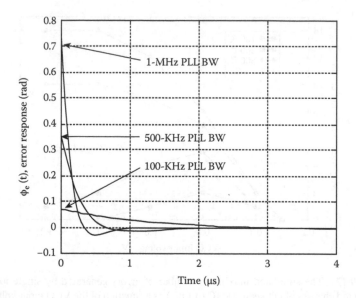

FIGURE 9.18 Transient PLL error response as a function of PLL bandwidth (BW). (From H. Chung, W. Chen, B. Bakkaloglu, H. J. Barnaby, B. Vermeire, and S. Kiaei, *IEEE Trans. Nucl. Sci.*, Vol. 53, No. 6, pp. 3539–3543, Dec. 2006.)

to SEU. Generally, reducing the window of vulnerability improves the SEU performance. Kauppila et al. use the concept to determine the subcircuits of a flash ADC vulnerable to single events within a single conversion cycle (Figure 9.19) [63]. The knowledge from this type of analysis was useful for targeted hardening solutions. Mikkola et al. expand on this concept to AMS designs through the implementation of an auto-zeroed CMOS comparator. By sampling and resetting the comparator's initial state each clock cycle, SET pulse widths were limited to the length of a single clock period [64]. The concept has also been employed to examine the relative SE sensitivities of SerDes and PLL subcircuits through what was termed phase-dependent single-event analysis. Asynchronous laser injections were used to deposit energy into the circuits at random time within the clock cycles. Figure 9.20 shows the number of errors versus signal phase for laser strikes in (a) a pre-emphasis amplifier in a 2-Gbps SerDes and (b) the output switches in the charge pump of a 200-MHz PLL. While errors occur during the entire clock period, they tend to be concentrated about the rising and falling clock edges [65].

9.3.7 Reduction of High-Impedance Nodes

The aforementioned circuit-level mitigation approaches function through the modification of characteristic circuit parameters such as gain, bandwidth, frequency, and drive strength. Each technique, though effective, may require special attention in compromising performance trade-offs (most AMS circuits already have stringent design requirements with little room for modification). One technique for reducing the nodal sensitivity of AMS circuits is to reduce or eliminate high impedance

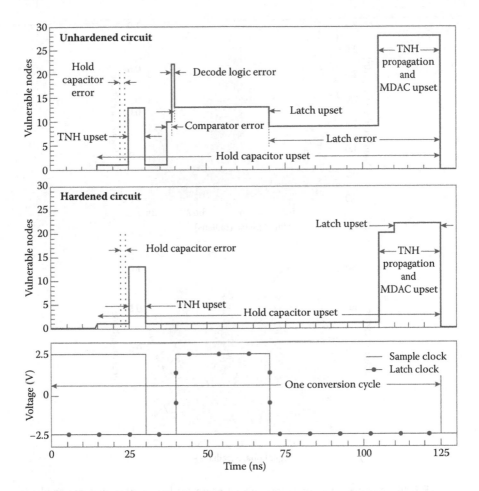

FIGURE 9.19 Windows of vulnerability for unhardened (top) and hardened (middle) 2-bit flash ADC circuits with respect to the clock (bottom) for one conversion cycle. (From J. S. Kauppila, L. W. Massengill, W. T. Holman, A. V. Kauppila, and S. Sanathanamurthy, *IEEE Trans. Nucl. Sci.*, Vol. 51, No. 6, pp. 3603–3608, Dec. 2004.)

nodes, thus improving the recovery time of the circuit following the ion strike [3,20,52,53,66]. This has shown to be applicable at the circuit [52,53,66] and device level [20].

Chen et al. accomplish this through circuit modification of a cross-coupled differential and voltage-controlled Colpitts oscillator. Through additional PMOS cross-coupled switching pairs at the oscillator output and decoupling the tail current source, output low impedance nodes were created and significantly improve the SET performance [66]. Lapuyade et al. implemented a similar approach in an injection-locked oscillator designed using a SiGe BiCMOS (integrated bipolar junction and CMOS transistors) process. First, a PMOS cross-coupled pair is utilized to increase the transconductance. Further, the length of ASETs is shown to decrease when operating in the injection-locked mode [67]. In general, free-running oscillators tend to

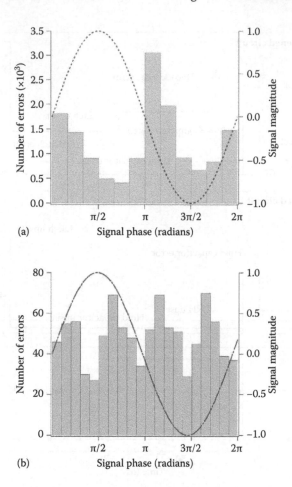

FIGURE 9.20 Number of errors versus signal phase for laser strikes in a pre-emphasis amplifier in a 2-Gbps SerDes (a) and for laser strikes in the output switches in the charge pump of a 200-MHz PLL (b). While errors occur during the entire clock period, they tend to be concentrated about the rising and falling clock edges. (From S. E. Armstrong, T. D. Loveless, J. R. Hicks, W. T. Holman, D. McMorrow, and L. W. Massengill, *IEEE Trans. Nucl. Sci.*, Vol. 58, No. 3, pp. 1066–1071, June 2011.)

exhibit poor SET performance when compared to synchronized oscillators such as the injection locked oscillator and VCO implemented in a PLL [32,47,53,67,68].

Sutton et al. provide a technique for creating a low impedance path within a SiGe HBT device designed to shunt charge away from the collector terminal. The path is realized by including an additional reverse biased p-n junction formed between the p-substrate and guard ring (n-ring), resulting in a secondary electric field [20].

9.3.8 Hardening via Charge Sharing

As previously shown, the DCC hardened layout technique takes advantage of charge sharing between adjacent transistors by using common-mode cancellation in

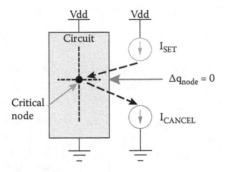

FIGURE 9.21   SNACC hardening technique. (From W. T. Holman, J. S. Kauppila, T. D. Loveless, L. W. Massengill, B. L. Bhuva, and A. F. Witulski, *Proc. 2013 GOMACTech Conf.*, 2013.)

differential signal paths. The concept of hardening via charge sharing (HCS) can be extended beyond layouts alone and exploited at the circuit level in AMS designs by using a combination of DCC layouts and appropriate analog circuit topologies [69].

Figure 9.21 illustrates one HCS technique: sensitive-node active charge cancellation (SNACC) [70]. This technique can be used to harden critical nodes that have a global effect on a larger circuit; that is, the output node of a bias circuit for an OA or the reference circuit of a data converter. SNACC works by balancing out collected charge with an equal but opposite amount of charge at a critical node with a net charge (ideally) of zero at the affected node.

Figure 9.22 shows a complementary folded cascade OA design with transistors M27 through M30 forming a bias circuit for the amplifier. Any SETs that perturb bias voltages *Vp*, *Vn*, or *Vnp* will affect the entire amplifier circuit. To prevent this, several additional transistors can be added to the bias circuit, as shown in Figure 9.23. Transistors S1 through S8 form the SNACC circuit. The following transistor pairs have DCC layouts: M27-S1, M28-S2, M29-S7, and M30-S8. Collected charge on a sensitive junction in M28 through M30 is shared with its DCC twin, resulting in an equal but opposite current pulse being coupled into the bias circuit nodes through the S3-S4 or S5-S6 current mirrors. The cancellation will not be perfect due to the different time constants of the two propagation paths, but the amplitude and duration of SETs can be reduced by more than 60% [70].

It is not necessary for the SNACC transistors to have the same width-to-length ratios as the transistors they protect; the sizes of S1 through S8 can be scaled proportionally to save area, with the shared charge amplified by the SNACC current mirrors. Furthermore, strikes on the SNACC transistors do not affect the bias circuit, since charge sharing and cancellation also take place in a path from the SNACC transistors to the bias circuit transistors. The SNACC transistors have a negligible effect on the circuit's power consumption because they are normally biased off. The main design penalty is increased layout area, but SNACC is only needed for a few critical nodes. In the case of the OA in Figure 9.22, the increase in area was only 7% [70].

In some circuits, the SNACC concept can be implemented without adding additional transistors. For example, consider the high-gain cascade output stage of the op

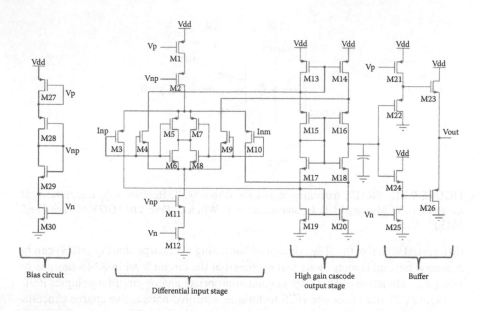

FIGURE 9.22 Complementary folded cascode OA that is suitable for hardening using SNACC. (From R. W. Blaine, S. E. Armstrong, J. S. Kauppila, N. M. Atkinson, B. D. Olson, W. T. Holman, and L. W. Massengill, *IEEE Trans. Nucl. Sci.*, Vol. 58, No. 6, pp. 3060–3066, Dec. 2011.)

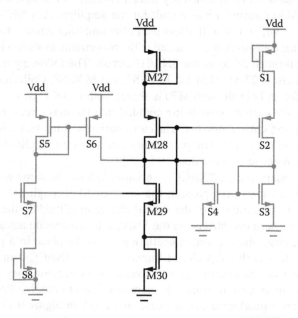

FIGURE 9.23 SNACC-hardened OA bias circuit with SNACC transistors highlighted. (From R. W. Blaine, S. E. Armstrong, J. S. Kauppila, N. M. Atkinson, B. D. Olson, W. T. Holman, and L. W. Massengill, *IEEE Trans. Nucl. Sci.*, Vol. 58, No. 6, pp. 3060–3066, Dec. 2011.)

amp in Figure 9.22. If transistors pairs M13-M14, M15-M16, M17-M18, or M19-M20 are laid out in DCC configuration, collected charge on any of these devices will be shared, mirrored, and cancelled at the output node (the drains of M16 and M18) of the stage, with a mitigation mechanism essentially identical to that of the SNACC bias circuit in Figure 9.23. As this example illustrates, analog and mixed-signal circuits with symmetrical subcircuit topologies are obvious candidates for the implementation of HCS techniques. The main drawback of hardening via charge sharing is that it cannot be used in integrated circuit (IC) fabrication processes such as SOI, where charge sharing is minimal or nonexistent. For bulk processes, however, HCS techniques can be very effective in mitigating SETs.

9.3.9 HARDENING VIA NODE SPLITTING

One of the most versatile and effective methodologies for hardening AMS circuits is that of node splitting. Hardening via node splitting (HNS) can be used in both discrete-time (e.g., switched capacitor) and continuous-time circuits for any integrated circuit process and in many cases has almost negligible impact on circuit performance, layout area, or power dissipation [69]. HNS hardening is a form of redundancy, as shown in Figure 9.24. The idea is to divide a circuit into two or more parallel signal paths, such that an ion strike on one path does not affect the other path(s). Ideally, the struck path will be disabled so that the remaining path(s) will maintain signal integrity. However, even if the struck path is not disabled, the remaining signal path(s) will still tend to mitigate the amplitude and duration of the SET. Although HNS techniques are well suited to differential circuit topologies, they can be applied to single-ended circuit topologies as well.

One HNS technique that can significantly reduce the SET vulnerability of differential switched-capacitor circuits is dual-path hardening [71]. The principle of the technique is to split the input nodes into separate parallel signal paths to provide significant immunity to a voltage perturbation on any single floating node of a switched-capacitor signal paths. This technique is applicable to any differential switched capacitor circuit and has been used with op amps and comparators in [71] and [72], respectively.

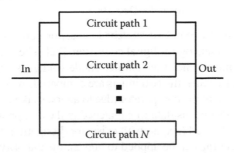

FIGURE 9.24 Illustration of the concept of HNS. (From W. T. Holman, J. S. Kauppila, T. D. Loveless, L. W. Massengill, B. L. Bhuva, and A. F. Witulski, *Proc. 2013 GOMACTech Conf.*, 2013.)

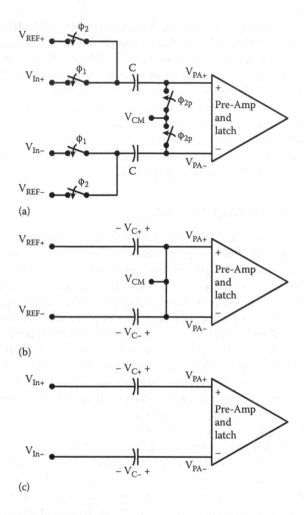

FIGURE 9.25 (a) Switched-capacitor comparator operates in two phases: (b) reset phase and (c) evaluation phase. The clocked switches are implanted by NMOS transistors. (From B. D. Olson, W. T. Holman, L. W. Massengill, B. L. Bhuva, and P. R. Fleming, *IEEE Trans. Nucl. Sci.*, Vol. 55, No. 6, pp. 3440–3446, Dec. 2008.)

Figure 9.25 illustrates a standard switched-capacitor comparator design as commonly used in pipelined analog-to-digital converters [72]. The comparator operates in two phases: the reset phase when the common-mode voltage is applied to both inputs, and the evaluation phase when the two inputs are compared. A voltage perturbation in the differential data path of the comparator (due to an ion strike on a NMOS switching transistor) may cause erroneous data to be latched at the comparator output if a floating capacitor node is affected (i.e., no path exists for dissipating the collected charge). Dual-signal path hardening can be applied to prevent the majority of such errors from generating an erroneous latched value. Figure 9.26 shows the comparator with dual inputs employed in the differential input stage. Input transistors M1 and M2 are each split into two identical transistors connected in parallel such that the width-to-length

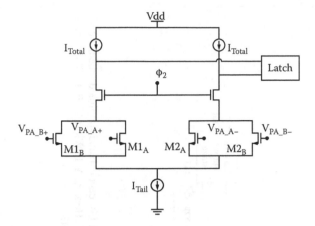

FIGURE 9.26 Simplified circuit schematic of the differential amplifier showing the split input paths. (From B. D. Olson, W. T. Holman, L. W. Massengill, B. L. Bhuva, and P. R. Fleming, *IEEE Trans. Nucl. Sci.*, Vol. 55, No. 6, pp. 3440–3446, Dec. 2008.)

ratio of each parallel device is one-half the width-to-length ratio of the original transistor. If the gates of M1A and M1B are shorted together, the configuration is identical to a standard differential amplifier. The switched-capacitor differential input network is also duplicated, as shown in Figure 9.27.

If one of the two signal paths is struck by an ion, the result is that the struck path is disabled, and the remaining path maintains signal integrity. Simulation and experimental results indicate significant improvement in single-event performance [72]. For the design depicted in Figures 9.26 and 9.27, the output perturbation was reduced to values correctable by standard digital error correction.

As shown in Figure 9.28, the main drawback of dual-path hardening is that the transistor switches must be the same type as the input transistors (i.e., NMOS switches with NMOS inputs) (or PMOS switches with PMOS inputs). This requirement

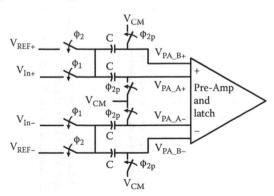

FIGURE 9.27 The switched-capacitor comparator with split differential amplifier input paths to harden the floating nodes against single-event upsets (From B. D. Olson, W. T. Holman, L. W. Massengill, B. L. Bhuva, and P. R. Fleming, *IEEE Trans. Nucl. Sci.*, Vol. 55, No. 6, pp. 3440–3446, Dec. 2008.)

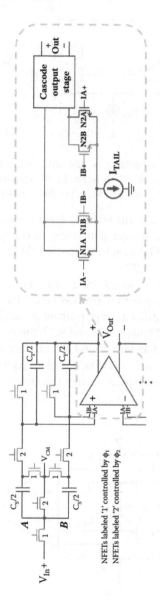

FIGURE 9.28 Schematic showing half of the dual-path sample and hold (S/H) amp topology (negative half is identical) and simplified schematic of a modified dual-input NFET OTA. (From N. M. Atkinson, W. T. Holman, J. S. Kauppila, T. D. Loveless, N. C. Hooten, A. F. Witulski, B. L. Bhuva, L. W. Massengill, E. X. Zhang, and J. H. Warner, *IEEE Trans. Nucl. Sci.*, Vol. 60, No. 6, pp. 4356–4361, Dec. 2013.)

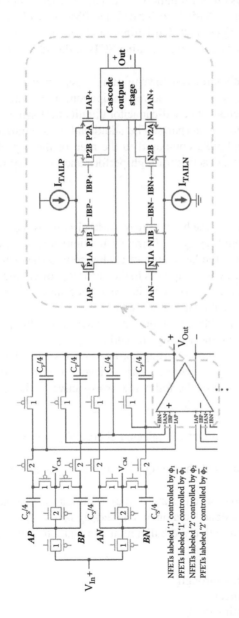

FIGURE 9.29 Schematic showing half of the quad-path sample and hold (S/H) amp topology (negative half is identical) and simplified schematic of modified quad-input complementary folded-cascode OTA. (From N. M. Atkinson, W. T. Holman, J. S. Kauppila, T. D. Loveless, N. C. Hooten, A. F. Witulski, B. L. Bhuva, L. W. Massengill, E. X. Zhang, and J. H. Warner, *IEEE Trans. Nucl. Sci.*, Vol. 60, No. 6, pp. 4356–4361, Dec. 2013.)

ensures that the struck path will be disabled (e.g., an ion strike on a NMOS switch pulls the gate of the NMOS input transistor to ground and shuts it off) but also limits dynamic input range, which makes dual-path hardening impractical for low-voltage circuits. To address this limitation, quad-path hardened designs have also been developed, with parallel NMOS-to-NMOS and PMOS-to-PMOS signal paths, as shown in Figure 9.29. Quad-path hardening has been proven to be even more effective than dual-path hardening in mitigating SETs at the added cost of increased layout complexity [73].

Dual-path and quad-path hardening have minimal design penalties beyond added layout complexity. In terms of noise, power dissipation, and frequency response, dual- and quad-path-hardened switched-capacitor circuits perform essentially identically to their unhardened counterparts. The additional area required by the split signal paths is almost negligible compared to the area of the capacitors, making these HNS techniques very attractive for high-performance AMS systems.

9.4 SUMMARY

There are unique challenges in the hardening of AMS circuits as the effect of a SE is dependent on the circuit topology, type of circuit (function), and the operating mode. As with digital components, the hardening of analog components can generally be accomplished through a brute force approach; that is, area, power, and/or bandwidth are sacrificed through the increase of capacitance, device size, and/or current drive, in order to increase the critical charge required to generate ASETs [2–4]. The challenge of radiation-hardened AMS circuit design is to develop techniques for mitigating ASETs while minimizing these design penalties.

SET mitigation, whether achieved through brute force or some other sophisticated design approach, can be implemented at various levels of abstraction and fundamentally involves one or both of the following, irrespective of the technology and abstraction level:

1. The reduction in the amount of collected charge (Q_{coll}) at a metallurgical junction [2]
2. The increase in the critical charge (Q_{crit}) required to generate an ASET [2]

This chapter overviewed basic and state-of-the-art mitigation approaches and provided some examples of hardened analog and mixed-signal circuits. The techniques presented are as follows (note that the bullet numbers and letters correspond to the chapter sections):

9.2 The reduction of Q_{coll} through
   9.2.1 Substrate engineering
   9.2.2 Layout-level mitigation
      9.2.2.1 Nodal separation and interleaved layout
      9.2.2.2 Differential design/differential charge cancellation layout
9.3 The reduction of Q_{crit} through
   9.3.1 Redundancy

REFERENCES

1. Y. Boulghassoul, L. W. Massengill, A. L. Sternberg, R. L. Pease, S. Buchner, J. W. Howard, D. McMorrow, M. W. Savage, and C. Poivey, Circuit Modeling of the LM124 Operational Amplifier for Analog Single-Event Transient Analysis, *IEEE Trans. Nucl. Sci.*, Vol. 49, No. 6, pp. 3090–3096, Dec. 2002.

2. S. Buchner and D. McMorrow, Single-Event Transients in Bipolar Linear Integrated Circuits, *IEEE Trans. Nucl. Sci.*, Vol. 53, No. 6, pp. 3079–3102, Dec. 2006.

3. J. Popp, Developing Radiation Hardened Complex System on Chip ASICs in Commercial Ultra Deep Submicron CMOS Processes, *2010 NSREC Short Course*, Denver, CO, July 2010.

4. Q. Zhou and K. Mohanram, Transistor Sizing for Radiation Hardening, *Proc. of 42nd IEEE IRPS*, pp. 310–315, Apr. 2004.

5. O. A. Amusan, A. F. Witulski, L. W. Massengill, B. L. Bhuva, P. R. Fleming, M. L. Alles, A. L. Sternberg, J. D. Black, and R. D. Schrimpf, Charge Collection and Charge Sharing in a 130 nm CMOS Technology, *IEEE Trans. Nucl. Sci.*, Vol. 53, No. 6, pp. 3253–3258, Dec. 2006.

6. O. A. Amusan, L. W. Massengill, M. P. Baze, B. L. Bhuva, A. F. Witulski, J. D. Black, A. Balasubramanian, M. C. Casey, D. A. Black, J. R. Ahlbin, R. A. Reed, and M. W. McCurdy, Mitigation Techniques for Single-Event-Induced Charge Sharing in a 90-nm Bulk CMOS Process, *IEEE Trans. Device Mater. Rel.*, Vol. 9, No. 2, pp. 468–472, June 2009.

7. O. A. Amusan, M. C. Casey, B. L. Bhuva, D. McMorrow, M. J. Gadlage, J. S. Melinger, and L. W. Massengill, Laser Verification of Charge Sharing in a 90 nm Bulk CMOS Process, *IEEE Trans. Nucl. Sci.*, Vol. 56, No. 6, pp. 3065–3070, Dec. 2009.

8. J. A. Pellish, R. A. Reed, R. D. Schrimpf et al., Substrate Engineering Concepts to Mitigate Charge Collection in Deep Trench Isolation Technologies, *IEEE Trans. Nucl. Sci.*, Vol. 53, No. 6, pp. 3298–3305, Dec. 2006.

9. J. R. Schwank, V. Ferlet-Cavrois, M. R. Shaneyfelt, P. Paillet, and P. E. Dodd, Radiation Effects in SOI Technologies, *IEEE Trans. Nucl. Sci.*, Vol. 50, No. 3, pp. 522–538, June 2003.

10. T. Poiroux, O. Faynot, C. Tabone, H. Tigelaar, H. Mogul, N. Bresson, and S. Cristoloveanu, Emerging Floating-Body Effects in Advanced Partially-Depleted SOI Devices, *IEEE International SOI Conf.*, pp. 99–100, 2002.

11. L. W. Massengill and P. W. Tuinenga, Single-Event Transient Pulse Propagation in Digital CMOS, *IEEE Trans. Nucl. Sci.*, Vol. 55, No. 6, pp. 2861–2871, Dec. 2008.

12. M. J. Gadlage, P. Gouker, B. L. Bhuva, B. Narasimham, and R. D. Schrimpf, Heavy-Ion-Induced Digital Single Event Transients in a 180 nm Fully Depleted SOI Process, *IEEE Trans. Nucl. Sci.*, Vol. 56, No. 6, pp. 3483–3488, Dec. 2009.

13. T. D. Loveless, J. S. Kauppila, S. Jagannathan, D. R. Ball, J. D. Rowe, N. J. Gaspard, N. M. Atkinson, R. W. Blaine, T. R. Reece, J. R. Ahlbin, T. D. Haeffner, M. L. Alles, W. T. Holman, B. L. Bhuva, and L. W. Massengill, On-Chip Measurement of

Single-Event Transients in a 45 nm Silicon-on-Insulator Technology, *IEEE Trans. Nucl. Sci.*, Vol. 59, No. 6, pp. 2748–2755, Dec. 2012.

14. J. A. Maharrey, R. C. Quinn, T. D. Loveless, J. S. Kauppila, S. Jagannathan, N. M. Atkinson, N. J. Gaspard, E. X. Zhang, M. L. Alles, B. L. Bhuva, W. T. Holman, and L. W. Massengill, Effect of Device Variants in 32 nm and 45 nm SOI on SET Pulse Distributions, *IEEE Trans. Nucl. Sci.*, Vol. 60, No. 6, pp. 4399–4404, Dec. 2013.

15. T. D. Loveless, J. S. Kauppila, J. A. Maharrey, R. C. Quinn, S. Jagannathan, M. L. Alles, B. L. Bhuva, W. T. Holman, and L. W. Massengill, Single-Event Transients in 45 nm and 32 nm Partially Depleted SOI Technologies, *Proc. 2014 GOMACTech Conf.*, 2014.

16. J. D. Black, A. L. Sternberg, M. L. Alles, A. F. Witulski, B. L. Bhuva, L. W. Massengill, J. M. Benedetto, M. P. Baze, J. L. Wert, and M. G. Hubert, HBD Layout Isolation Techniques for Multiple Node Charge Collection Mitigation, *IEEE Trans. Nucl. Sci.*, Vol. 52, No. 6, pp. 2536–2541, Dec. 2005.

17. B. D. Olson, O. A. Amusan, S. DasGupta, L. W. Massengill, A. F. Witulski, B. L. Bhuva, M. L. Alles, K. M. Warren, and D. R. Ball, Analysis of Parasitic PNP Bipolar Transistor Mitigation Using Well Contacts in 130 nm and 90 nm CMOS Technology, *IEEE Trans. Nucl. Sci.*, Vol. 54, No. 4, pp. 894–897, Aug. 2007.

18. B. Narasimham, R. L. Shuler, J. D. Black, B. L. Bhuva, R. D. Schrimpf, A. F. Witulski, W. T. Holman, and L. W. Massengill, Quantifying the Reduction in Collected Charge and Soft Errors in the Presence of Guard Rings, *IEEE Trans. Device Mater. Rel.*, Vol. 8, No. 1, pp. 203–209, Mar. 2008.

19. B. Narasimham, J. W. Gambles, R. L. Schuler, B. L. Bhuva, and L. W. Massengill, Quantifying the Effect of Guard Rings and Guard Drains in Mitigating Charge Collection and Charge Spread, *IEEE Trans. Nucl. Sci.*, Vol. 55, No. 6, pp. 3456–3460, Dec. 2008.

20. A. K. Sutton, M. Bellini, J. D. Cressler, J. A. Pellish, R. A. Reed, P. W. Marshall, G. Niu, G. Vizkelethy, M. Turowski, and A. Raman, An Evaluation of Transistor-Layout RHBD Techniques for SEE Mitigation in SiGe HBTs, *IEEE TNS*, Vol. 54, No. 6, pp. 2044–2052, Dec. 2007.

21. R. R. Troutman, *Latchup in CMOS Technology: The Problem and Its Cure*. Norwell, MA: Kluwer, 1986.

22. A. Hastings, *The Art of Analog Layout*, 2nd ed. New York: Prentice-Hall, 2005, Ch. 4, 7.

23. B. Mossawir, I. R. Linscott, U. S. Inan, J. L. Roeder, J. V. Osborn, S. C. Witczak, E. E. King, and S. D. LaLumondiere, A TID and SEE Radiation-Hardened, Wideband, Low-Noise Amplifier, *IEEE Trans. Nucl. Sci.*, Vol. 53, No. 6, pp. 3439–3448, Dec. 2006.

24. M. Varadharajaoerumal, G. Niu, X. Wei, T. Zhang, J. D. Cressler, R. A. Reed, and P. W. Marshall, 3-D Simulation of SEU Hardening of SiGe HBTs Using Shared Dummy Collector, *IEEE Trans. Nucl. Sci.*, Vol. 54, No. 6, pp. 2330–2337, Dec. 2007.

25. O. A. Amusan, L. W. Massengill, B. L. Bhuva, S. DasGupta, A. F. Witulski, and J. R. Ahlbin, Design Techniques to Reduce SET Pulse Widths in Deep-Submicron Combinational Logic, *IEEE Trans. Nucl. Sci.*, Vol. 54, No. 6, pp. 2060–2064, Dec. 2007.

26. B. Narasimham, B. L. Bhuva, R. D. Schrimpf et al., Characterization of Digital Single Event Transient Pulse-Widths in 130-nm and 90-nm CMOS Technologies, *IEEE Trans. Nucl. Sci.*, Vol. 54, No. 6, pp. 2506–2511, Dec. 2007.

27. M. J. Gadlage, J. R. Ahlbin, B. Narasimham, B. L. Bhuva, L. W. Massengill, R. A. Reed, R. D. Schrimpf, and G. Vizkelethy, Scaling Trends in SET Pulse Widths in Sub-100 nm Bulk CMOS Processes, *IEEE Trans. Nucl. Sci.*, Vol. 57, No. 1, pp. 3336–3341, Dec. 2010.

28. M. L. Alles, R. D. Schrimpf, R. A. Reed, and L. W. Massengill, Radiation Hardness of FDSOI and FinFET Technologies, *Proc. 2011 IEEE Int. SOI Conference*, Tempe, AZ, Oct. 2011.

29. P. Roche and G. Gasiot, SEE on Advanced CMOS BULK, FinFET, and UTTB SOI Technologies, *2014 IEEE NSREC Short Course*, 2014.

30. D. R. Ball, M. L. Alles, R. D. Schrimpf, and S. Cristoloveanu, Comparing Single Event Upset Sensitivity of Bulk vs. SOI based FinFET SRAM Cells Using TCAD Simulations, *Proc. 2010 IEEE Int. SOI Conference*, San Diego, CA, Oct. 2010.

31. T. D. Loveless, M. L. Alles, D. R. Ball, K. M. Warren, and L. W. Massengill, Parametric Variability Affecting 45 nm SOI SRAM Single Event Upset Cross-Sections, *IEEE Trans. Nucl. Sci.*, Vol. 57, No. 6, pp. 3228–3233, Dec. 2010.

32. Y. Boulghassoul, L. W. Massengill, A. L. Sternberg, B. L. Bhuva, and W. T. Holman, Towards SET Mitigation in RF Digital PLLs: From Error Characterization to Radiation Hardening Considerations, *IEEE Trans. Nucl. Sci.*, Vol. 53, No. 3, pp. 2047–2053, Aug. 2006.

33. T. Calin, M. Nicolaidis, and R. Velazco, Upset Hardened Memory Design for Submicron CMOS Technology, *IEEE Trans. Nucl. Sci.*, Vol. 43, No. 6, pp. 2874–2878, Dec. 1996.

34. K. Lilja, M. Bounasser, S. Wen, R. Wong, J. Holst, N. Gaspard, S. Jagannathan, D. Loveless, and B. Bhuva, Single-Event Performance and Layout Optimization of Flip-Flops Ion a 28 nm Bulk Technology, *IEEE Trans. Nucl. Sci.*, Vol. 60, No. 4, pp. 2782–2788, Aug. 2013.

35. M. P. Baze, B. Hughlock, J. Wert, J. Tostenrude, L. W. Massengill, O. Amusan, R. Lacoe, K. Lilja, and M. Johnson, Angular Dependence of Single Event Sensitivity in Hardened Flip-Flop Designs, *IEEE Trans. Nucl. Sci.*, Vol. 55, No. 6, pp. 3295–3301, Dec. 2008.

36. E. Cannon, M. Cabanas-Holmen, T. McKay, M. Carson, A. Kleinosowksi, S. Rabaa, and J. Wert, On-Edge Irradiation of SOI Logic Cells Hardened by Spatial Redundancy, *Proc. Single-Event Effects Symp.*, La Jolla, CA, Apr. 2013.

37. M. Cabanas-Holmen, E. H. Cannon, S. Rabaa, T. Amort, J. Ballast, M. Carson, D. Lam, and R. Brees, Robust SEU Mitigation of 32 nm Dual Redundant Flip-Flops Through Interleaving and Sensitive Node-Pair Spacing, *IEEE Trans. Nucl. Sci.*, Vol. 60, No. 6, pp. 4374–4380, Dec. 2013.

38. J. E. Knudsen and L. T. Clark, An Area and Power Efficient Radiation Hardened by Design Flip-Flop, *IEEE Trans. Nucl. Sci.*, Vol. 53, No. 6, pp. 3392–3399, Dec. 2006.

39. B. I. Matush, T. J. Mozdzen, L. T. Clark, and J. E. Knudsen, Area-Efficient Temporally Hardened by Design Flip-Flop Circuits, *IEEE Trans. Nucl. Sci.*, Vol. 57, No. 6, pp. 3588–3593, Dec. 2010.

40. A. T. Kelly, P. R. Fleming, W. T. Holman, A. F. Witulski, B. L. Bhuva, and L. W. Massengill, Differential Analog Layout for Improved ASET Tolerance, *IEEE Trans. Nucl. Sci.*, Vol. 54, No. 6, pp. 2053–2059, Dec. 2007.

41. S. E. Armstrong, B. D. Olson, W. T. Holman, J. Warner, D. McMorrow, and L. W. Massengill, Demonstration of a Differential Layout Solution for Improved ASET Tolerance in CMOS AMS Circuits, *IEEE Trans. Nucl. Sci.*, Vol. 57, No. 6, pp. 3615–3619, Dec. 2010.

42. J. R. Ahlbin, L. W. Massengill, B. L. Bhuva, B. Narasimham, M. J. Gadlage, and P. H. Eaton, Single-Event Transient Pulse Quenching in Advanced CMOS Logic Circuits, *IEEE Trans. Nucl. Sci.*, Vol. 56, No. 6, pp. 3050–3056, Dec. 2009.

43. N. Seifert, V. Ambrose, B. Gill, Q. Shi, R. Allmon, C. Recchia, S. Mukherjee, N. Nassif, J. Krause, J. Pickholtz, and A. Balasubramanian, On the Radiation-Induced Soft Error Performance of Hardened Sequential Elements in Advanced Bulk CMOS Technologies, *Proc. 2010 IEEE Int. Reliability Physics Symp.*, pp. 188–197, May 2010.

44. N. W. van Vonno and B. R. Doyle, Design Considerations and Verification Testing of an SEE-Hardened Quad Comparator, *IEEE Trans. Nucl. Sci.*, Vol. 48, No. 6, pp. 1859–1864, Aug. 2002.
45. B. D. Olson, W. T. Holman, L. W. Massengill, and B. L. Bhuva, Evaluation of Radiation-Hardened Design Techniques Using Frequency Domain Analysis, *IEEE Trans. Nucl. Sci.*, Vol. 55, No. 6, pp. 2957–2961, Dec. 2008.
46. T. D. Loveless, L. W. Massengill, B. L. Bhuva, W. T. Holman, M. C. Casey, R. A. Reed, S. A. Nation, D. McMorrow, and J. S. Melinger, A Probabilistic Analysis Technique Applied to a Radiation-Hardened-by-Design Voltage-Controlled Oscillator for Mixed-Signal Phase-Locked Loops, *IEEE Trans. Nucl. Sci.*, Vol. 55, No. 6, pp. 3447–3455, Dec. 2008.
47. T. D. Loveless, L. W. Massengill, W. T. Holman and B. L. Bhuva, Modeling and Mitigating SETs in Voltage-Controlled Oscillators, *IEEE Trans. Nucl. Sci.*, Vol. 54, No. 6, pp. 2561–2567, Dec. 2007.
48. R. Kumar, V. Karkala, R. Garg, T. Jindal, and S. P. Khatri, A Radiation Tolerant Phase Locked Loop Design for Digital Electronics, *Proc. of IEEE ICCD*, pp. 505–510, Oct. 2009.
49. J. L. Andrews, J. E. Schroeder, B. L. Gingerich, W. A. Kolasinski, R. Koga, and S. E. Diehl, Single Event Error Immune CMOS RAM, *IEEE Trans. Nucl. Sci.*, Vol. 29, No. 6, pp. 2040–2043, Dec. 1982.
50. S. E. Diehl, A. Ochoa, P. V. Dressendorfer, R. Koga, and W. A. Kolasinski, Error Analysis and Prevention of Cosmic Ion-Induced Soft Errors in Static CMOS RAMs, *IEEE Trans. Nucl. Sci.*, Vol. 29, No. 6, pp. 2032–2039, Dec. 1982.
51. A. L. Sternberg, L. W. Massengill, M. Hale, and B. Blalock, Single-Event Sensitivity and Hardening of a Pipelined Analog-to-Digital Converter, *IEEE Trans. Nucl. Sci.*, Vol. 53, No. 6, pp. 3532–3538, Dec. 2006.
52. T. D. Loveless, L. W. Massengill, B. L. Bhuva, W. T. Holman, A. F. Witulski, and Y. Boulghassoul, A Hardened-by-Design Technique for RF Digital Phase-Locked-Loops, *IEEE Trans. Nucl. Sci.*, Vol. 53, No. 6, pp. 3432–3438, Dec. 2006.
53. T. D. Loveless, L. W. Massengill, B. L. Bhuva, W. T. Holman, R. A. Reed, D. McMorrow, J. S. Melinger, and P. Jenkins, A Single-Event-Hardened Phase-Locked Loop Fabricated in 130 nm CMOS, *IEEE Trans. Nucl. Sci.*, Vol. 54, No. 6, pp. 2012–2020, Dec. 2007.
54. D. McMorrow, S. Buchner, W. T. Lotshaw, J. S. Melinger, M. Maher, and M. W. Savage, Demonstration of Single-Event Effects Induced by Through-Wafer Two-Photon Absorption, *IEEE Trans. Nucl. Sci.*, Vol. 51, No. 6, pp. 3553–3557, Dec. 2004.
55. Y. Boulghassoul, P. C. Adell, J. D. Rowe, L. W. Massengill, R. D. Schrimpf, and A. L. Sternberg, System-Level Design Hardening Based on Worst Case ASET Simulations, *IEEE Trans. Nucl. Sci.*, Vol. 51, No. 5, pp. 2787–2793, Oct. 2004.
56. S. E. Armstrong, B. D. Olson, J. Popp, J. Braatz, T. D. Loveless, W. T. Holman, D. McMorrow, and L. W. Massengill, Single-Event Transient Error Characterization of a Radiation-Hardened by Design 90 nm SerDes Transmitter Driver, *IEEE Trans. Nucl. Sci.*, Vol. 56, No. 6, pp. 3463–3468, Dec. 2009.
57. T. Uemura, R. Tanabe, Y. Tosaka, and S. Satoh, Using Low Pass Filters in Mitigation Techniques against Single-Event Transients in 45nm technology LSIs, *14th IEEE Int. On-line Testing Symposium*, pp. 117–122, July 2008.
58. A. L. Sternberg, L. W. Massengill, R. D. Schrimpf, Y. Boulghassoul, H. J. Barnaby, S. Buchner, R. L. Pease, and J. W. Howard, Effect of Amplifier Parameters on Single-Event Transients in an Inverting Operational Amplifier, *IEEE Trans. Nucl. Sci.*, Vol. 49, No. 3, pp. 1496–1501, June 2002.
59. H. Chung, W. Chen, B. Bakkaloglu, H. J. Barnaby, B. Vermeire, and S. Kiaei, Analysis of Single Event Effects on Monolithic PLL Frequency Synthesizers, *IEEE Trans. Nucl. Sci.*, Vol. 53, No. 6, pp. 3539–3543, Dec. 2006.

60. T. D. Loveless, L. W. Massengill, W. T. Holman, B. L. Bhuva, D. McMorrow, and J. H. Warner, A Generalized Linear Model for Single Event Transient Propagation in Phase-Locked Loops, *IEEE Trans. Nucl. Sci.*, Vol. 57, No. 5, pp. 2933–2947, Oct. 2010.

61. M. J. Gadlage, P. H. Eaton, J. M. Benedetto, M. Carts, V. Zhu, and T. L. Turflinger, Digital Device Error Rate Trends in Advanced CMOS Technologies, *IEEE Trans. Nucl. Sci.*, Vol. 53, No. 6, Dec. 2006.

62. Y. Boulghassoul et al., Effects of Technology Scaling on the SET Sensitivity of RF CMOS Voltage-Controlled Oscillators, *IEEE Trans. Nucl. Sci.*, Vol. 52, No. 6, pp. 3466–3471, Dec. 2005.

63. J. S. Kauppila, L. W. Massengill, W. T. Holman, A. V. Kauppila, and S. Sanathanamurthy, Single Event Simulation Methodology for Analog/Mixed Signal Design Hardening, *IEEE Trans. Nucl. Sci.*, Vol. 51, No. 6, pp. 3603–3608, Dec. 2004.

64. E. Mikkola, B. Vermeire, H. J. Barnaby, H. G. Parks, and K. Borhani, SET Tolerant CMOS Comparator, *IEEE Trans. Nucl. Sci.*, Vol. 51, No. 6, pp. 3609–3614, Dec. 2004.

65. S. E. Armstrong, T. D. Loveless, J. R. Hicks, W. T. Holman, D. McMorrow, and L. W. Massengill, Phase-Dependent Single-Event Sensitivity Analysis of High-Speed A/MS Circuits Extracted from Asynchronous Measurements, *IEEE Trans. Nucl. Sci.*, Vol. 58, No. 3, pp. 1066–1071, June 2011.

66. W. Chen, V. Pouget, G. K. Gentry, H. J. Barnaby, B. Vermeire, B. Bakkaloglu, S. Kiaei, K. E. Holbert, and P. Fouillat, Radiation Hardened by Design RF Circuits Implemented in 0.13 um CMOS Technology, *IEEE Trans. Nucl. Sci.*, Vol. 53, No. 6, pp. 3449–3454, Dec. 2006.

67. H. Lapuyade, V. Pouget, J.-B. Begueret, P. Hellmuth, T. Taris, O. Mazouffre, P. Fouillat, and Y. Deval, A Radiation-Hardened Injection Locked Oscillator Devoted to Radio-Frequency Applications, *IEEE Trans. Nucl. Sci.*, Vol. 53, No. 4, pp. 2040–2046, Aug. 2006.

68. H. Lapuyade, O. Mazouffre, B. Goumballa, M. Pignol, F. Malou, C. Neveu, V. Pouget, Y. Deval, and J.-B. Begueret, A Heavy-Ion Tolerant Clock and Data Recovery Circuit for Satellite Embedded High-Speed Data Links, *IEEE Trans. Nucl. Sci.*, Vol. 54, No. 6, pp. 2080–2085, Dec. 2007.

69. W. T. Holman, J. S. Kauppila, T. D. Loveless, L. W. Massengill, B. L. Bhuva, and A. F. Witulski, Low-Penalty Radiation-Hardened-by-Design Concepts for High-Performance Analog, Mixed-Signal, and RF Circuits, *Proc. 2013 GOMACTech Conf.*, 2013.

70. R. W. Blaine, S. E. Armstrong, J. S. Kauppila, N. M. Atkinson, B. D. Olson, W. T. Holman, and L. W. Massengill, RHBD Bias Circuits Utilizing Sensitive Node Active Charge Cancellation (SNACC) Designs, *IEEE Trans. Nucl. Sci.*, Vol. 58, No. 6, pp. 3060–3066, Dec. 2011.

71. P. R Fleming, B. D. Olson, W. T. Holman, B. L. Bhuva, and L. W. Massengill, Design Technique for Mitigation of Soft Errors in Differential Switched-Capacitor Circuits, *IEEE Trans. Circuits Syst. II, Exp. Briefs*, Vol. 55, No. 9, pp. 838–842, Sept. 2008.

72. B. D. Olson, W. T. Holman, L. W. Massengill, B. L. Bhuva, and P. R. Fleming, Single-Event Effect Mitigation in Switched-Capacitor Comparator Designs, *IEEE Trans. Nucl. Sci.*, Vol. 55, No. 6, pp. 3440–3446, Dec. 2008.

73. N. M. Atkinson, W. T. Holman, J. S. Kauppila, T. D. Loveless, N. C. Hooten, A. F. Witulski, B. L. Bhuva, L. W. Massengill, E. X. Zhang, and J. H. Warner, The Quad-Path Hardening Technique for Switched-Capacitor Circuits, *IEEE Trans. Nucl. Sci.*, Vol. 60, No. 6, pp. 4356–4361, Dec. 2013.

10 CMOS Monolithic Sensors with Hybrid Pixel-Like, Time-Invariant Front-End Electronics

TID Effects and Bulk Damage Study

Lodovico Ratti, Luigi Gaioni, Massimo Manghisoni, Valerio Re, Gianluca Traversi, Stefano Bettarini, Francesco Forti, Fabio Morsani, Giuliana Rizzo, Luciano Bosisio, and Irina Rashevskaya

CONTENTS

10.1   INTRODUCTION

Electronic circuits and systems are employed in a number of different fields where some degree of radiation tolerance is required. These fields include, to mention but a few, space and avionic applications, high-energy physics experiments, nuclear and (still at an exploratory stage) thermonuclear power plants, and medical diagnostic imaging and therapy. When operated in these environments, electronic systems may be directly struck by photons, electrons, nucleons or heavier particles, with a subsequent alteration of their electrical properties. Depending on the type and characteristics of the impinging radiation and on the fabrication technology of the circuit, different effects, either irreversible or (partially or totally) reversible, may arise. Knowledge of the mechanisms underlying the behavior under irradiation of electronic devices and circuits is of paramount importance for

- devising hardness assurance methodologies to guarantee that they can work reliably in the target environment,
- developing rad-hard circuits and design techniques to improve their tolerance to specific radiation effects in specific applications.

This chapter focuses on the cumulative effects of ionizing radiation and neutrons on complementary metal-oxide semiconductor (CMOS) monolithic sensors, also known as monolithic active pixel sensors (MAPSs) or as CMOS image sensors (CISs), developed for particle tracking applications in high-energy physics (HEP) experiments. The analysis will focus on a couple of case studies, where the analog front-end electronics (the analog electronic circuits integrated in each of the elementary cells making up the sensor array) is more complicated than the typical three-transistor structure of CMOS MAPS and actually has the same architecture as the circuits used in hybrid pixel sensors (HPDs) for the readout of pixel sensors in a high-resistivity substrate. For this reason, the results discussed here can be of interest for the radiation-tolerant design not only of CMOS MAPS, but also more generally of circuits for the readout of capacitive detectors broadly used in fundamental physics experiments. The purpose is to draw the attention to the most critical points, as far as radiation tolerance is concerned, in the design of classical analog blocks for radiation detection applications. The chapter is organized as follows. After a short section about the development and use of CMOS MAPS in charged particle tracking applications, total ionizing dose (TID) effects will be discussed for two different MAPS devices: one fabricated in a 130-nm, triple-well CMOS technology, and the other in a 180-nm, quadruple-well CMOS technology. While CMOS circuits are known to be significantly affected by ionizing radiation, they are mostly insensitive to bulk damage. On the other hand, the collecting electrode of monolithic sensors, and in particular, its capability of collecting charge from the substrate to detect the transit of a particle, may be sizably degraded by neutron-induced increase in bulk trap density. This subject will be investigated for the same two devices, the 130-nm and 180-nm CMOS MAPS sensors, in the last section of the chapter.

10.2 CMOS MONOLITHIC SENSORS FOR CHARGED PARTICLE TRACKING

The need for charged particle trackers with low material budget for precise momentum measurements at the future high luminosity colliders (B-factories, International Linear Collider [1,2]), has prompted several research groups in the particle physics community to explore solutions involving the use of monolithic active pixel sensors in CMOS technology [3–5]. Although the spatial resolution of CMOS sensors for imaging applications (the area where they first emerged and thrive [6]) is much higher than that required by particle trackers, they can provide an appealing solution to the design of multilayer thin detectors with improved momentum resolution. In HEP applications, depending on the experiment characteristics, they may be required to withstand TIDs ranging from a few hundreds of krad(SiO_2) to a few tens of Mrad(SiO_2) and 1-MeV neutron equivalent fluences from 10^{11} to a few 10^{13} cm^{-2} [7,8]. In a monolithic sensor, the readout electronics is fabricated in the same substrate as the detector, which is typically an N-well diffusion. Also, the substrate itself can be thinned down to a few tens of micrometers with no significant signal loss, since the operating principle of MAPS is based on the collection of charge diffusing in the mostly undepleted device bulk, and therefore the sensitive volume is limited to the very first silicon layer just beneath the collecting N-well diffusion (the charge released deeper in the substrate recombines before reaching the sensor). Standard MAPS are laid out with a minimum size N-type collecting electrode for noise minimization purposes. Therefore, any P-type MOS (PMOS), in the surroundings would steal a significant amount of charge from the sensor with its N-well, unacceptably impairing the device collection efficiency. This is the reason why only N-type metal oxide semiconductor (NMOS) transistors (and generally, a small number of them, in a source follower configuration [6]) are allowed in the elementary cell of conventional MAPS, therefore preventing the designer from taking full advantage of the features of CMOS processes. Attempts to overcome these limitations by incorporating PMOS transistors in the design of the electronics at the pixel level include both design/layout strategies, as in the case of deep N-well (DNW) MAPS [9], and technological solutions, as in the case of monolithic sensors in quadruple-well CMOS processes [10]. The availability of both PMOS and NMOS transistors makes it possible to include high-performance amplifiers and low-power digital blocks in the MAPS front-end cell. This in turn can be exploited to implement sparsified readout architectures (which are typical of HPDs) capable of complying with the intense data rates foreseen for tracking applications at future high-energy physics facilities [11]. The two different approaches mentioned above have been exploited in the development of the MAPS devices discussed in the following sections. It is worth noting here that DNW and quadruple-well MAPS are not the only options to make CMOS processes fully available in the design of monolithic sensors. Complete freedom to use PMOS transistors in MAPS front-end circuits is also guaranteed by silicon-on-insulator (SOI) [12–14] and high-voltage (HV) processes [15].

10.3   DNW MAPS IN A TRIPLE-WELL, 130-nm CMOS TECHNOLOGY

The DNW MAPS approach takes advantage of the properties of triple-well structures to implement a collecting electrode with a relatively large area (as compared
to standard three-transistor MAPS [6]) and at the same time sufficiently low pitch
to meet the demanding resolution requirements set by future high-energy physics
experiments. This solution does not require any nonstandard processing steps, as
triple-well structures are typically available in modern commercial CMOS technologies to isolate NMOS transistors from the P-type substrate (to avoid, for instance,
possible coupling to digital signals). In a DNW MAPS sensor, the signal from the
collecting electrode is read out by a classical optimal chain for capacitive detectors,
including a charge preamplifier whose charge sensitivity is independent of the collecting electrode capacitance. Use of a sensing electrode with a large area makes it
possible to integrate PMOS devices in the elementary cell as long as its area significantly exceeds the area covered by the N-wells used for those PMOS field-effect
transistors, or PMOSFETs. Also, NMOS parts of the analog front-end (which may
include a shaping stage) can be laid out inside the DNW sensor to optimize the usage
of the elementary cell area.

10.3.1   DESCRIPTION OF THE DEVICES UNDER TEST AND IRRADIATION PROCEDURE

A number of different test structures fabricated in a 130-nm CMOS technology and
implementing different solutions both for the front-end electronics and for the collecting electrode layout have been fabricated taking advantage of the DNW MAPS
design approach. In particular, the radiation tolerance test results discussed in this
section are relevant to a test chip called Apsel2T. The chip includes a set of stand-
alone channels (MAPS where the readout channel is actually not connected to the
sensor) and small, 3 × 3 arrays of pixels with a 50-μm pitch and different layouts
for the collecting electrodes. Here we will be exclusively interested in the effects of
ionizing radiation on the readout channel, which is the same for all the structures
and is shown in Figure 10.1. It includes a charge preamplifier and an RC-CR (or
semi-Gaussian unipolar) shaper. As already mentioned, the NMOS transistors of the
analog front-end are integrated inside the deep N-well collecting electrode, although
this is not explicitly indicated in the figure. The charge preamplifier features a folded
cascode scheme [16] where M_{in} is the input device and M_{cs} is the PMOS current
source in the circuit input branch. C_F is the preamplifier feedback capacitor (nominally 8 fF) continuously reset by means of the M_F NMOS transistor operated in the
deep subthreshold region. The aspect ratio of this device, W/L = 0.18 μm/10 μm, has
been chosen for the purpose of maximizing the equivalent feedback resistance, and
therefore the feedback time constant, while minimizing the area. The preamplifier
output stage is represented by means of an ideal buffer. In the shaping filter, C_1 is the
differentiating input capacitance while C_2 is the capacitance in the shaper feedback
network, which also includes a transconductor (transistors MN1, MN2, MP1, and
MP2). Again, the output stage of the shaper is depicted as an ideal buffer. By acting
(through slow control bits b0 and b1) on the capacitance C_z in the high-impedance

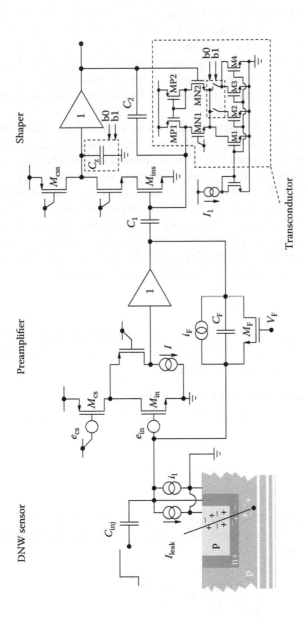

FIGURE 10.1 Schematic diagram of the analog readout channel integrated in the DNW MAPS. A conceptual cross-sectional view of the DNW sensor is also shown. (After L. Ratti, C. Andreoli, L. Gaioni, M. Manghisoni, E. Pozzati, V. Re et al., *IEEE Trans. Nucl. Sci.*, Vol. 56, No. 4, pp. 2124–2131, Aug. 2009.)

node of the shaper gain stage and on the transconductor current sources (M1 to M4, all with identical gate dimensions), the peaking time t_{p0} can be set to 0.5, 1, or 2 μs [17,18]. Figure 10.1 also includes the main preamplifier noise sources (to be described and discussed later on in this section). The capacitor C_{inj} is used to inject charge into the channel input terminal and test its response. The overall power dissipation is about 30 μW per channel at a V_{DD} of 1.2 V. Besides the DNW MAPS, a set of test structures fabricated with the same technology and including NMOS transistors was characterized from the standpoint of ionizing radiation. The purpose was to improve the understanding of the relationship between the radiation-induced degradation of the front-end performance and the fundamental damage mechanisms in elementary devices.

The devices under test (DUTs) were exposed to γ-rays from a [60]Co (cobalt 60) source featuring a dose rate of 12 rad/s(SiO$_2$). The final integrated dose, 1100 krad(SiO$_2$), was reached through one or more intermediate irradiation steps. The DUTs were also subjected to a 100°C annealing cycle for 168 hours. All the DUTs were biased during both the irradiation and the annealing cycles. In particular, DNW MAPS were biased as they would be in actual applications and NMOS transistors had their gate terminal at V_{DD} while keeping all the other terminals grounded. It is worth noting here that no specific radiation hardening techniques (e.g., use of enclosed layout transistors or of thin oxides over the junctions) were applied in the design of either the collecting electrode or the processing electronics.

10.3.2   TID Effects

The effect of ionizing radiation is studied in terms of its impact on the main parameters and features of the analog channel (i.e., the charge sensitivity, the shape of the output signal and the equivalent noise charge).

10.3.2.1   Charge Sensitivity and Response Shape

In a front-end channel for charge measurement, the charge sensitivity G_Q is defined as

$$G_Q = \frac{dV_{peak}}{dQ},$$

(10.1)

where $V_{peak}(Q)$ is the peak value of the signal at the shaper output as a function of the input charge Q. If the system is linear, like in the case under examination, then V_{peak} is proportional to Q and

$$\frac{dV_{peak}}{dQ} = \frac{V_{peak}(Q)}{Q}.$$

(10.2)

The charge sensitivity in linear systems is generally measured by interpolating the V_{peak} versus Q characteristic with a straight line, with the purpose of averaging out possible small nonlinearities. If a direct access to the front-end input is available through an injection capacitor, then the V_{peak} versus Q characteristic can be easily measured by injecting a varying, known amount of charge and measuring the amplitude of the relevant channel response. The transfer function of the shaper in Figure 10.1 is given by

$$T_s(s) = \frac{2\pi s \dfrac{C_1 GBP_s}{(C_1 + C_2)}}{s^2 + 2\pi s \dfrac{C_2 GBP_s}{(C_1 + C_2)} + 2\pi \dfrac{G_m GBP_s}{(C_1 + C_2)}}. \tag{10.3}$$

In the previous equation

$$G_m = k_1 G_{m0}, \quad G_{m0} = g_{m0,MN1} = \frac{q I_{t0}}{2 n k_B T}, \tag{10.4}$$

where G_m is the transconductance of the shaper transconductor (equal to its preirradiation value, G_{m0}, when the parameter k_1 is equal to 1), $g_{m0,MN1}$ is the preirradiation value of the transconductance of transistor MN1 (and of MN2), I_{t0} is the preirradiation value of the current flowing through M1 to M4, q is the elementary charge, n is the subthreshold slope coefficient, k_B is the Boltzmann's constant and T is the absolute temperature. It is assumed here that the NMOS transistors in the transconductor differential pair are operated in the weak inversion region. Again in Equation 10.3,

$$GBP_s = k_2 GBP_{s0}, \quad GBP_{s0} = k_2 \frac{g_{m0,ins}}{C_z}, \tag{10.5}$$

where GBP_s is the gain-bandwidth product of the shaper gain stage (the same as its preirradiation value, GBP_{s0}, when $k_2 = 1$) and $g_{m0,ins}$ is the preirradiation value of the transconductance of the shaper input transistor M_{ins}. The parameters k_1 and k_2, both equal to 1 in the fresh circuits, will be used in the following to account for radiation effects in the shaping stage. The transconductor and the gain stage of the shaper have been designed in such a way that the equality

$$G_{m0} = \frac{\pi}{2} \frac{GBP_{s0} C_2^2}{(C_1 + C_2)} \tag{10.6}$$

holds and a semi-Gaussian, first-order unipolar shaping is achieved with an overall response of the channel to an input charge pulse Q given by [9],

$$v_{out,s}(t) = -2 \frac{C_1 Q}{C_2 C_F} e^{-\frac{t}{t_{p0}}} \frac{t}{t_{p0}}, \quad t_{p0} = \frac{C_2}{2 G_{m0}}, \tag{10.7}$$

with the signal reaching its peak at t_{p0}. The ratio between G_{m0} and GBP_{s0} is maintained when the peaking time is changed through the control bits b0 and b1. In the case of an ideal charge preamplifier, providing a step signal as the response to a delta-shaped input pulse (i.e., acting as a pure integrator), the charge sensitivity can be shown to be given by

$$G_Q = \frac{V_{\text{peak}}}{Q} = 2\frac{C_1}{eC_FC_2}, \tag{10.8}$$

where e is the Neper's constant. In a real system, the charge sensitivity is affected by the finite bandwidth of the charge preamplifier and by the finite value of the time constant in the preamplifier feedback network, $\tau_F = C_F/g_{\text{ds,F}}$, with $g_{\text{ds,F}}$ the output conductance of the MOSFET M_F. Let us assume a single-pole approximation for the open-loop transfer function of the charge preamplifier, with DC gain A_0 and time constant τ. If $\tau_F \gg \tau$ and A_0C_F is much larger than the total capacitance C_T shunting the preamplifier input (including the detector capacitance C_D, the input capacitance C_{in}, the feedback capacitance C_F, and the injection capacitance C_{inj}, which, it is worth recalling, has been included in the design just for test purposes), then the charge sensitivity in Equation 10.8 is modified by a factor

$$k_F \cong \left(\frac{\tau_r}{\tau_F}\right)^{\frac{\tau_r}{\tau_F}} - \frac{\tau_r}{\tau_F}. \tag{10.9}$$

In Equation 10.9, $\tau_r = \tau C_T/(A_0 C_F)$ is the rise time in the charge preamplifier response. Typically, $\tau_r/\tau_F \ll 1$.

The charge sensitivity is remarkably affected by exposure to ionizing radiation. Figure 10.2 shows the charge sensitivity as a function of the TID and after annealing in an Apsel2T readout channel not connected to the collecting DNW electrode for the three different possible values of the peaking time. A steady decrease of G_Q with the dose can be observed, followed by a partial recovery after annealing. The degradation induced by irradiation on G_Q can be appropriately studied by analyzing the effects taking place in the two main front-end blocks, namely

- a change in the equivalent preamplifier feedback resistance (i.e., the output conductance of the transistor M_F) due to a threshold voltage shift,
- a change in the feedback transconductance and in the gain-bandwidth product in the shaping stage.

In a NMOS transistor operated in the deep subthreshold region, such as M_F, the drain current is given by [19]

$$I_D = I_{D0}\frac{W}{L}e^{\frac{q(V_{GS}-V_{\text{th,F}})}{nk_BT}}\left(1 - e^{-\frac{qV_{DS}}{k_BT}}\right), \tag{10.10}$$

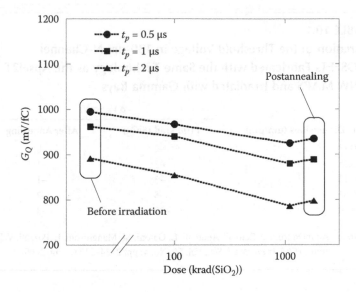

FIGURE 10.2 Charge sensitivity as a function of the total ionizing dose and after the annealing cycle for a readout channel not connected to the DNW collecting electrode at the three available peaking time values. (From L. Ratti, C. Andreoli, L. Gaioni, M. Manghisoni, E. Pozzati, V. Re et al., *IEEE Trans. Nucl. Sci.*, Vol. 56, No. 4, pp. 2124–2131, Aug. 2009.)

where $I_{D0} \approx 6\mu_{n0}C_{OX}(k_B T/q)^2$, μ_{n0} is the zero-bulk-bias electron mobility, C_{OX} is the gate oxide capacitance per unit area, W and L are the transistor channel width and length, respectively, V_{GS} is the gate-to-source voltage, $V_{th,F}$ is the threshold voltage, and V_{DS} is the drain-to-source voltage. From Equation 10.10, a change in the threshold voltage determines a variation in the output conductance; in particular

$$\frac{\partial g_{ds,F}}{\partial V_{th,F}} = \frac{\partial}{\partial V_{th,F}} \frac{\partial I_D}{\partial V_{DS}} < 0. \tag{10.11}$$

It can be calculated as well that, if $\tau_r/\tau_F \ll 1$

$$\frac{dk_F}{d\tau_F} > 0. \tag{10.12}$$

Therefore, a decrease in $V_{th,F}$ would result in an increase in the output conductance of M_F, and in turn, in a decrease of τ_F. Eventually, this would yield a reduction in the charge sensitivity through a reduction of the k_F factor. Although deep submicron CMOS technologies are widely known to feature a high degree of radiation tolerance [20], narrow channel devices ($W < 1$ μm) belonging to the same CMOS generation as the technology used for the Apsel2T DNW MAPS were found to undergo the so-called radiation-induced narrow channel (RINC) effect originating from charge

TABLE 10.1

Variation of the Threshold Voltage in 130-nm N-Channel MOSFETs Fabricated with the Same Technology as the Apsel2T DNW MAPS and Irradiated with Gamma Rays

Gate Dimensions (μm/μm)	ΔV_{th} (mV)	
	1100 krad (SiO$_2$)	After Annealing
20/0.13	−2	0
20/0.35	−3	−1
10/0.13	−3	−1
10/0.35	−5	−2
0.18/10	−20	−3

Source: Adapted from L. Ratti, C. Andreoli, L. Gaioni, M. Manghisoni, E. Pozzati, V. Re et al., *IEEE Trans. Nucl. Sci.*, Vol. 56, No. 4, pp. 2124–2131, Aug. 2009.

trapping in the shallow trench isolation oxides and proven to affect both N- and P-channel devices [21]. A very similar behavior was actually found in irradiated devices from the same 130-nm process as the DNW MAPS. This is shown in Table 10.1, displaying the threshold voltage shift $\Delta V_{th,F}$ for a set of devices with different gate dimensions after exposure to a 1100-krad(SiO$_2$) γ-ray TID and after annealing. The threshold voltage shift, (a few millivolts in all the devices with gate width no smaller than 10 μm), is significantly larger in the narrow channel device, which, by the way, features the same gate dimensions as M_F in the circuit of Figure 10.1. After the 100°C/168 h annealing cycle, the starting conditions are almost completely recovered in the irradiated devices, including the narrow channel one. As a first conclusion, a change in the threshold voltage of the preamplifier feedback MOSFET associated with RINC effects may contribute to the observed change in the charge sensitivity. As anticipated, other contributions may come from the shaping stage, where again the RINC effect may be responsible for charge sensitivity degradation. In particular, the NMOS current sources (M1 to M4) in the transconductor, featuring a channel width of 0.25 μm, and the PMOS current source in the input stage, M_{css}, with $W = 0.18$ μm, may suffer a nonnegligible shift in their threshold voltage due to charge trapping in the shallow trench isolation oxide. With the help of circuit simulations, it can be verified that if the threshold voltage of transistors M1 to M4 is reduced, then an increase in the current sunk by the sources follows. In the case of the NMOS current sources, a further contribution, again due to charge accumulation in the shallow trench isolations, may be provided by the increase in the leakage current. The two effects in turn result in an increase in G_m, which can be accounted for by assuming $k_1 > 1$ in Equation 10.4. On the other hand, a decrease of the threshold voltage in M_{css} may lead to a decrease in the gain-bandwidth product of the shaper through a decrease of the transconductance of its input device. This may be accounted for by a value smaller than 1 for the k_2 parameter. Therefore, after irradiation, Equation 10.7 will no longer be true in general and the shape of

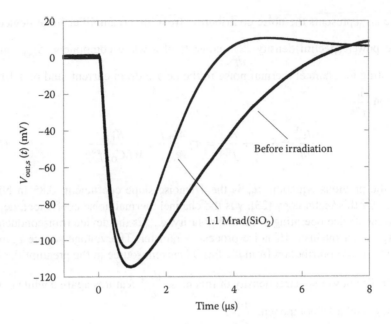

FIGURE 10.3 Response of the readout circuit of Figure 10.1 (not connected to a DNW collecting electrode) before irradiation and after exposure to a total ionizing dose of 1.1 Mrad(SiO$_2$).

the response will change. In particular, once G_m and GBP$_s$ have been replaced in Equation 10.3 with their expressions given in Equations 10.4 and 10.5, respectively, in the approximation that the response of the charge preamplifier is an ideal step and given $k_2/k_1 < 1$, the overall channel response to an input charge pulse Q can be shown to take the form

$$v_{out,s}(t) = -2 \frac{C_1 Q}{C_F C_2} \frac{e^{-k_2 \frac{t}{t_{p0}}}}{\sqrt{\left(\frac{k_1}{k_2} - 1\right)}} \sin\left(\frac{t}{\tau_1}\right), \quad \tau_1 = \frac{t_{p0}}{\sqrt{k_1 k_2 - k_2^2}}, \quad (10.13)$$

resulting in a charge sensitivity decrease both for k_1 increasing and for k_2 decreasing. The effect on the shape of the signal is shown in Figure 10.3. In particular, an increase in k_1 (i.e., an increase in G_m) can be shown to be responsible for the faster return of the signal to the baseline and for the more marked overshoot.

10.3.2.2 Equivalent Noise Charge

Equivalent noise charge (ENC) is a figure of merit for charge-sensitive amplifiers. It is defined as the charge that has to be injected at the input of the charge measuring system in order to find a unit signal-to-noise ratio at its output [22]. The main noise sources in the DNW MAPS under test are indicated in Figure 10.1. The voltage source e_{in} represents the noise contribution from the preamplifier input device M_{in}, whose power spectral density $\dfrac{\overline{de_{in}^2}}{df}$ consists of a white component, $S_{ws,in}$, mainly accounting for channel thermal noise in the device drain current, and of a 1/f contribution $\dfrac{A_{f,in}}{f^{\alpha_{fn}}}$:

$$\frac{\overline{de_{in}^2}}{df} = S_{ws,in} + \frac{A_{f,in}}{f^{\alpha_{fn}}} = n\gamma\frac{4k_BT}{g_{m,in}} + \frac{K_{f,in}}{WLC_{OX}f^{\alpha_{fn}}}. \tag{10.14}$$

In the previous equation, α_{fn} is the 1/f noise slope coefficient, 0.85 in NMOS devices for this technology [23], γ is the channel thermal noise coefficient, depending on the device operating point and polarity, $g_{m,in}$ is the device transconductance, and $K_{f,in}$ is an intrinsic 1/f noise process parameter. The voltage source e_{cs} represents the noise contribution from the PMOS current source in the preamplifier input branch. The power spectral density of this noise $\dfrac{\overline{de_{cs}^2}}{df}$ features again a white component $S_{ws,cs}$ and a 1/f component $\dfrac{A_{f,cs}}{f^{\alpha_{fp}}}$,

$$\frac{\overline{de_{cs}^2}}{df} = S_{ws,cs} + \frac{A_{f,cs}}{f^{\alpha_{fp}}} = n\gamma\frac{4k_BT}{g_{m,cs}} + \frac{K_{f,cs}}{WLC_{OX}f^{\alpha_{fp}}}, \tag{10.15}$$

where α_{fp} is the 1/f noise slope coefficient, around 1.1 for PMOSFETs in this 130 nm CMOS technology, and $K_{f,cs}$ is the intrinsic 1/f noise process parameter depending on overdrive voltage in PMOS transistors [23]. In Equations 10.14 and 10.15 n is again the subthreshold slope coefficient, as defined in Equation 10.4. The preamplifier feedback MOSFET M_F also provides a noise contribution, represented by the current source i_F, whose power spectral density $\dfrac{\overline{di_F^2}}{df}$, in the case of NMOS transistors operated in the deep subthreshold region, can be modeled as [24]

$$\frac{\overline{di_F^2}}{df} = S_{wp,F} = 2qI_D = 2qI_{D0}\frac{W}{L}e^{\frac{q(V_{GS}-V_{th,F})}{nk_BT}}\left(1+e^{-\frac{qV_{DS}}{k_BT}}\right). \tag{10.16}$$

The above three noise terms yield the following ENC expression:

$$\mathrm{ENC} = \left\{C_T^2\left[\frac{A_1S_{ws,in}}{t_p} + (2\pi)^{\alpha_{fn}}A_2A_{f,in}t_p^{\alpha_{fn}-1} + \left(\frac{A_1S_{ws,cs}}{t_p} + (2\pi)^{\alpha_{fp}}A_2A_{f,cs}t_p^{\alpha_{fp}-1}\right)\frac{g_{m,cs}^2}{g_{m,in}^2}\right]\right.$$

$$\left.+ A_3S_{wp,F}t_p\right\}^{\frac{1}{2}}, \tag{10.17}$$

where A_1, A_2, and A_3 are shaping coefficients. Some of the terms appearing in Equation 10.17 exhibit a certain degree of sensitivity to ionizing radiation and are expected to feature a nonnegligible increase after exposure to γ-rays. Figure 10.4 provides a typical example of the effects of a 1100-krad(SiO_2) TID on an NMOS transistor belonging to the same technology as the DNW MAPS readout channel under test. The low-frequency portion of the spectrum is clearly affected by irradiation, whereas white noise does not seem to undergo any sizeable degradation, with only a partial recovery after the 100°C/168 h annealing procedure. This is actually found to be correlated to an incomplete recovery in the static characteristics, pointing to the fact that parasitic lateral transistors are still present. Flicker noise performance degradation can be explained by means of the 1/f noise contributions coming from lateral parasitic devices turned on by positive charge buildup in the shallow trench isolation oxides (STI) [25,26]. The effect is larger in transistors with many fingers and operated at small current densities, where the parasitic lateral devices have a greater impact on the behavior of the main transistor. A minor, further contribution may be ascribed to border trap density increase in the main device gate oxide [27] although a large fraction of the holes trapped in the gate oxide after irradiation, including in particular near-interfacial ones, is likely to undergo fast direct tunneling annealing [28]. The 1/f component of the noise voltage spectrum after irradiation, $S_{f,post}$, is therefore obtained by summing the contributions from the

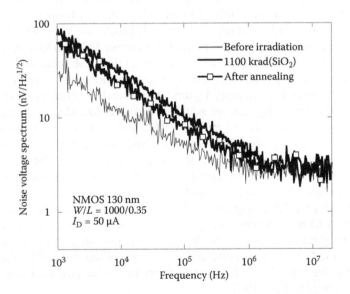

FIGURE 10.4 Noise voltage spectra before irradiation, after exposure to a 1.1 Mrad(SiO_2) total ionizing dose and after an annealing cycle (100°C/168 h) for a NMOS transistor with $W/L = 1000$ μm/0.35 μm and operated at $I_D = 50$ μA. The transistor belongs to the same technology as the DNW MAPS under test. (From L. Ratti, C. Andreoli, L. Gaioni, M. Manghisoni, E. Pozzati, V. Re et al., *IEEE Trans. Nucl. Sci.*, Vol. 56, No. 4, pp. 2124–2131, Aug. 2009.)

main transistor and from the equivalent parasitic transistor representing the effects of sidewall leakage in all the device fingers:

$$S_{f,post} = \frac{\dfrac{K_{f,main}}{WLC_{OX}} g_{m,main}^2 + \dfrac{K_{f,lat}}{W_{lat}LC_{OX,lat}} g_{m,lat}^2}{(g_{m,main} + g_{m,lat})^2 f^{\alpha_{fn}}}. \tag{10.18}$$

In the previous equation, $K_{f,main}$ and $K_{f,lat}$ are postirradiation intrinsic process parameters for 1/f noise in the main device channel and in the equivalent parasitic transistor, featuring a transconductance $g_{m,main}$ and $g_{m,lat}$, respectively. For each transistor finger, two parasitic devices get switched on by radiation, each featuring a channel width $W_{lat,f}$ and contributing to an equivalent parasitic transistor with channel width $W_{lat} = 2n_f W_{lat,f}$, n_f being the number of fingers in the main device. The $C_{OX,lat}W_{lat,f}$ product, where $C_{OX,lat}$ is the effective oxide capacitance of the parasitic STI sidewall transistor, can be extracted from static I_D–V_{GS} curves [25]. Extraction of the geometrical, electrical, and process parameters appearing in Equation 10.18 from the characterization of single transistors with different gate dimensions and operated at different drain currents makes it possible to estimate ionizing radiation damage in the preamplifier input device. In particular, a $C_{OX,lat}W_{lat,f}$ product of 1.39×10^{-10} fF/μm, a $K_{f,main}$ of 20×10^{-25} J Hz$^{-0.15}$ and a $K_{f,lat}$ of 11.9×10^{-25} J Hz$^{-0.15}$ were obtained at a TID of 1.1 Mrad(SiO$_2$) for the considered CMOS process. The flicker noise slope coefficient, α_{fn}, is not significantly affected by irradiation. Changes in the threshold voltage of the preamplifier feedback NMOSFET, which have already been discussed in the previous section, may be responsible for further ENC performance degradation in the readout channel. In order to determine the noise contribution from M_F, its quiescent operating conditions (drain current and drain-to-source voltage) were extracted from circuit simulations. Figure 10.5 shows the equivalent noise charge as a function of the peaking time for a DNW MAPS readout channel before irradiation, after exposure to a 1100 krad(SiO$_2$) TID, and after annealing. Experimental data is compared to a theoretical evaluation based on Equation 10.17 and on the parameters extracted from single device characterization. The ENC increase after irradiation can be satisfactorily explained with the 1/f noise contribution from the preamplifier input device and the parallel noise contribution from the preamplifier feedback MOSFET. ENC degradation is larger at short peaking times (about 25%), mostly due to 1/f noise increase in M_{in}. M_F starts playing a significant role at $t_p = 2$ μs. Postannealing partial recovery is likely to be correlated both to the observed annealing induced reduction in 1/f noise and to the shift of the threshold voltage of the preamplifier feedback MOSFET toward its original value (see the behavior of the narrowest device in Table 10.1).

10.4 MAPS IN A QUADRUPLE-WELL, 180-nm CMOS TECHNOLOGY

A different approach, as compared to DNW MAPS, was used for the design of the Apsel4well monolithic sensor, also implementing a classical readout chain for

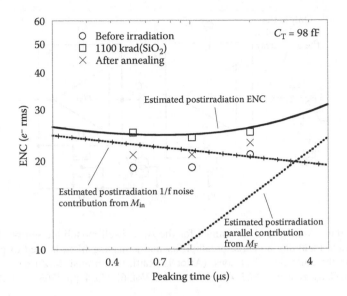

FIGURE 10.5 Measured values of the ENC (markers) as a function of the peaking time for a readout channel not connected to the collecting electrode before irradiation, after exposure to a 1100 krad(SiO$_2$) TID and after annealing. Experimental, postirradiation data is compared to the ENC estimation (continuous curve) based on Equation 10.17. The estimated postirradiation contributions to the ENC coming from the preamplifier input device and from the preamplifier feedback MOSFET (dotted curves) are also shown. (Adapted from L. Ratti, C. Andreoli, L. Gaioni, M. Manghisoni, E. Pozzati, V. Re et al., *IEEE Trans. Nucl. Sci.*, Vol. 56, No. 4, pp. 2124–2131, Aug. 2009.)

continuous time signal processing in the elementary cell, with a charge preamplifier and a shaper. The approach is based on a 180-nm CMOS planar technology with quadruple-well option, called INMAPS, in which a deep P-well is implanted beneath the PMOS N-wells [29]. Apart from this additional step, the one discussed here is a standard CMOS process. The deep P-well provides a potential barrier for the charge diffusing in the epitaxial layer, preventing the carriers from being collected by the positively biased N-wells hosting the PMOS transistors of the in-pixel readout circuits. The NMOS transistors are designed in heavily doped P-wells located over a lightly P-doped epitaxial layer about 12 µm thick that has been grown on a substrate featuring a doping concentration slightly lower than 10^{19} atoms/cm^{-3}. The epitaxial layer, featuring a resistivity higher than both the deep P-well and the substrate, plays an important role in improving the charge collection properties. The presence of two potential barriers (deep P-well/epitaxial layer or P-well/epitaxial layer on one side and epitaxial layer/substrate on the other) prevents the carriers from diffusing through the substrate.

10.4.1 Description of the DUTs and Irradiation Procedure

The Apsel4well chip includes different test structures, all laid out in the form of 3 × 3 arrays of elementary cells with a pitch of 50 µm. Chips with different epitaxial layer

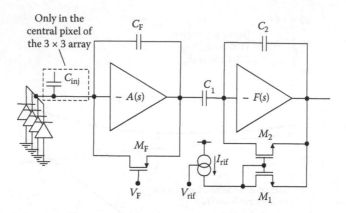

FIGURE 10.6   Analog readout channel for the Apsel4well monolithic sensor. A classical readout chain, including a charge preamplifier and a shaping filter, is used to process the signal from the collecting electrodes. (After L. Ratti, G. Traversi, S. Zucca, S. Bettarini, F. Morsani, G. Rizzo et al., *IEEE Trans. Nucl. Sci.*, Vol. 61, No. 4, pp. 1763–1771, Aug. 2014.)

resistivity, 10 Ω·cm and 1 kΩ·cm (in the following also referred to as standard and high resistivity, respectively), were tested [30]. In the Apsel4well monolithic sensor, the analog channel reads out the charge signal collected by four electrodes (each consisting of a simple N-well over P-epitaxial layer junction with no buried layer) laid out with the minimum size allowed by the technology (1.5 μm side). The block diagram of the in-pixel analog circuit, with a few transistor level details, is shown in Figure 10.6. The signal from the sensor (the four collecting electrodes in parallel) is processed by a charge preamplifier, where the charge across the feedback capacitor C_F is continuously reset by a P-channel MOSFET biased by the voltage V_F in the deep subthreshold region. The shaping stage is AC-coupled to the charge preamplifier through the C_1 capacitor. The voltage V_{rif} sets the current used to discharge the feedback capacitor C_2 with a constant slope by means of a current mirror structure. The peaking time is about 300 ns when V_{rif} is adjusted for a return-to-baseline time of 2 μs. The overall power dissipation is about 18 μW per channel at a V_{DD} of 1.8 V. The output of the shaping stage is accessible through dedicated pads for all of the nine pixels in the 3 × 3 arrays. The readout channel of the central pixel is provided with an injection capacitor C_{inj} (with a nominal capacitance of 30 fF) used to feed the circuit input with a known amount of charge and emulate the signal from the collecting electrodes. The input device of the charge preamplifier is an enclosed layout NMOS transistor in all of the structures integrated in the test chip, with the exception of a single 3 × 3 array, where an open, interdigitated transistor was used instead as the input device (all the other monolithic sensor features remaining the same). The layout used for the transistor is a variation of the typical annular enclosed geometry [31]. The polysilicon gate, under which thin gate oxide is placed, is closed around the drain terminal. This is supposed to prevent thick STI oxides from running along and close to the channel between the source and the drain terminals, therefore minimizing the effects of STI charge trapping on the device operation.

The following discussion refers to four Apsel4well samples, all with a high resistivity epitaxial layer, that were exposed to γ-rays from a cobalt 60 source, featuring a dose rate of about 6 rad(SiO₂)/s. The DUTs were irradiated up to a final dose of 10 Mrad(SiO₂) (chips 44 and 48 in the following) or 11.5 Mrad(SiO₂) (chips 45 and 47 in the following), which were reached in three steps, with intermediate integrated doses of 1 and 3 Mrad(SiO₂). During irradiation, the DUTs were biased in the same conditions as during measurements. Of the four 3 × 3 arrays under test, two have the input device of the charge preamplifier laid out with an enclosed technique while the other two use standard interdigitated MOSFETs. One purpose of the tests was actually that of understanding whether any advantage in terms of radiation hardness, in particular from the standpoint of noise performance, could be obtained from using enclosed layout techniques in this quadruple-well process.

10.4.2 TID Effects

The effect of ionizing radiation is studied again, as in the case of the DNW MAPS, in terms of the degradation induced on the main parameters of the front-end channel (i.e., the charge sensitivity and the equivalent noise charge).

10.4.2.1 Charge Sensitivity

The effects of ionizing radiation on the charge sensitivity were studied by using a ^{55}Fe source, which provides a convenient reference for charge sensitivity calibration in radiation detection systems. ^{55}Fe X-rays release their entire energy in the substrate through photoelectric interaction. Photons from the ^{55}Fe 5.9-keV line generate about $Q_{Fe} = 1640$ electron/hole pairs each. In the case of monolithic sensors, when photons release their energy into the junction depleted region (relevant to one of the four collecting diffusions), the corresponding charge is virtually entirely collected by the sensor, yielding the peak in the ^{55}Fe spectrum. Therefore, the front-end channel charge sensitivity G_Q can be calculated as

$$G_Q = \frac{V_{peak}(Q_{Fe})}{Q_{Fe}},$$ (10.19)

$V_{peak}(Q_{Fe})$ being the shaper output amplitude corresponding to the 5.9-keV peak of the ^{55}Fe source. The tail at lower amplitudes is due to the charge released in the epitaxial layer out of the electrode depleted region and only partially collected. Events above the spectrum peak (like some of those located at slightly smaller energies) are instead due to the noise in the front-end electronics. Figure 10.7 shows ^{55}Fe spectra detected by the central pixel of chip 44 before irradiation and after exposure to different doses of γ-rays. After a shift of about 20% toward lower amplitude values following the first irradiation step, the spectrum peak position does not change significantly as a result of the exposure to higher ionizing doses. Further confirmation of this result is provided by charge sensitivity measurements performed by direct charge injection at the DUT input through the injection capacitance C_{inj} (see Figure 10.6). Figure 10.8 shows the charge sensitivity, normalized with respect to the

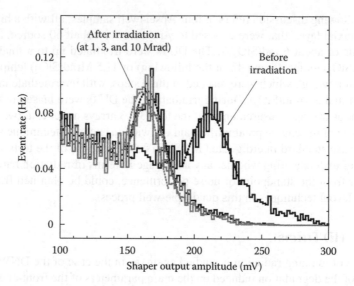

FIGURE 10.7 Spectra of a ^{55}Fe source detected by the central pixel of a 3 × 3 array of Apsel4well MAPS before irradiation and after exposure to γ-rays. (From L. Ratti, G. Traversi, S. Zucca, S. Bettarini, F. Morsani, G. Rizzo et al., *IEEE Trans. Nucl. Sci.*, Vol. 61, No. 4, pp. 1763–1771, Aug. 2014.)

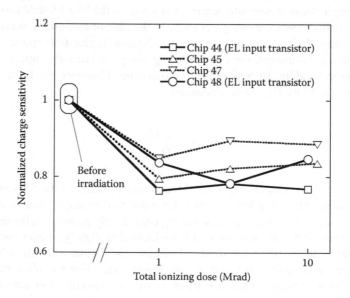

FIGURE 10.8 Charge sensitivity normalized to the preirradiation value as a function of the total ionizing dose. The charge sensitivity was measured in the central pixels of the four arrays under test through charge injection techniques. (After L. Ratti, G. Traversi, S. Zucca, S. Bettarini, F. Morsani, G. Rizzo et al., *IEEE Trans. Nucl. Sci.*, Vol. 61, No. 4, pp. 1763–1771, Aug. 2014.)

preirradiation value, as a function of the TID for the central pixels of the four 3 × 3 arrays. A decrease in the 15% to 25% range can be detected after the first step, at an integrated dose of 1 Mrad(SiO_2). Exposure to larger γ-ray doses does not result in any further decrease of the charge sensitivity. Instead, a slight recovery is observed in some of the samples. Analysis of the waveforms at the shaper output may help understand the phenomena underlying the behavior of charge sensitivity. Figure 10.9a shows the channel response to a 750-electron input pulse in an Apsel4well monolithic sensor (i.e., the central pixel of the 3 × 3 array from chip 48) before irradiation and after exposure to different values of the TID. The biasing conditions indicated in the figure, $V_F = 0.16$ V and $V_{rif} = 1.40$ V, are needed to have a 2-μs return-to-baseline time before irradiation and were used while acquiring each of the waveforms in the figure. The specification on the return-to-baseline time is no longer satisfied after irradiation with integrated doses of 1 and 3 Mrad(SiO_2). The slope is almost restored to its preirradiation state with the final ionizing dose step. A modification of the peaking time, correlated with the slope change in all of the irradiation steps (with the peaking time decreasing when the slope increases and vice versa), can also be detected. Together with the change in the slope (and in the peaking time), a significant reduction of the peak amplitude of the signal can be noticed after the first irradiation step with a further slight decrease after the second step, and a partial recovery, likely to be correlated with the recovery in slope, after the 10 Mrad(SiO_2) step. The preirradiation response and the response after a 1 Mrad(SiO_2) ionizing dose for the same device as in Figure 10.9a are shown again in Figure 10.9b. Here, a comparison is made with the response of the circuit after the 1 Mrad(SiO_2) step obtained when the bias conditions (i.e., the values of V_F and V_{rif}) are modified in such a way that the response is as close as possible to the preirradiation one. In particular, the signal amplitude was restored virtually to the preirradiation value mostly by suitably adjusting the bias voltage V_F at the gate of the preamplifier feedback MOSFET, whereas the return-to-baseline slope could be brought back to the earlier state by acting on the V_{rif} voltage. Note that acting on both bias points is needed to restore the channel operation from the standpoint of both charge sensitivity and signal shape. This points to the fact that ionizing radiation affects not only the charge preamplifier, but also the shaper behavior.

In the case of the charge preamplifier, one possible degradation source is represented by the radiation-induced increase in the sensor leakage current, which may be responsible for an increase in the feedback transconductance, bringing about a faster discharge of the C_F capacitor and a gain decrease. This may be actually proven by noting that if a single pole approximation is assumed for the open-loop gain of the charge preamplifier, $A(s) = A_0/(1 + s\tau_0)$, with DC gain A_0 and time constant τ_0, then the signal V_p at the preamplifier output as a response to an input charge Q, in the Laplace domain, is given by

$$V_p(s) = \frac{A_0 Q}{s^2 C_T \tau + s(C_F A_0 + C_T) + g_{m,F} A_0},$$

(10.20)

where $g_{m,F}$ is the transconductance of the PMOS transistor M_F in the preamplifier feedback network. Now it can be demonstrated that in the time domain, the peak

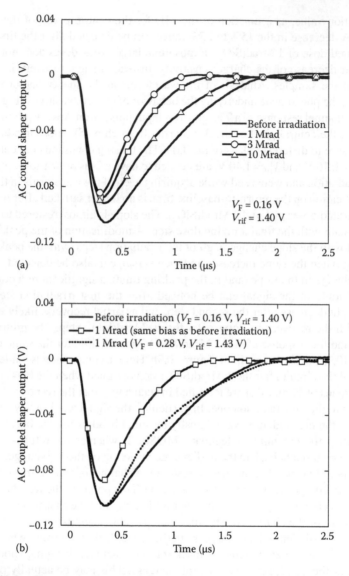

FIGURE 10.9 Channel response to a 750-electron input pulse in the central pixel of the 3 × 3 array from chip 48 (a) before irradiation and after exposure to different values of total ionizing dose and (b) before irradiation and after exposure to a 1-Mrad(SiO₂) dose for two different bias conditions in the preamplifier and shaper restoring networks. (From L. Ratti, G. Traversi, S. Zucca, S. Bettarini, F. Morsani, G. Rizzo et al., *IEEE Trans. Nucl. Sci.*, Vol. 61, No. 4, pp. 1763–1771, Aug. 2014.)

value $V_{p,MAX}$ of the response is such that $\dfrac{\partial V_{p,MAX}}{\partial g_{m,F}} < 0$. Since in a MOSFET the channel transconductance increases if the drain current increases, and since, in the case of M_F, the drain current corresponds to the sensor leakage current I_{leak}, then

$$\frac{dV_{p,MAX}}{dI_{leak}} = \frac{\partial V_{p,MAX}}{\partial g_{m,F}} \frac{dg_{m,F}}{dI_{leak}} < 0. \tag{10.21}$$

Therefore, an increase in the sensor leakage current will lead to a decrease in the charge sensitivity. As far as the shaping stage is concerned, a shift in the threshold voltage of the N-type transistor M_2 in the current mirror feedback network may determine a change in the mirror ratio, therefore explaining the faster return to baseline of the signal detected after irradiation with 1-Mrad(SiO_2) and 3-Mrad(SiO_2) γ-ray doses. A change in the leakage current in the same transistor would actually produce the same effect. Both phenomena (the change in the threshold voltage and the increase in the leakage current) may result from RINC effects [21], in particular from the accumulation of positive charge in the STIs. This may actually alter the behavior of M_2, which features an aspect ratio $W/L = 0.4$ μm/1.5 μm (the other transistor in the mirror has $W/L = 8$ μm/0.25 μm and is not expected to undergo any RINC effect). At larger doses, negative charge trapped in interface states may start competing with the positive charge trapped in the bulk STI oxide, therefore shifting the threshold voltage and the leakage current of M_2 back to the preirradiation values. This is likely the reason for the recovery detected in the peaking time and in the return-to-baseline slope after the last irradiation step.

10.4.2.2 Equivalent Noise Charge

Equivalent noise charge in quadruple-well MAPS is also found to be affected by ionizing radiation. Figure 10.10 shows the ENC normalized to the preirradiation value as a function of the TID. Before irradiation, the ENC was found to be between 30 and 40 electrons in the different samples. Each of the four plots results from the average over the nine pixels making up each of the four tested 3 × 3 arrays. After the first irradiation step, a change of less than 20% can be observed. A slight, further increase is detected at an integrated dose of 3 Mrad(SiO_2). After the last step, three out of four curves exhibit a slight reduction in the ENC, with the fourth one instead showing an increase. No significant difference can be found in the device behavior between pixels using an enclosed layout transistor as the input device of the charge preamplifier and pixels using a standard design. This points to the fact that no sizeable contribution, in terms of noise, emerges as the result of charge trapping in the STI in the preamplifier input device. The reason may lie in a very high doping concentration of the P-wells where NMOS transistors are laid out, preventing the P-type polarity of the silicon beneath the STIs from being inverted by the holes trapped in the oxide. The observed change in the ENC may instead be ascribed to an increase of the flicker noise contribution from the same transistor, resulting from the increase in the border (or near-interfacial) trap density due to hole-trapping in the device gate

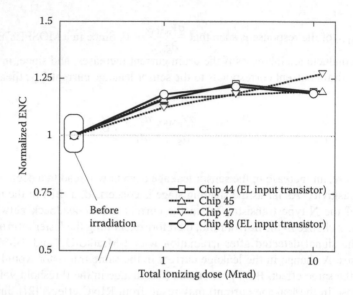

FIGURE 10.10 ENC normalized to the preirradiation value as a function of the TID. Each of the four plots results from the average over the nine pixels making up each of the four tested arrays. (After L. Ratti, G. Traversi, S. Zucca, S. Bettarini, F. Morsani, G. Rizzo et al., *IEEE Trans. Nucl. Sci.*, Vol. 61, No. 4, pp. 1763–1771, Aug. 2014.)

oxide. This is supported by literature data discussing γ-ray effects on low-frequency noise in NMOS transistors belonging to the 180-nm node [32].

10.5 BULK DAMAGE IN HYBRID PIXEL-LIKE MAPS

Exposure to neutrons is not expected to significantly degrade the performance of a CMOS circuit. Actually, MOSFET transistors, whose operation is based on the drift of majority carriers at the device channel surface, are known to be largely insensitive to bulk damage, at least in a neutron fluence range not exceeding 10^{15} 1-MeV-neutron equivalent/cm^2 [33,34]. On the other hand, the charge collection properties of CMOS monolithic sensors may be impaired to a nonnegligible extent by exposure to nonionizing radiation. In order to evaluate bulk damage effects on quadruple-well and DNW MAPS performance, four sets of Apsel4well chips, each including two samples with a resistivity of 10 Ω cm and two samples with a resistivity of 1 kΩ cm, have been irradiated with neutrons from a Triga MARK II nuclear reactor. Each set was exposed to a different fluence, namely 2×10^{12}, 7.4×10^{12}, 2.7×10^{13}, and 10^{14} 1-MeV-neutron equivalent/cm^2, and characterized only after irradiation. The DUTs were irradiated as naked dice before being bonded to the test board to minimize the quarantine period required for neutron-induced activation to die down. A fifth set of four chips (again two with low-resistivity and two with high-resistivity epitaxial layers) was characterized to study the performance of nonirradiated DUTs. Also DNW MAPS (whose sensitive layer features a resistivity of 10 Ω cm, the same as in the case of the standard resistivity epitaxial layer in the quadruple-well

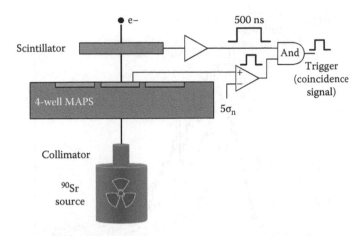

FIGURE 10.11 Measurement setup used for the characterization of MAPS collection properties with a ^{90}Sr/^{90}Y radioactive source. (From L. Ratti, G. Traversi, S. Zucca, S. Bettarini, F. Morsani, G. Rizzo et al., *IEEE Trans. Nucl. Sci.*, Vol. 61, No. 4, pp. 1763–1771, Aug. 2014.)

technology), very similar to those discussed in Section 10.3, were irradiated with neutrons from the same source, but with different fluence steps, namely 2×10^{11}, 7×10^{11}, 1.7×10^{12}, and 6.7×10^{12} 1-MeV-neutron equivalent/cm^2. All the DUTs were left unbiased during irradiation.

Charge collection properties of the monolithic sensors under test were evaluated by means of a ^{90}Sr/^{90}Y source. The measurement setup, including a scintillator and a few logic blocks for coincidence measurements as well as the DUT, is schematically represented in Figure 10.11. Electrons released by the source through beta decay have a broad continuous spectrum, with end-point energy beyond 2 MeV. In the measurement system, upon detection of an electron, the pulse from the scintillator is used to generate a 500-ns gate signal. If, within this time interval, the signal in the central pixel (the cluster seed) of the 3×3 array exceeds $5\sigma_n$ (σ_n being the root mean square [rms] noise at the shaper output), a trigger pulse is issued and the waveforms at the output of the nine channels are stored with predefined pretrigger and posttrigger intervals (1.1 and 4 μs, respectively, in the cases discussed below) for off-line processing. For each event, the amplitudes of the signals from the nine pixels taken at a time t_p after the scintillator pulse are summed and the sum is stored as a measurement of the charge collected by the cluster. Figure 10.12 shows the results from tests performed with a ^{90}Sr/^{90}Y source on a nonirradiated 3×3 Apsel4well array and on two arrays irradiated with different neutron fluences. The 3×3 cluster signal amplitude (in volts) was converted to the relevant amount of collected charge (this is needed to compare the event rates at different fluences) by using the measured charge sensitivity. Results from the irradiation campaign show that after exposure to the maximum fluence, 10^{14} n/cm^2, the most probable value (MPV) of the event rate curves is reduced by slightly less than 50%. Charge collection degradation can be ascribed to an increase in the recombination probability of the carriers released in the P-type epitaxial layer, which in turn is related to the increased defect concentration in the silicon lattice.

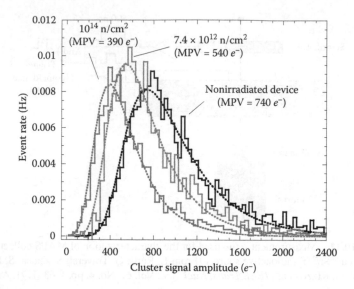

FIGURE 10.12 Event rate in a nonirradiated Apsel4well array and in arrays exposed to different neutron fluences tested with a $^{90}Sr/^{90}Y$ source. The samples feature a high-resistivity epitaxial layer. (After L. Ratti, G. Traversi, S. Zucca, S. Bettarini, F. Morsani, G. Rizzo et al., *IEEE Trans. Nucl. Sci.*, Vol. 61, No. 4, pp. 1763–1771, Aug. 2014.)

Figure 10.13 compares charge collection degradation in quadruple-well MAPS with high-resistivity and standard-resistivity epitaxial layer evaluated again through tests with a $^{90}Sr/^{90}Y$ source. Data relevant to DNW MAPS devices has also been included in the same figure. The names M1 and M2 given to DNW MAPS refer to sensors using different layouts for the collecting electrode. In the figure, the MPV of the event rate histogram, normalized with respect to the MPV as detected in nonirradiated devices, is plotted as a function of the fluence. As far as DNW devices are concerned, each data point refers to a different sample, exposed to a specific fluence. In the case of quadruple-well MAPS data, each point results from the average over a set of two samples irradiated with that particular neutron fluence. Quadruple-well monolithic sensors fabricated with the high-resistivity epi-layer option are found to offer the best performance in terms of tolerance to bulk damage. Actually, in the neutral epitaxial region, doping concentration plays a role in determining the equilibrium Fermi level, which in turn influences the effectiveness of neutron-induced defects as recombination centers [33]. If two substrates with different doping concentration are considered, the Fermi level is higher in the one with smaller concentration, and fewer centers (roughly those located in the silicon bandgap between the Fermi level itself and the conduction band) can take active part in the recombination process. This provides an explanation for the better behavior of DUTs with a high-resistivity sensitive layer as compared to DUTs with a small resistivity sensitive layer (whether they are of the quadruple-well or of the DNW type). As far as the sensor size is concerned, at relatively small fluences, up to a value slightly in excess of 10^{12} n/cm^2, use of large collecting electrodes seems to provide, for a given resistivity of the sensitive layer, some benefits in terms of charge collection capabilities. Note

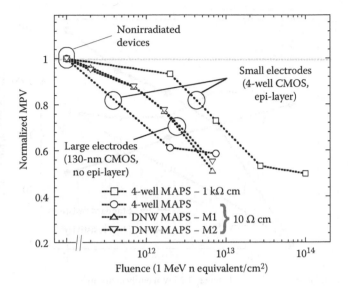

FIGURE 10.13 Most probable value of the $^{90}Sr/^{90}Y$ event rate histograms normalized to the preirradiation value as a function of the fluence in the case of Apsel4well 3 × 3 arrays in high-resistivity and standard-resistivity epitaxial layer. A comparison is made with results from the characterization of DNW MAPS devices also irradiated with neutrons from the same source. (From L. Ratti, G. Traversi, S. Zucca, S. Bettarini, F. Morsani, G. Rizzo et al., *IEEE Trans. Nucl. Sci.*, Vol. 61, No. 4, pp. 1763–1771, Aug. 2014.)

that the two curves relevant to quadruple-well MAPS seem to flatten out after the last considered irradiation step (7.4 × 10^{12} n/cm² for MAPS with standard resistivity epitaxial layer, 10^{14} n/cm² for MAPS with high resistivity epi-layer). The explanation for such a result may not lie as much in the actual behavior of the DUT as in a limited sensitivity of the adopted measurement method.

While no direct effect of bulk damage is expected on the front-end electronics, as anticipated at the beginning of this section, in the case of MAPS devices, a change in the collecting electrode properties may affect the readout channel performance. Figure 10.14 shows the charge sensitivity as a function of the fluence for Apsel4well samples with different epitaxial layer resistivity. The charge sensitivity was normalized with respect to the value detected in nonirradiated devices. While no particular trend can be observed in the case of the set of devices laid out in a standard-resistivity epitaxial layer, in MAPS fabricated with the high-resistivity epi-layer option, the charge sensitivity is found to feature a nonnegligible decrease with increasing neutron fluence. As already observed in Section 10.4.2.1, a change in the charge sensitivity may be ascribed to a variation in the sensor leakage current. The different behavior detected in Figure 10.14 between the samples with high-resistivity epi-layer and those with standard-resistivity epi-layer would imply a different behavior also in terms of leakage current response to radiation. Actually, the leakage current increase can be reasonably expected to be larger in the case of the higher-resistivity epi-layer, where, at a given reverse voltage, the depleted volume under the collecting electrodes is larger, resulting in a larger increase in the radiation-induced contribution of

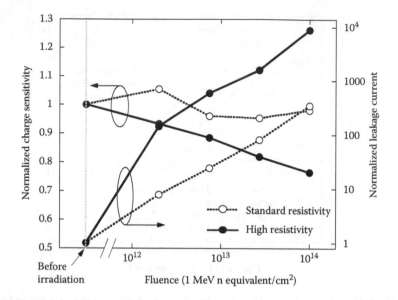

FIGURE 10.14 Charge sensitivity for the readout channel and leakage current in N-well/P-epi diodes as a function of the fluence for samples with different epitaxial layer resistivity. The two parameters are normalized to the value detected in nonirradiated devices. Note that the leakage current scale is logarithmic while the charge sensitivity scale is linear.

generation current. This consideration is supported by leakage current measurements in irradiated N-well/P-epi diodes featuring epitaxial layers with different resistivity and included as supplementary test structures in the Apsel4well chip. The behavior of a pair of samples, representative of a larger set of devices, is shown again in Figure 10.14. Here the leakage current as a function of the fluence has been normalized to the value detected in nonirradiated diodes. The current in the diodes with high-resistivity P-side, at the different fluence steps, was found to be almost two orders of magnitude larger than in the lower resistivity case. This result seems to be compatible with the above hypothesis about the effects of the leakage current on the charge sensitivity in the DUTs. The increase in the leakage current may be responsible for a decrease in the charge sensitivity for the devices with high-resistivity epitaxial layer, while in the standard-resistivity samples, the leakage current increase, two orders of magnitude smaller, is insufficient to produce a detectable effect.

10.6 CONCLUSION

Full-custom radiation-aware design of microelectronic circuits is of paramount importance in applications for particle detection systems, where the front-end electronics, as well as the detector, may be subjected to sizeable levels of both ionizing and nonionizing radiation. In the case of pixel detectors for the next generation, high luminosity colliders, radiation-tolerant design of integrated circuits is further constrained by the limited area and power budget available for the front-end electronics. Results from the characterization of monolithic sensors with hybrid pixel-like, continuous-time front-end channels have been presented and discussed in this chapter. The purpose was to provide an insight into some crucial points of front-end electronics rad-hard design. Given the particular nature of the monolithic sensors readout chains discussed here, which are very similar in their architecture to those used for the readout of hybrid pixel detectors, the obtained results can be extended to a broader class of analog circuits for radiation measurements. The study included devices fabricated in different CMOS processes with a few different technology options to emphasize the dependence of radiation effects on the process characteristics.

ACKNOWLEDGMENTS

The research activity leading to the results discussed in this work has been partially funded by the European Commission under the FP7 Research Infrastructures project AIDA, by the Italian Ministry for Education, University and Research through PRIN projects, and by the Italian Institute for Nuclear Physics (INFN) in the frame of the SLIM5 and VIPIX projects. The authors are indebted to Vladimir Cindro, Jozef Stefan Institute, Ljubljana, Slovenia, for performing the neutron irradiations on the devices tested in this work. They wish to thank Professor Armando Buttafava, University of Pavia, Italy, for making the ^{60}Co source available for ionization damage tests. They also acknowledge the work of Enrico Pozzati, Claudio Andreoli and Stefano Zucca, who designed and characterized the DNW and quadruple-well MAPS chips.

REFERENCES

1. M. Friedl, T. Bergauer, P. Dolejschi, A. Frankenberger, I. Gfall, C. Irmler et al., The Belle II Silicon Vertex Detector, *Physics Procedia*, Vol. 37, pp. 867–873, 2012.
2. T. Behnke, The International Linear Collider, *Fortschritte der Physik*, Vol. 58, No. 7–9, pp. 622–627, Jul. 2010.
3. Y. Degerli, M. Besanon, A. Besson, G. Claus, G. Deptuch, W. Dulinski et al., Performance of a Fast Binary Readout CMOS Active Pixel Sensor Chip Designed for Charged Particle Detection, *IEEE Trans. Nucl. Sci.*, Vol. 53, No. 6, pp. 3949–3955, Dec. 2006.
4. M. Barbero, G. Varner, A. Bozek, T. Browder, F. Fang, M. Hazumi et al., Development of a B-Factory Monolithic Active Pixel Detector—The Continuous-Acquisition Pixel Prototypes, *IEEE Trans. Nucl. Sci.*, Vol. 52, No. 4, pp. 1187–1191, Aug. 2005.
5. D. Contarato, J.-M. Bussat, P. Denes, L. Greiner, T. Kim, T. Stezelberger et al., CMOS Monolithic Pixel Sensors Research and Development at LBNL, *PRANAMA Journal of Physics*, Vol. 69, No. 6, pp. 963–967, Dec. 2007.
6. E.R. Fossum, CMOS Image Sensors: Electronic Camera on-a-Chip, *IEEE Trans. El. Dev.*, Vol. 44, No. 10, pp. 1689–1698, Oct. 1997.
7. A. Besson, G. Claus, C. Colledani, Y. Degerli, G. Deptuch, M. Deveaux et al., A Vertex Detector for the International Linear Collider Based on CMOS Sensors, *Nucl. Instrum. Methods*, Vol. A568, pp. 233–239, 2006.
8. SuperB, a High-Luminosity Asymmetric e+e− Super Flavour Factory. Conceptual Design Report Online. Available: http://www.pi.infn.it/SuperB/CDR.
9. L. Ratti, M. Manghisoni, V. Re, V. Speziali, G. Traversi, S. Bettarini et al., Monolithic Pixel Detectors in a 0.13 μm CMOS Technology with Sensor Level Continuous Time Charge Amplification and Shaping, *Nucl. Instrum. Methods*, Vol. A568, pp. 159–166, 2006.

10. J. P. Crooks, J. A. Ballin, P. D. Dauncey, A.-M. Magnan, Y. Mikami, O. Miller et al., A Novel CMOS Monolithic Active Pixel Sensor with Analog Signal Processing and 100% Fill Factor, *2007 Nuclear Science Symposium Conference Record*, Vol. 2, pp. 931–935, Oct. 26–Nov. 3, 2007.

11. A. Gabrielli, G. Batignani, S. Bettarini, F. Bosi, G. Calderini, R. Cenci et al., On-Chip Fast Data Sparsification for a Monolithic 4096-Pixel Device, *IEEE Trans. Nucl. Sci.*, Vol. 56, No. 3, pp. 1159–1162, Jun. 2009.

12. Y. Arai, Y. Ikegami, Y. Unno, T. Tsuboyama, S. Terada, M. Hazumi et al., SOI Pixel Developments in a 0.15 μm Technology, *2007 Nuclear Science Symposium Conference Record*, Vol. 2, pp. 1040–1046, Oct. 26–Nov. 3, 2007.

13. W. Kucewicz, A. Bulgheroni, M. Caccia, P. Grabiec, J. Marczewski, H. Niemiec, Development of Monolithic Active Pixel Detector in SOI Technology, *Nucl. Instrum. Methods*, Vol. A541, pp. 172–177, 2005.

14. M. Battaglia, D. Bisello, D. Contarato, P. Denes, P. Giubilato, L. Glesener et al., A Monolithic Pixel Sensor in 0.15 μm Fully Depleted SOI Technology, *Nucl. Instrum. Methods*, Vol. A583, pp. 526–528, 2007.

15. I. Peric, A Novel Monolithic Pixel Detector Implemented in High-Voltage CMOS Technology, *2007 Nuclear Science Symposium Conference Record*, Vol. 2, pp. 1033–1039, Oct. 26–Nov. 3, 2007.

16. H. Spieler, *Semiconductor Detector Systems*, Oxford University Press, New York, 2005.

17. L. Ratti, C. Andreoli, L. Gaioni, M. Manghisoni, E. Pozzati, V. Re et al., TID Effects in Deep N-Well CMOS Monolithic Active Pixel Sensors, *IEEE Trans. Nucl. Sci.*, Vol. 56, No. 4, pp. 2124–2131, Aug. 2009.

18. L. Ratti, Continuous Time Charge Amplification and Shaping in CMOS Monolithic Sensors for Particle Tracking, *IEEE Trans. Nucl. Sci.*, Vol. 53, No. 6, pp. 3918–3928, Dec. 2006.

19. R. J. Baker, H. W. Li, D. E. Boyce, *CMOS Circuit Design, Layout and Simulation*. The Institute of Electrical and Electronics Engineers, New York, 1998.

20. V. Re, M. Manghisoni, L. Ratti, V. Speziali, G. Traversi, Total Ionizing Dose Effects on the Noise Performances of a 0.13 μm CMOS Technology, *IEEE Trans. Nucl. Sci.*, Vol. 53, No. 3, pp. 1599–1606, Jun. 2006.

21. F. Faccio, G. Cervelli, Radiation-Induced Edge Effects in Deep Submicron CMOS Transistors, *IEEE Trans. Nucl. Sci.*, Vol. 52, No. 6, pp. 2413–2420, Dec. 2005.

22. E. Gatti, P. F. Manfredi, Processing the Signals from Solid-State Detectors in Elementary Particle Physics, *La Rivista del Nuovo Cimento*, Vol. 9, pp. 1–147, 1986.

23. L. Ratti, M. Manghisoni, V. Re, G. Traversi, Design Optimization of Charge Preamplifiers with CMOS Processes in the 100 nm Gate Length Regime, *IEEE Trans. Nucl. Sci.*, Vol. 56, No. 1, pp. 235–242, Feb. 2009.

24. Y. Tsividis, *Operation and Modeling of the MOS Transistor*, McGraw-Hill, Boston, 1999.

25. V. Re, L. Gaioni, M. Manghisoni, L. Ratti, V. Speziali, G. Traversi, Impact of Lateral Isolation Oxides on Radiation-Induced Noise Degradation in CMOS Technologies in the 100-nm Regime, *IEEE Trans. Nucl. Sci.*, Vol. 54, No. 6, pp. 2218–2216, Dec. 2007.

26. L. Ratti, L. Gaioni, M. Manghisoni, G. Traversi, D. Pantano, Investigating Degradation Mechanisms in 130 nm and 90 nm Commercial CMOS Technologies under Extreme Radiation Conditions, *IEEE Trans. Nucl. Sci.*, Vol. 55, No. 4, pp. 1992–2000, Aug. 2008.

27. D. M. Fleetwood, P. S. Winokur, R. A. Reber, Jr., T. L. Meisenheimer, J. R. Schwank, M. R. Shaneyfelt et al., Effects of Oxide Traps, Interface Traps, and Border Traps on Metal-Oxide-Semiconductor Devices, *J, Appl. Phys.*, Vol. 73, No. 10, pp. 5058–5074, May 1993.

28. J. M. Benedetto, H. E. Boesch, Jr., F. B. McLean, J. P. Mize, Hole Removal in Thin-Gate MOSFET's by Tunneling, *IEEE Trans. Nucl. Sci.*, Vol. 32, pp. 3916–3920, Dec. 1985.

29. J. A. Ballin, J. P. Crooks, P. D. Dauncey, A. M. Magnan, Y. Mikami, O. D. Miller et al., Monolithic Active Pixel Sensors (MAPS) in a Quadruple Well Technology for Nearly 100% Fill Factor and Full CMOS Pixels, *Sensors*, 2008, 5336–5351; doi: 10.3390 /s8085336.

30. L. Ratti, G. Traversi, S. Zucca, S. Bettarini, F. Morsani, G. Rizzo et al., Quadruple Well CMOS MAPS with Time-Invariant Processor Exposed to Ionizing Radiation and Neutrons, *IEEE Trans. Nucl. Sci.*, Vol. 61, No. 4, pp. 1763–1771, Aug. 2014.

31. W. J. Snoeys, T. A. P. Gutierrez, G. Anelli, A New NMOS Layout Structure for Radiation Tolerance, *IEEE Trans. Nucl. Sci.*, Vol. 49, No. 4, pp. 1829–1833, Aug. 2002.

32. M. Manghisoni, L. Ratti, V. Re, V. Speziali, G. Traversi, A. Candelori, Comparison of Ionizing Radiation Effects in 0.18 and 0.25 μm CMOS Technologies for Analog Applications, *IEEE Trans. Nucl. Sci.*, Vol. 50, No. 6, pp. 1827–1833, Dec. 2003.

33. G. C. Messenger, A Summary Review of Displacement Damage from High Energy Radiation in Silicon Semiconductors and Semiconductor Devices, *IEEE Trans. Nucl. Sci.*, Vol. 39, No. 3, pp. 468–473, Jun. 1992.

34. T. P. Ma, P. V. Dressendrofer, *Ionizing Radiation Effects in MOS Devices and Circuits*, John Wiley & Sons, New York, 1989.

11 Radiation Effects on CMOS Active Pixel Image Sensors

Vincent Goiffon

CONTENTS

11.1 INTRODUCTION

11.1.1 CONTEXT

Today, complementary metal-oxide semiconductor (CMOS) image sensors (CISs) [1–4], also called active pixel sensors (APSs), are the most popular imager technology with several billion manufactured every year [5,6]. They represent about 90% of the imager market and should exceed 95% in a few years [5]. Compared to the main alternative imager technology, the charge-coupled device (CCD), CISs have several major benefits such as low-power consumption, high-integration, high speed, and the capacity to integrate advanced CMOS functions on-chip (and even inside the pixel). Thanks to the latest technology innovations, CISs are now matching the performances of CCDs in terms of image quality and sensitivity, placing them at the forefront even in high-end applications such as digital single-lens reflex, scientific instruments, and machine vision. Thanks to these advantages, CISs are also used in harsh radiation environment for applications such as space applications, X-ray medical imaging, electron microscopy, nuclear facility monitoring and remote handling (nuclear power plants, nuclear waste repositories, nuclear physics facilities, etc.), particle detection and imaging, and military applications. Designing, hardening, and testing a sensor for such applications require the understanding of the CIS behavior when exposed to radiation sources. Understanding and improving further the intrinsically good radiation hardness of APS has been a topic of interest since its invention [7–13]. This interest has been recently growing with the coming of new behaviors brought by the profound evolution of CIS technologies (as discussed throughout this chapter) compared to the older-generation mainstream CMOS processes used in early work.

The aim of this chapter is to give an overview of the parasitic effects that can degrade a modern CIS when it is exposed to a high-energy particle radiation field.

11.1.2 APS, CIS, AND MONOLITHIC ACTIVE PIXEL-SENSORS

APS, CIS and monolithic active pixel sensors (MAPS) [14,15] designate the same type of CMOS integrated circuit (IC): a pixel array with a photodetector and an amplifier inside each pixel [1,2]. Depending on the community, one of these names may be used preferentially. APS is the generic term, CIS is mainly used for imaging applications, whereas MAPS is the main term used in the particle detection community to emphasize the monolithic nature of the device compared to hybrid detectors. In most cases, a CIS is an APS manufactured using a CMOS process optimized for imaging applications (called CIS process) whereas MAPS are generally manufactured using standard, or high-voltage CMOS processes and their main purpose is not optical imaging but high-energy particle detection (and imaging). From the radiation effect point of view, there is qualitatively no major difference between MAPS and CIS if the photodetector technology is the same. It means that despite the fact this chapter focuses on CIS, most of the discussions developed here apply to both families of sensors.

11.1.3 BASIC KNOWLEDGE OF RADIATION EFFECTS

The following radiation effect concepts are used in this chapter to describe the influence of high-energy particles on CIS. The reader is invited to look at the first chapter of this book or at the references given in this section to have the details of the origin and limitations of these definitions, mechanisms, and properties.

When passing through the layers of the materials that constitute an IC, ionizing particles (such as high-energy photons [X- and γ-rays] and charged particles [electrons, protons, heavy ions, etc.]) lose most of their energy by generating electron-hole pairs. This excess of charge carriers can disturb or damage ICs by inducing single-event effects (SEE) (see [16] and references therein) or total ionizing dose (TID) effects. SEE occurs when the electron-hole pairs generated by a single particle are sufficient to disturb or damage the IC whereas TID effects are the result of the cumulative exposure to ionizing radiation.

The TID (or absorbed dose) represents the mean energy imparted to matter per unit mass by ionizing interaction and it is expressed here in $Gy(SiO_2)$ (i.e., 1J of energy per kg of SiO_2).*† The ionizing radiation dose absorbed by electronic circuits in medical and space applications are generally below 100 Gy–1 kGy whereas the MGy range can be reached in electron microscopes or nuclear and particle physics experiments. Throughout this chapter the reader should keep in mind that the absorbed TID leads to the buildup of trapped positive charge in the dielectrics, to the buildup of interface states at the Si/oxide interfaces and that these defect densities increase with TID. A detailed review of TID effects can be found in [17–22].

High-energy particles can also lose their energy in matter through nonionizing interactions. These interactions can be summarized as direct interactions with atomic nucleus and they generally result in the displacement of this nucleus. Contrary to TID effects that are mainly a concern in dielectrics, atomic displacement is mainly an issue in the crystalline silicon part of the circuit. The effects linked to radiation-induced atomic displacements are called displacement damage effects and the mean energy imparted to matter per unit mass by nonionizing interaction is called displacement damage dose (D_d) (generally expressed in eV/g(Si)). It is important to note that the D_d leads to the creation of defects in silicon lattice that can act as Shockley-Read-Hall (SRH) generation/recombination centers or SRH carrier traps. These defects can take the form of point defects in the lattice or to clusters of defects (also called amorphous inclusions). Reviews of displacement damage effects that discuss the origin and the limitation of the D_d concept (and especially the nonionizing energy loss [NIEL] concept) can be found in [23–26].

11.2 INTRODUCTION TO CISs

11.2.1 OVERVIEW OF CIS TECHNOLOGY

The basic working principle of CMOS APSs can be found in [2,3,27–30]. As with any APS, CISs are constituted by [2] a pixel array, addressing circuits to access the pixels (the address decoders), and an analog signal-processing circuit (often called readout circuit). This basic architecture common to nearly every APS IC (including MAPS) is presented in Figure 11.1a. In addition to these necessary building blocks, modern CIS products [29–32] often integrate on-chip one or more of the following functions: one analog-to-digital converter (ADC) per column (see [32–34] and

---

* $Gy(SiO_2)$ is used here instead of Gy(Si) because TID effects are due to the absorbed dose in the dielectrics (mainly constituted by SiO_2) and not to the absorbed dose in the silicon.
† 1 Gy = 100 rad.

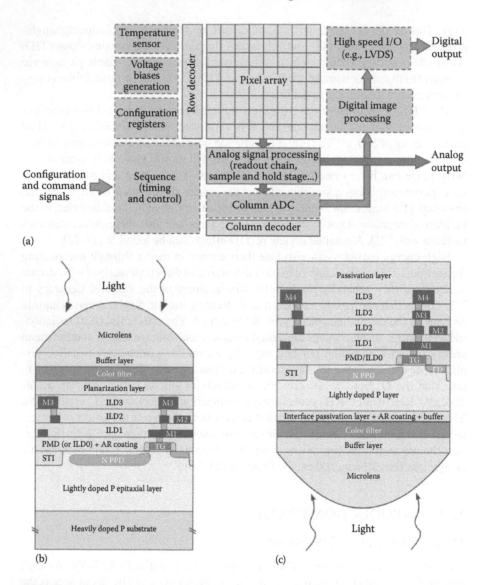

FIGURE 11.1 Overview of CIS technology: (a) Typical CIS IC architecture (the dashed blocks are optional, and usually only one type of output is available in a CIS [digital or analog]). Example of FSI (b) and BSI (c) CIS cross-sectional views. (Inspired from the cross-sectional view shown in R. Fontaine, *IEEE Trans. Semicond. Manuf.* Vol. 26, no. 1, pp. 11–16, Feb. 2013.)

references therein for ADC architectures used in CISs), a sequencer, a digital-image-processing unit, high-speed input/output (I/O) interfaces, configuration registers, and so forth.

Unlike MAPS that are generally manufactured using standard commercial CMOS processes [35] (standard mixed-mode or high-voltage processes, sometimes slightly customized), most of CIS ICs are produced by dedicated CIS processes optimized for visible light detection. Figure 11.1b and c present simplified cross-sectional views of typical modern CIS technologies [36–39]. The base of a CIS process is similar to a standard Deep SubMicron (DSM) CMOS technology [40]: outside the pixel array, metal-oxide semiconductor field-effect-transistors (MOSFETs) are most often the same as the ones used in the mixed-mode version of the process (i.e., non-CIS) with the use of classical source/drain implants, N- and P wells, shallow trench isolation (STI), polysilicon gates, and the typical dielectric stack (constituted by the interlayer dielectrics [ILD]) on top of the semiconductor devices to insure the isolation between the interconnect layers. The first ILD between the first level of metal and the active silicon or the polysilicon layer is often called the premetal dielectric (PMD). However, compared to mainstream CMOS ICs, CISs have several unique features to improve the light collection:

- A reduced number of interconnection metal levels
- Dedicated dielectrics such as antireflection (AR) coatings
- Filters for color imaging
- Microlenses and light guides [41,42], etc.

Several improvements are also made at the device level to optimize the photogenerated charge collection while reducing the dark signal and the noise:

- Dedicated photodiode and in-pixel isolation doping profiles (P-wells, trench sidewall passivation, etc.)
- Dedicated pixel devices (optimized in-pixel MOSFETs with specific threshold voltages, dedicated MOSFET devices for in-pixel charge transfer, etc.)
- A lightly doped epitaxial layer with a thickness optimized for the targeted wavelength range
- Dedicated in-pixel trench isolations to minimize crosstalk [43], such as deep trench isolation (DTI), etc.

In addition to these special features, CIS can be front-side illuminated (FSI) or backside illuminated (BSI), as illustrated in Figure 11.1b and c. BSI technologies allow to collect more light (leading to higher External Quantum Efficiency (EQE) [31]) for a given fill factor but they require the thinning of the sensitive layer down to a few micrometers and the use of backside passivation techniques to reduce signal charge recombination and dark current generation at the back interface [39,44–46].

Several active pixel architectures have been proposed [2] but the vast majority of modern device pixels are based on these two basic designs: the 3T pixel based on a conventional photodiode (Figure 11.2a) and the 4T pixel based on a dedicated buried photodiode (Figure 11.2b) called pinned photodiode (PPD) [47–50]. Because of its low noise, high quantum efficiency, and low dark current [50], the PPD is used in almost all consumer applications. However, this photodetector is only available in CIS processes and the conventional photodiode used in 3T pixel may exhibit some advantages in niche applications (e.g., where large pixel pitches or high full well capacity are required). Therefore conventional photodiodes are still used in most of

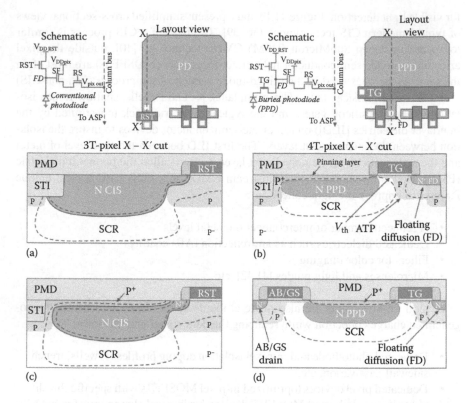

FIGURE 11.2 Typical schematic, layout, and cross-sectional views of (a) a typical 3T pixel and of (b) a typical 4T PPD pixel. Cross-sectional views of (c) a 3T partially pinned photodiode pixel and (d) a 5T pinned photodiode pixel are also presented in this figure. SCR = space charge region, ATP = antipunchthrough implant, V_{th} = threshold voltage implant.

MAPS, in some specific CISs that do not require the use of PPD, and in other APSs not manufactured with CIS processes. As shown in Figure 11.2, both basic pixel architectures share the same three transistors:

- The reset (RST) MOSFET used to reset the floating diffusion (FD) (also called sense node [SN]) that performs the charge-to-voltage conversion thanks to its intrinsic capacitance. In the case of 3T pixel, the photodiode is the FD.
- The source follower (SF) MOSFET used to perform the in-pixel amplification.
- The row select (RS) switch MOSFET used to connect the pixel to the column sample-and-hold stage.

In PPD-based pixels, another MOSFET is necessary to transfer the charge collected during integration to the FD (and also to empty the PPD potential well); this additional transistor is called the transfer gate (TG).

The cross-sectional view in Figure 11.2a shows that the conventional CIS photo-diode used in 3T pixels is typically a deep N-CIS implant (similar to N-well implants but optimized for photodetection) on P-epitaxial layer and surrounded by a P-well. Depending on the design, the STI can cover the whole N-CIS implant or be recessed from the N-CIS region. In any case, the depletion region in conventional photo-diodes reaches an oxide interface (generally the STI bottom, as illustrated in Figure 11.2a, or the PMD/Si interface if the STI is recessed) all over its perimeter. In other types of APS (such as MAPS), this conventional photodiode can be made using the N-MOSFET source/drain N+ implant or the P-MOSFET N-well implant. Some APSs even use triple- or quadruple-well technologies to realize deeper photodiodes [35,51,52]. It is also possible to reverse the doping types and to use P on N substrate photodiodes.

Contrary to the conventional CIS photodiode, the PPD is a buried N-PPD implant surrounded by a P-well (or STI passivation P doping) and protected from the PMD interface by a P+ pinning implant on top of it. This pinning layer is also used to insure the full depletion of the PPD N region after a complete charge transfer. If the TG is completely turned OFF (i.e., biased in accumulation regime, generally with the use of negative gate voltage), the PPD depletion region does not reach any oxide interface because it is protected from the STI by the P-well (or P-STI) doping, from the PMD by the pinning layer, and from the TG channel by the TG accumulation layer (as illustrated in Figure 11.2b).

For discussing the radiation effects on CIS, two variations have to be presented: the partially pinned photodiode (Figure 11.2c) and the 5T-PPD pixel (Figure 11.2d). The first is similar to the 3T pixel conventional photodiode except it is covered by a P+ pinning implant. In the case of partially pinned-photodiode, the pinning implant's sole purpose is to reduce the dark current by reducing the contact area between the photodiode depletion region and the surrounding oxides (STI and PMD). In order to connect this photodiode to the SF and RST MOSFETs, the P+ pinning layer has to be opened somewhere to let the N region reach the surface. Therefore, contrary to PPDs, partially pinned photodiode depletion region is in contact with oxide inter-faces in the vicinity of the SF/RST contacts. This is illustrated in Figure 11.2c where one can see that the contact between the (SCR) and the oxide interface (PMD here) is near the RST MOSFET. The total area of this depleted oxide interface is smaller than in a 3T pixel conventional diode design where the depleted interface runs all along the photodiode perimeter (indeed the contact between the SCR and the oxide is located below the peripheral STI in Figure 11.2a). Since the dark current rises with the total depleted oxide interface area in the photodiode (as explained in the next section), the dark current in a partially pinned photodiode is much higher than in a PPD but it is lower than in a conventional photodiode.

The other interesting variation from the radiation effect point of view is the 5T PPD pixel (Figure 11.2d) in which an additional TG is added in a PPD-based pixel to perform an antiblooming (AB) or global shutter (GS) function (or both) [53]. Any more complex pixels with even more transistors (such as those found in so-called smart sensors [29,54–56]) will be based on one of these building blocks (the 3T pixel, the 4T PPD pixel, the 3T partially pinned photodiode pixel, or the 5T PPD pixel) and thus, to understand the radiation effect on any CIS, the first step is to understand the radiation effects on these elementary pixel structures. The discussions presented in this chapter can easily be transposed to more integrated pixel architecture.

11.2.2 Selected Important CIS Concepts for Radiation Effect Discussions

This section provides some details about a few selected CIS concepts that are necessary to discuss the radiation effects. More information about generic solid-state imager parameter definitions, such as EQE, charge-to-voltage conversion factor (CVF), charge transfer efficiency (CTE), charge transfer inefficiency (CTI), maximum output voltage swing (MOVS), and dynamic range (DR), can be found in [2,3,27–30] or [57–59].

11.2.2.1 Full-Well Capacity and Pinning Voltage

In 3T active pixels the saturation level is given by the saturation of the readout chain (or the ADC) and it is thus not related to the photodiode maximum charge (called the full well capacity [FWC] [58]). In a 4T pixel during integration (Figure 11.3a), the photo-generated electrons are collected in the PPD potential well that is isolated from the FD by turning the TG OFF (here, this TG OFF voltage is referred to as V_{LOTG}). At the end of the integration time t_{int}, the TG is turned ON (Figure 11.3b) and the collected charge is transferred to the FD for being readout. Since the collecting well (i.e., the PPD) is separated from the readout node (the FD), the saturation charge of PPD (the FWC) can be lower than the saturation charge of the FD. In this case, the output saturation level is given by the photodiode FWC.

An important parameter specific to PPD pixels is the pinning voltage V_{pin} of the buried photodiode. The pinning voltage represents the bottom of the PPD potential

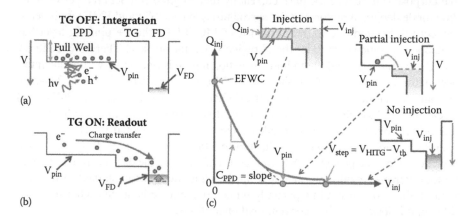

(a)

(b)

(c)

FIGURE 11.3 PPD-TG structure operation and V_{pin} measurement illustrations. (a) During integration, the TG is OFF and the photo-generated electrons are collected in the PPD potential well. (b) To read out the collected charge, the TG is turned ON and the electrons are transferred to the FD. (c) Pinning voltage characteristic with the PPD-TG physical parameters that can be extracted. C_{PPD} = PPD capacitance, EFWC = equilibrium full-well capacity, V_{pin} = pinning voltage, and V_{th} = TG threshold voltage. (From V. Goiffon, M. Estribeau et al., *IEEE J Electron Devices Soc* 2, 2014.)

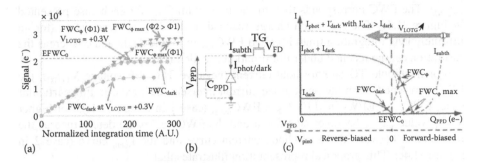

FIGURE 11.4 Illustrations of the different FWC definitions in PPD CIS. (a) Measurement of the different FWCs on a PPD CIS. (Modified from A. Pelamatti, J.-M. Belloir et al., *IEEE Trans. Electron Devices* 2015.) (b) Pixel electrical schematic used to determine the FWC. (c) Graphical representation of the Pelamatti et al. model. (From A. Pelamatti, V. Goiffon et al., *IEEE Electron Device Lett* 34, 2013, 900.)

well (as illustrated in Figure 11.3). More precisely, this potential corresponds to the maximum PPD channel potential [50,60,61]. The higher the V_{pin}, the higher the FWC is (but the pinning potential must stay low enough to ensure a good transfer). It is possible to measure the pinning voltage at the sensor output [62] from the pinning voltage characteristic presented in Figure 11.3c. This technique can also be used to extract several important physical parameters of the PPD-TG structure [60]. Its basic principle (presented in Figure 11.3c) is detailed in [60,62]. It consists in injecting electrons (charge Q_{inj}) in the PPD during the integration phase by applying the injection voltage V_{inj} on the FD (with TG and RST MOSFETs turned ON).

As explained in detail in [60], the pinning voltage corresponds to the boundary voltage between the injection and partial injection regime of Figure 11.3c, the *Y*-intersect of the characteristic provides the equilibrium full well capacity (EFWC),* the slope in the injection regime gives the PPD capacitance (C_{PPD}) and the step corresponding to the beginning of the partial injection regime allows to estimate the TG threshold voltage.

At a given temperature, the saturation charge of a PPD depends on the photon flux and on the TG bias (as shown in Figure 11.4a). As a consequence, several FWCs can be defined in a PPD CIS [63,64]. These different saturation levels can be explained by the TG-PPD electrical schematic presented in Figure 11.4b [63,65]. The saturation charge of the PPD capacitance is reached during integration when the current flowing through the diode (the photonic current I_{phot} and the dark current I_{dark}) is compensated by the TG subthreshold current (I_{subth}). A graphical representation of this model is presented in Figure 11.4c. It shows a part of the classical PN junction I-V characteristic [66] without illumination (I_{dark} curve) and with illumination ($I_{phot} + I_{dark}$ curve). The PPD charge is directly related to the PPD potential (through the PPD capacitance C_{PPD}). Hence, the voltage *x*-axis can be graduated in stored charge values

---

* The EFWC is the charge stored in the PPD at equilibrium [63,64], which corresponds to a PPD potential equal to 0V.

(Q_{PPD}). The FWC corresponds then to the maximum Q_{PPD} for each case presented in Figure 11.4c. The saturation charge reached in the dark at steady state (for an infinitely long integration time) is called FWC_{dark} in Figure 11.4a and c whereas the saturation charge reached under illumination is called FWC_Φ.

For negligible TG subthreshold current (negative TG OFF voltage V_{LOTG}), the FWC is reached when the photodiode current intersect the x-axis. In this particular case, $FWC_{dark} \approx EFWC$ and $FWC_\Phi = FWC_{\Phi max}$ (case 1 in Figure 11.4c). For higher V_{LOTG} values, I_{subth} becomes significant and the FWC values are determined as the intersection between the photodiode current curve and the I_{subth} curve (case 2 in Figure 11.4c). This graphical representation illustrates that

- FWC_{dark} is always lower or equal to EFWC.
- FWC_Φ is always greater than FWC_{dark}.
- FWC_Φ increases when the photonic current curve is shifted upward (i.e., when the photon flux increases) whereas FWC_{dark} stays unchanged.
- Both FWC_{dark} and FWC_Φ decrease when V_{LOTG} decreases.
- An increase of dark current shifts FWC_Φ toward EFWC, which leads to a decrease of FWC_Φ in most of the cases (i.e., when $FWC_\Phi > EFWC$).*

11.2.2.2 Dark Current Sources

At the beginning of the integration time (right after the reset phase), the CIS photodiode is reverse-biased to empty the collecting well of previously integrated charge carriers. This reverse bias increases the depletion volume and places the collecting well under nonequilibrium condition (i.e., $p \cdot n < n_i^2$ in the depletion region). The SRH recombination/generation process [67,68] induces a parasitic reverse current that fills the well by discharging the potential to return to equilibrium (i.e., 0V photodiode bias and $p \cdot n = n_i^2$). Since $p \cdot n < n_i^2$, the net recombination/generation rate U is negative and the dominating SRH mechanism is the electron-hole pair generation [3]. In solid-state imagers, this SRH generation induced parasitic current is called the dark current and it limits the sensor dynamic range. It can take several forms depending on its origin inside the pixel:

- *Interface state generation dark current:* If the photodiode depletion region is in contact with a Si-oxide interface, the high density of interface states (N_{it}) leads to an intense generation contribution [69] that generally hides the other dark current sources and that can be rewritten as

$$I_{itgen} = K_1 \exp\left(-\frac{E_g}{2kT}\right) A_{itdep} N_{it} \tag{11.1}$$

with K_1 a proportionality factor that includes several physical and technological constants, A_{itdep} the area of depleted Si/oxide interface, N_{it} the

---

* For high V_{LOTG} values or high I_{subth}, FWC_Φ can be lower than EFWC.

interface state density, k the Boltzmann constant, E_g the bandgap energy, and T the temperature.

- *Bulk generation dark current:* A bulk defect with energy E_T in the bandgap will generate the following dark current if it is located in the depletion region of the photodiode (under nonequilibrium) [69,70]:

$$I_{bkgen} = K_2 \frac{n_i V_{dep} N_t}{2\cosh\left(|E_i - E_t|/kT\right)} = \begin{cases} K_2' \dfrac{V_{dep} N_t}{2} \exp\left(-\dfrac{E_g}{2kT}\right) & \text{if } E_i = E_t \\[2em] K_2'' V_{dep} N_t \exp\left(-\dfrac{E_g/2 + |E_i - E_t|}{kT}\right) & \text{if } |E_i - E_t| \gg kT \end{cases}$$

$$(11.2)$$

with K_2, K_2' and K_2'' proportionality constants, V_{dep} the depleted volume, and N_t the defect concentration.

- *Interface state diffusion dark current:* if the generation currents originating from the depletion region (I_{itgen} and I_{bkgen}) are low enough (as in state-of-the-art PPDs), the dark current contribution coming from the generation of undepleted interfaces can be visible. Minority carriers generated at the interface (outside the depletion region) diffuse toward the photodiode depletion region, leading to a generation-induced diffusion current often simply referred to as diffusion current [69]. Under certain simplifying assumptions (realistic for CIS), this interface state diffusion dark current contribution can be expressed [71,72]:

$$I_{itdif} = K_3 \exp\left(-\frac{E_g}{kT}\right) \frac{A_{it}}{N_{A,D}} \frac{N_{it}}{1 + K'N_{it}} \approx K_3 \exp\left(-\frac{E_g}{kT}\right) \frac{A_{it}N_{it}}{N_{A,D}} \qquad (11.3)^*$$

with K_3 and K' as two proportionality factors, A_{it} as the area of the considered Si/oxide interface outside the depletion region, and $N_{A,D}$ as the P or N doping concentration at the interface. This contribution decreases when the distance between the photodiode depletion region and the Si/oxide interface increases (as shown by Equation 11.5 in [72]). Therefore, only the nearest Si/oxide interfaces generally bring a significant contribution (PMD or STI interfaces right above or beside the PPD).

- *Bulk diffusion dark current:* The dark current coming from the generation of minority carriers in the quasi-neutral region that diffuse toward the depletion region can be visible if all the other sources are weak enough. As the previous source, this current is also called a diffusion current despite

---

* Because $K' \times N_{it} \ll 1$ in general except in heavily irradiated devices where $K' \times N_{it} > 1$ leading to a saturation of this current contribution [71,72].

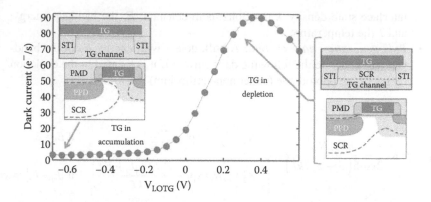

FIGURE 11.5 Dark current evolution with TG OFF bias (i.e., V_{LOTG}) in a PPD CIS. When the TG channel is accumulated (for negative V_{LOTG}) the PPD depletion regions does not reach any oxide interface and the dark current is minimum. If the TG channel is depleted, the PPD depletion region is in contact with the oxides in the PPD-TG transition region and the dark current is maximum. SCR= space charge region. (Data from V. Goiffon, M. Estribeau et al., *IEEE Trans. Nucl. Sci.* 61, 2014.)

the fact it also comes from a SRH generation process [69]. Its general form can be approximated by [58]:

$$I_{bkdif} = K_4 \exp\left(-\frac{E_g}{kT}\right)\frac{\sqrt{N_t}}{N_{A,D}} \qquad (11.4)$$

where K_4 is a proportionality constant* and $N_{A,D}$ the P or N doping concentration in this region.

Each of these contributions can dominate the dark current observed at a sensor output. In unirradiated 3T conventional and partially pinned photodiodes, Equation 11.1 dominates.[†] For these two photodiodes, the apparent activation energy[‡] of the dark current will be close to $E_g/2$ (as shown in Equation 11.1).[§]

In a state-of-the-art PPD, the depletion region is not supposed to reach any Si/ oxide interface when the TG is placed into accumulation (with a negative V_{LOTG} bias, typically lower than –0.5 V) as shown in Figure 11.5. In this case I_{itdif} and/or I_{bkdif} contributions (Equations 11.3 and 11.4) dominate and the apparent activation energy is close to E_g. If V_{LOTG} is increased, the PPD depletion region reaches the Si/

---

* K_4 decreases when the quasi-neutral region thickness is reduced or when the diffusion length increases (see Equation 7.21 in [58]).
† As discussed previously, the dark current in a partially pinned photodiode pixel is lower than the one of a conventional diode pixel simply because the depleted area A_{itdep} is smaller in the case of the partially pinned photodiode.
‡ If the dark current evolution with temperature is fitted using the Arrhenius law $A = K \exp(-E_a/kT)$ [73], the value of E_a is the activation energy of the dark current.
§ In reality, a slightly higher value than $E_g/2$ is measured (typically 0.63 eV) because of the temperature dependence of the terms included in the proportionality factor K_1.

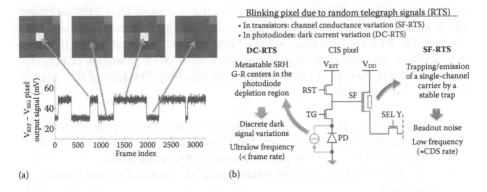

FIGURE 11.6 Random telegraph signal overview: (a) Typical effect of an RTS pixel on dark frames and (b) summary of the two main RTS phenomena in CIS: dark current RTS (DC-RTS) and source follower RTS (SF-RTS).

oxide interface near the TG sidewall spacer. The interface state generation current coming from this depleted interface (Equation 11.1) becomes the dominant current contribution (and it is much higher than the diffusion dark current, as illustrated in Figure 11.5).

Other dark current sources, such as metallic contamination [74,75] or electric field enhancement (EFE) are very infrequent in modern CIS where the manufacturing process and the operating conditions are optimized to mitigate these unwanted sources. Such effects may however happen in other types of APS because of the nondedicated process or because of the high voltages used in some of these devices.

11.2.2.3 Random Telegraph Signal Noises: DC-RTS and SF-RTS

A random telegraph signal (RTS) is a random process that switches randomly between two or more discrete levels [76,77], as illustrated in Figure 11.6a. This phenomenon has been observed in many electronic devices [77] and can have several different physical origins. Several names are used to describe it: RTS, random telegraph noise (RTN), burst noise, popcorn noise, variable junction leakage (VJL) [78], and some application-specific names, such as blinking pixels in CIS or variable retention time (VRT) in dynamic random access memories (DRAMs) [79–81]. In CISs, RTSs lead to bright pixels that seem to be turned on and off randomly (Figure 11.6a). Such pixels are generally called blinking pixels. Because of the constant progress in the reduction of dark current and noises, this parasitic behavior is becoming the limiting factor in more and more high-end applications. Two kinds of RTS have been observed in CIS: one due to the discrete fluctuation of the photodetector dark current, called the dark-current RTS (DC-RTS) and the other one due to the discrete switching of the in-pixel SF MOSFET channel resistance (called the SF-RTS).

DC-RTS was first reported [82–84] and analyzed [85–90] in irradiated CCDs and this phenomenon has been attributed to bulk metastable SRH generation centers*

---

* In this chapter, "metastable generation centers" means a generation center (whatever its physical origin: a point defect, an interface state, a cluster, etc.) that exhibits a generation rate (i.e., a dark current) with two or more metastable states (whatever the physical origin of this apparent signal metastability).

located in a depleted region (with $p \cdot n < n_i^2$, such as a reverse-biased PN junction), which are able to instantaneously switch between two generation rates. Since then, this bulk DC-RTS has been observed in CMOS APS [12,91,92], in MAPS [93], and in CIS [94]. It has been recently demonstrated that DC-RTS generation centers can also be located at the depleted Si-oxide interface [95–100]. This kind of RTS has the following characteristics:

- The amplitudes of the dark signal discrete fluctuations are proportional to the integration time (as illustrated in Figure 11.7a) and they can be much higher than what would be expected from a single-point defect.
- The DC-RTS fluctuations are instantaneous (no transient regime between two states).
- The RTS behavior can directly be observed at the sensor output on the dark signal evolution with time of a single RTS pixel.
- The intertransition times are exponentially distributed and the time constants of DC-RTS are not limited to a particular range (time constants ranging from milliseconds to hours have been reported).
- The maximum transition amplitudes are also exponentially distributed [94,96].
- The time constants and the amplitudes are thermally activated (typically with an activation energy of about 0.6 eV for the amplitude in modern CIS).
- One possible explanation for the observed metastable generation rate is that DC-RTS is due to a change of configuration of an SRH generation center

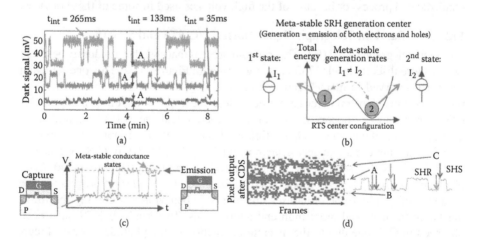

FIGURE 11.7 DC RTS and SF RTS illustrations. (a) One RTS pixel output dark signal versus time (frames) for three different integration durations (t_{int}). (b) Illustration of a possible DC-RTS mechanism. (c) Illustration of the SF RTS physical mechanism (assuming fixed VG, VD, and IDS as in real pixel operation conditions). (d) Effect of SF-RTS on the sensor output signal: a discrete dark voltage variation is seen only if the source follower RTS state has changed between the two samples (SHS and SHR). Contrary to DC-RTS, SF-RTS metastable states are not visible at the sensor output and the SF-RTS amplitude is not proportional to integration time.

that has a different generation rate in each configuration (as illustrated in Figure 11.7b). There is, however, no clear evidence that these different configurations are linked to the trapping and the emission of charge carriers (contrary to SF-RTS) and the real nature of the DC-RTS centers is not yet fully understood.

- The wide majority of DC-RTS pixels observed in unirradiated CIS is coming from metastable generation centers located at Si-oxide interfaces (i.e., not from bulk DC-RTS centers) [95,98].

SF-RTS in CIS is due to the well-known MOSFET gate oxide trapping RTS [101,102]. It is due to the random trapping and emission of inversion channel carriers ([77,103] and references therein). When an electron is trapped by a channel interface state (at the gate oxide or STI oxide interface [103]), the channel conductance is reduced, leading to a low source potential state (as shown in Figure 11.7c). When the electron is released, the channel conductivity is increased back to its initial value and the MOSFET source voltage is then in a high potential state.

Both phenomena are different physical mechanisms: SF-RTS is a discrete change of channel resistance whereas DC-RTS is a discrete fluctuation of the dark current. They also have different signatures at the CIS output and can thus be easily discriminated. For example, SF-RTS leads to an increase of the sensor temporal noise but the RTS behavior is hardly visible at the output and no stable RTS state can be seen in this case (Figure 11.7d), contrary to DC-RTS (Figure 11.7a).

11.3 SINGLE EVENT EFFECTS

As any CMOS IC, APS, and CIS are in theory sensitive to all kinds of single event effects (SEEs) [16] and this sensitivity strongly depends on the design and the technology of the tested sensor. However, the peripheral circuits of most of the CIS tested in literature do not exhibit high SEE sensitivity (for example, no effect has been observed in [104]) even in extreme conditions such as inertial confinement fusion (ICF) radiation environment [105]. Single-event latchups (SELs) have been observed in CIS digital circuits [106] whereas simple analog readout circuits are generally immune to SEL (because they usually do not have N and P-MOSFETs close to each other [107]). Single-event upsets (SEUs) have been observed in CIS [108,109] but only in peripheral circuits (such as on-chip sequencers) that embed digital memories, latches or flip-flops. Some single-event functional interrupts (SEFIs) [106,110] and some single-event transients (SETs) [107] have also been reported. Most of these SEEs occurring in peripheral circuits can be mitigated by using the classical radiation-hardening-by-design (RHBD) techniques [16,111–114] and SEEs in peripheral circuits are not likely to be a problem in radiation-hardened CIS.

As with most of radiation effects in CIS, the most frequent SEEs occur in the pixel array. Incoming particles generate electron-hole pairs in the sensitive silicon volume through direct (for charged particles) or indirect ionization (for neutrons). These high densities of parasitic carriers are collected like photo-generated signal carriers and they lead to the transient saturation of the collecting pixels. Such pixel SETs can saturate large clusters of pixels [107,115,116] or generate secondary recoil ion tracks [105,117] and several modeling approaches have been proposed in the literature to predict their effect on image quality, their occurrence, or shape of their track [108,118–120]. The same single-particle-induced parasitic charge collection mechanisms are also actively studied and modeled in the MAPS community, but in this case, the collected charge is seen as the signal, not as a parasitic SET (see [121] for example). Pixel SETs do not last more than a frame and antiblooming techniques can help reduce the size of the particle track (in PPD pixels [106] but not in 3T pixels [107]). Anticrosstalk DTI [43] can also limit the spreading of parasitic charge over several pixels. By thinning the sensor active volume (typically the epitaxial layer in CISs), the effect of SET can be reduced but the signal sensitivity is most often also decreased. In some particular applications, hardening-by-system techniques can be used to get rid of the unwanted parasitic charge [122,123]. Other kinds of SEEs are not likely in simple 3T or 4T pixels but can be an issue in smart sensors with in-pixel integrated functions.

11.4 CUMULATIVE RADIATION EFFECTS ON PERIPHERAL CIRCUITS

In most cases, radiation effects on peripheral circuits such as address decoders and analog readout circuits are limited to the degradation of the MOSFET characteristics. Due to the reduced gate oxide thickness of DSM CMOS technologies, the radiation-induced gate oxide degradation is usually negligible in CIS (even after high levels of TID [124]) and the main MOSFETs degradations come from the radiation-induced positive charge trapped in the lateral STI (on the channel edges) [19,112,125]. This remains true despite the fact that CISs use, in the analog parts of the circuit (especially inside the pixel array), high-voltage* MOSFETs (also called I/O or GO2 MOSFETs) with thicker gate oxide (typically 7 nm) than the core low-voltage MOSFETs. This trapped charge leads to three main effects in MOSFETs:

- The creation of a parasitic drain to source leakage path on the STI sidewalls (in N-channel MOSFET only)
- A threshold voltage variation in narrow N and P-MOSFETs called radiation-induced narrow-channel effect (RINCE) [126,127]
- The creation of leakages paths between N-doped regions called interdevice leakages

For low and moderate TID levels (below 1–10 kGy), these effects are generally not a problem in CIS digital and analog circuits outside the pixel array and radiation hardness above 10 kGy(SiO$_2$) have been demonstrated without the use of RHBD techniques [60,70,99]. By using these hardening techniques [112–114], such as enclosed layout transistors (ELT), CIS and APS peripheral circuits can even handle TID levels beyond the 100 kGy(SiO$_2$) range [8,10,13,128,129]. Some specific parts of mixed-signal CMOS circuits used in some CISs may exhibit a lower radiation

---

* This high voltage is typically between 2.5 V and 5 V, with a popular 3.3 V value, which is higher than the operating voltage of core MOSFETs (typically around 1 V).

hardness (and require specific hardening technique), such as sampling nodes (which can suffer from retention time reduction), voltage references, or parasitic bipolar transistors [130,131].

Despite the fact that CISs with integrated sequencers, registers, and ADCs have been exposed to ionizing radiation and tested [12,13,109,132–138], it has not been demonstrated so far that such peripheral circuits are limiting the TID hardness of CISs. The part of CIS circuits that is the most sensitive to TID is the pixel itself, especially before the charge to voltage conversion (i.e., from the photodetector to the SF MOSFET gate electrode in Figure 11.2) where any leakage current can impact the sensor performance.

As a surface device, MOSFETs are known to be almost immune to displacement damage effects whereas, as discussed in Section 11.5.2, pixels are very sensitive to D_d. Therefore pixel arrays generally become unusable before any displacement damage effect can be observed on the peripheral CMOS circuits.

11.5 CUMULATIVE RADIATION EFFECTS ON PIXEL PERFORMANCES

11.5.1 TOTAL IONIZING DOSE EFFECTS

11.5.1.1 Degradation Mechanism Overview and Common Effects

In DSM CIS processes, the total ionizing dose effects (TID) induced degradations of in-pixel MOSFETs are generally secondary [139,140] compared to the effects on the photodiode (and the associated TGs in PPD pixels). Moreover, in-pixel MOSFETs can be easily radiation-hardened by using the RHBD techniques [112–114] to extend their radiation hardness much beyond the 10 kGy range. This section mainly focuses on the radiation effects on the photodiode (and the TG in PPD pixels).

An overview of the TID-induced degradation of three selected pixels is presented in Figure 11.8. The left column (Figure 11.8a) represents a 4T PPD pixel with an accumulated TG whereas the center column shows a 4T PPD pixel with a TG in depletion. The right column shows 3T pixel cross sections. As mentioned in Section 11.1.3, ionizing radiation induces the buildup of trapped positive charge in the dielectrics (here the STI and the PMD) and interface states at the Si/oxide interfaces. These defects are distributed all over the dielectrics but only the ones that play a significant role in the radiation-induced degradation are represented in Figure 11.8. Despite the fact that the defect distributions are similar in the three illustrated pixel cross sections, the most limiting defects are not the same in the three pixels. A first look at Figure 11.8 allows us to realize that in the accumulated TG PPD pixel, the most active defects are located on top of the PPD (in the PMD or at its interface) whereas TG channel interface states play the main role at low and moderate TID in a PPD pixel where a TG is in depletion. In a 3T pixel and in any APS pixel based on a conventional or partially pinned photodiode, the most active defects are located on the photodiode perimeter where the depletion region reaches the oxide.

During TID irradiation, the most degraded parameter in every CIS, APS, and MAPS pixel is the dark current (shown in Figure 11.9a and b). In PPD based pixels with accumulated TG, the dark current starts to rise with TID at low dose because

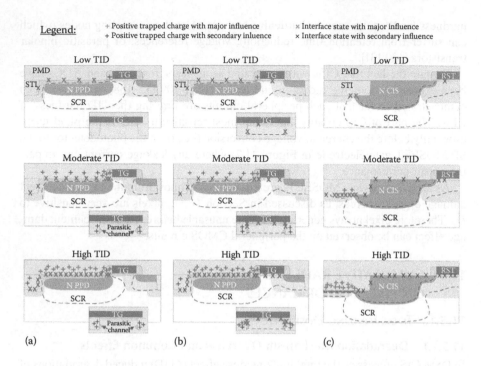

FIGURE 11.8 Pixel degradation mechanism illustration in (a) a 4T pinned photodiode pixel with accumulated TG, (b) a 4T pinned photodiode pixel with depleted TG, and (c) a 3T pixel with a conventional photodiode. Partially pinned photodiodes behave similarly to 3T pixel conventional photodiodes except that some effects may be delayed due to the P+ pinning layer. The low, moderate, and high TID levels depend on the manufacturing process but they roughly correspond to the following ranges, respectively: below 0.5–1 kGy(SiO$_2$), in the 0.5–10 kGy(SiO$_2$) range, and beyond 1–50 kGy(SiO$_2$).

of the interface state buildup at the nearest Si/oxide interfaces [71,72]. In most of the cases, the generating interface is the Si/PMD interface on top of the PPD [72] but in some particular cases* the main contribution can come from the Si/STI (or Si/ DTI) interface [71]. This contribution is an interface-state-diffusion dark current proportional to the PPD area or perimeter (depending on the dominating oxide) and is described by Equation 11.3. At higher TID, the Si/PMD interface becomes slightly depleted in the area where it is nearest to the PPD depletion region [72,99,141]. This particular area is the region in the vicinity of the TG sidewall spacer,† as illustrated in Figures 11.2b and 11.8a. In this case, the main contribution is an interface-state-generation dark current (Equation 11.1) coming from this depleted interface (that is proportional to the TG width [141]). At even higher TID the PPD depletion region

---

* Such as a very small pixel pitch, a reduced STI-PPD distance, a high P+ pinning doping concentration, or a low P doping concentration between the STI and the PPD.
† The P doping concentration in the spacer vicinity is lower than the P+ pinning layer doping concentration to reduce the potential barrier between the PPD and the TG when it is turned ON.

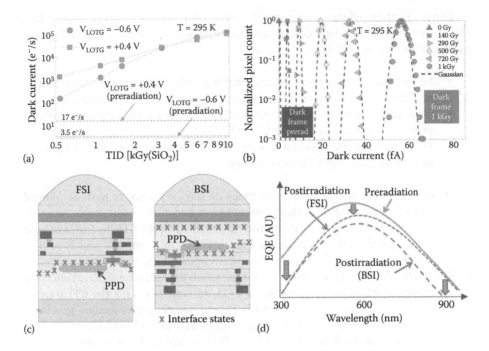

FIGURE 11.9 Common TID-induced degradations. (a) Typical mean dark current increase measured on a 7-μm pitch 4T PPD CIS. (Modified from V. Goiffon, M. Estribeau et al., *IEEE Trans. Nucl. Sci.* 61, 2014.) (b) Typical dark current distribution evolution with TID measured on a 10-μm pitch 3T image sensor. (Modified from V. Goiffon, C. Virmontois et al., in 2010, pp. 78261S–78261S–12.) Both sensors were exposed to ^{60}Co gamma rays. (c) Localization of TID-induced interface recombination centers in the FSI and BSI cases. (d) Illustration of the TID-induced EQE degradation.

reaches the PMD interface,* the PPD is not pinned anymore, and the interface-state-generation dark current (Equation 11.1) coming from the PMD interface dominates (this contribution is proportional to the PPD area) [99].

In the case of PPD pixels in which a TG is not properly accumulated,† the PPD depletion region reaches the depleted TG channel that is in contact with the gate oxide and channel STI interfaces. The dominant contribution at low and moderate TID becomes the interface-state-generation dark current (Equation 11.1) coming from the gate oxide and TG channel STI sidewalls. It leads to a higher radiation-induced dark current increase in depleted TG PPD pixels [62,72] compared to the one with accumulated TG, as shown in Figure 11.9a. At high TID (3 kGy in Figure 11.9a), the main contribution comes from the depleted PMD interface and the additional contribution of the depleted TG channel is not visible anymore.

---

* Possibly the STI interface, too, in some extreme cases.
† This is generally the case for the second TG used for AB of GS operation in a 5T PPD pixel such as the one tested in [138]. It also happens in 4T PPD pixels where a positive voltage is applied on the TG to use the FD as an AB drain.

Other radiation-induced dark current sources that are not likely to appear in a PPD pixel with an optimized design manufactured using a mature CIS process have been reported in the literature:

- An interface-state-generation dark current coming from the peripheral STI interface [142] if this interface is in contact with the PPD depletion region*
- Tunneling currents and other high electric field effects (and contamination-related dark current issues) [62,142,143]
- An unidentified dark current source in [72] that could be explained by an underestimation of the TG contribution[†]

In 3T pixels (and in any active pixel based on a conventional or partially pinned photodiode) the main source of dark current before and after irradiation is the interface-state-generation dark current coming from the depleted STI (or PMD in case of partially pinned diode, as shown in Figure 11.2c). In a conventional N-well or N^+-based photodiode, this contribution is proportional to the photodiode perimeter [7,10,139,144]. At low TID the dark current rises with the radiation-induced interface state buildup (as expected from Equation 11.1). At moderate TID, the STI trapped charge extends the depletion region on the photodiode perimeter leading to a large dark current enhancement (by increasing A_{itdep} in Equation 11.1) [8,13,145,146]. At high TID, the field oxide interface (STI or DTI) is inverted, the photodiodes are shorted, the CVF drops, and the pixel array is no longer functional (see [70] for example). The same degradation mechanism appears in partially pinned photodiodes except that the depleted interface area is smaller before irradiation than in a conventional diode and except that the effects might be delayed to higher TID if the pinning implant concentration is much higher than the one at the bottom of the STI in a conventional diode.

In all the pixel types, this TID-induced dark current is pretty uniform as illustrated by the nice Gaussian distribution (from [147]) shown in Figure 11.9b that leads to a uniform increase of the gray level in the dark frame.

The EQE is also degraded by ionizing radiation. The radiation-induced interface states (blue crosses in Figure 11.9c) act as recombination centers for the excess minority carriers generated by the incoming light. The signal carriers generated close to the interfaces are lost by recombination at this interface leading to a decrease of EQE for the shortest wavelengths in FSI devices [140,142] (as shown in Figure 11.9d). In the case of a BSI sensor, a Si-oxide interface exists on both CIS sides (front and back sides). Therefore, the EQE drop may appear at both short and long wavelengths, as illustrated in Figure 11.9d. For the same reasons, charge collection efficiency also drops in TID irradiated MAPS [148]. With regard to the effect of ionizing radiation

---

* It should not happen if the PPD-STI distance is sufficient and if the P-doping concentration between the PPD and the STI is sufficiently high.
[†] In light of recent conclusions on the effect of the EFWC concept on dark current nonlinearities [141], it seems possible that the most intense dark current contributions (the TG-related ones) have been underestimated in [72]. Since I_0 is determined in [72] by subtracting all the other contributions (PPD area, PPD perimeter, TG channel, etc.) to the measured dark current, the reason why I_0 is not null could be a consequence of such underestimation.

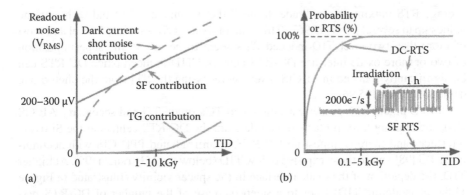

FIGURE 11.10 Radiation effects on CIS noises. (a) Typical evolution with TID of the main temporal noise contributions. This illustration represents a possible dark current shot noise increase assuming a linear dark current evolution with TID. (b) Typical evolution with TID of the number of pixels exhibiting DC-RTS or SF-RTS behaviors. The range at which nearly 100% of the pixel population exhibit DC-RTS can be below 100 Gy for 3T pixels and in the kGy range for PPD pixels.

on the optical properties of the dielectric stack (ILDs, passivation and buffer layers, microlenses, etc.), no clear evidence of degradation has been reported in APS/CIS and it seems that up to 3 kGy(SiO$_2$) microlense properties are not degraded [140]. Similar conclusions have been drawn on the radiation hardness of color filters (but in a more limited TID range [149]).

Ionizing radiation also enhances the noise sources. The SF contribution increases* with TID [71,151] and generally dominates at low TID[†] but the dark current shot noise contribution (that rises as the square root of the dark signal) can quickly become the dominant temporal noise[‡] source at higher dose (as illustrated in Figure 11.10a). Comparable noise degradation mechanisms are observed in MAPSs [146], but the dominant radiation-induced noise sources may be different because MAPS readout circuit architecture differ generally from classical CIS ones (this can also be true in smart-pixel CISs). In PPD CISs, the charge transfer noise [153,154] can start to have a visible contribution at moderate or high TID [71] when the CTE is strongly degraded (as discussed in the next section).

The number of RTS pixels rises as well with the absorbed ionizing radiation dose, as illustrated in Figure 11.10b. Contrary to what was observed in the first DC-RTS studies [85,86,92], TID creates DC-RTS centers[§] very effectively in DSM CIS [95], and nearly 100% of 3T CIS pixels can exhibit an oxide DC-RTS behavior after a TID as low as 100 Gy [96] because of the interface state buildup on the photodiode-depleted perimeter (same origin as the mean dark current increase). As discussed in Section 11.2.2.3, the amplitude of DC-RTS are exponentially distributed (with an

---

* As in any irradiated MOSFET (see for example [150]).
† The SF contribution can be reduced by using a buried channel SF MOSFET [152].
‡ Depending on the integration time and the CVF value.
§ That is, metastable generation centers located at the Si-oxide interface.

average RTS maximum amplitude in the 100 e⁻/s range at 22°C and with extreme values up to several ke⁻/s [95,96]). This type of oxide RTS is mainly bilevel and most of more-than-two-level TID-induced RTS seem to be due to the superimposition of two or more oxide interface DC-RTS centers.* TID-induced oxide DC-RTS can be significantly reduced in 3T CIS pixels by recessing the STI from the photodiode depletion region [95,96].

In PPD pixels with properly accumulated TGs (main TG and secondary AB/GS TG), the pinning layer protects the diode from the DC-RTS centers at the Si-oxide interfaces and there is almost no DC-RTS in unirradiated PPD CIS with accumulated TG [98] and in those exposed to low TID (below the kGy range) [99]. At higher TID, the depletion of the oxide interface in the spacer vicinity (illustrated in Figure 11.8a, at moderate TID) leads to a steep increase of the number of DC-RTS pixels and most of the pixel array suffers from oxide RTS when the absorbed TID approaches the kGy range [99] (despite the TG accumulation). If at least one of the TGs is not accumulated, oxide DC-RTS pixels exist before irradiation [98] and their number starts to rise significantly at a much lower TID than in the accumulated TG case [99,100] because the PPD is not protected from the DC-RTS centers located at the TG channel oxide interfaces [98,99]. With a nonaccumulated TG, a PPD CIS can be saturated by oxide DC-RTS pixels well below the kGy range (as in 3T pixel).

Since SF-RTS is linked to the presence of interface states in the MOSFET channel (at the gate-oxide or STI interface), ionizing radiation should have a strong influence on the number of pixels exhibiting SF-RTS. The few results that can be found on TID-induced SF-RTS in literature show that the SF-RTS pixel creation is hardly noticeable even after 10–20 kGy of TID (no increase observed in [155] and less than 0.3% of SF-RTS pixels created by the radiation exposure in [151]). The very good radiation hardness of DSM-MOSFET gate oxides could explain why very few SF-RTS are created by TID but the reason why SF-RTS coming from the SF channel STI is not visible is still unclear.

11.5.1.2 Pinned Photodiode-Specific Effects

In addition to the cumulative ionizing radiation effects common to all type of APS/CIS, some specific radiation-induced degradations have been observed in PPD-based CISs. The first effect is a right shift of the V_{pin} characteristic, illustrated in Figure 11.11a, caused by the PMD-positive trapped charge [99,141] (represented by the plus sign in the PMD in Figure 11.8a) that acts as a positively biased CCD gate on top of the PPD. This right shift indicates an increase of pinning voltage and EFWC. At high TID (above ≈ 10 kGy(SiO₂)), the pinning layer effective doping is largely reduced by the PMD-positive trapped charge, leading to a PPD capacitance decrease (visible on the V_{pin} curve as a slope diminution) and eventually to the complete depletion of the Si/PMD interface on top of the PPD (as shown in Figure 11.8a) [99,156]. Despite a large degradation of the PPD structure in the kGy(SiO₂) range, the charge partition step (at V_{step}) linked to the TG threshold voltage is not shifted, indicating that the TG threshold voltage is not affected at this TID level. A subthreshold leakage may,

---

* However, some complex multilevel oxide DC-RTS that cannot be explained by the simple sum of independent DC-RTS center contributions have also been reported [95,98].

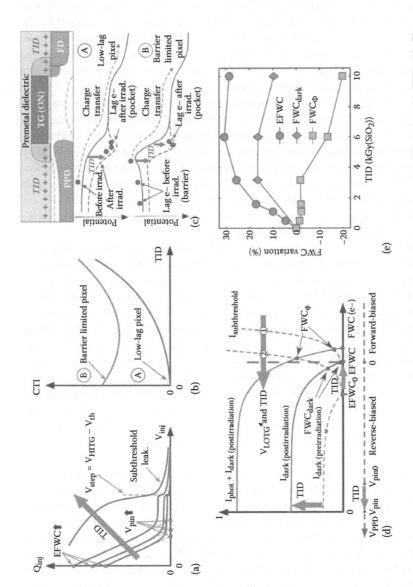

FIGURE 11.11 Pinned photodiode-specific radiation effects. (a) Evolution of the pinning voltage characteristic with TID. (b) Charge transfer inefficiency (CTI = 1-CTE) evolutions with TID of a barrier-limited pixel and a low-lag pixel. (c) Illustration of the CTE degradation mechanism. (Modified from V. Goiffon, M. Estribeau et al., *IEEE Trans. Nucl. Sci.* 61, no. 6, 2014.) (d) Illustration of the FWC degradation mechanism. (e) Measured relative FWC degradations on a 7-μm-pitch 4T PPD CIS. (Data from V. Goiffon, M. Estribeau et al., *IEEE Trans. Nucl. Sci.* 61, no. 6, 2014.)

however, appear on the characteristic at moderate/high TID, as presented in Figure 11.11a.

With regard to the transfer efficiency, several mechanisms can limit the CTE in a CIS: the existence of a potential barrier or pocket in the transition region from the PPD to the TG [157] or a high density of interface states in the channel that can trap the signal carriers [158]. The TID-induced surface potential increase caused by the PMD-positive trapped charge has an influence on the CTE (illustrated in Figure 11.11b and c). The most sensitive part of the pixel for the transfer is right below the TG sidewall spacer because the N-doped PPD region comes closer to the interface (partly because the surface P doping concentration is lower than in the rest of the pinning layer) to ensure a good CTE. In this region, the PMD/spacer positive trapped charge has a much stronger influence on the PPD channel than in the rest of the PPD. In low-lag pixels (i.e., with no limiting potential barrier or pocket before irradiation), this local potential augmentation creates a potential pocket that degrades the CTE (case A in Figure 11b and c). In PPD pixels limited by a potential barrier before irradiation, the exposure to ionizing radiation reduces the potential barrier and leads to a CTE improvement at low TID (case B in Figure 11.11b and c). At high TID, the potential pocket expansion outweighs the benefit of the potential barrier lowering and the CTE is degraded with TID as in case A.

The last PPD-specific effect concerns the FWC variations. We have just discussed that the PMD trapped charge increases the EFWC. Since all the FWC definitions are linked to the EFWC (as mentioned in Section 11.2.2.1 and as can be seen in Figures 11.4c and 11.11d), it should lead to an increase of both the FWC_Φ and the FWC_{dark}. In practice, other mechanisms are competing with the EFWC growth, as depicted in Figure 11.11d. The first one is the increase of TG subthreshold current [143] (mainly due to the TID-induced parasitic STI sidewall channel pictured in Figure 11.8.a and b) that can be represented as a left shift of the dashed I_{subth} curve in Figure 11.11d. It leads to a left shift of the intersections between the I_{subth} curve and the other photo-diode current curves. Since these intersections define the FWC values (as discussed in Section 11.2.2.1), it means that the TG I_{subth} reinforcement causes a reduction of the FWC_Φ and FWC_{dark} values that can compensate some of the EFWC augmentation. It is the main reason why the relative FWC_{dark} increase in Figure 11.11e is weaker than the EFWC growth.

The second phenomenon is the dark current enhancement with TID that leads to a much steeper slope* of the reverse current exponential I-V characteristics (I_{dark} and $I_{phot} + I_{dark}$) near I = 0. Because of this steeper slope, the FWC_Φ intersection is shifted toward the EFWC point (as depicted in Figure 11.4c). Since in most of the cases the FWC_Φ is greater than the EFWC, it means that the large dark current rise with TID leads to an important reduction of the FWC_Φ that can be larger than the EFWC increase and thus to an overall decrease of FWC_Φ with TID. This mechanism seems to be the main cause of the FWC_Φ drop with TID shown in Figure 11.11e and the subthreshold current contribution appears to be secondary in the few experimental conditions (technology, design, and operating conditions) tested in the

---

* That is, a higher first derivative absolute value of the photocurrent (I_{FW} in [63]) near the x-intersect point.

literature [141]. All these graphical interpretations are supported by the analytical model developed in [63,64], and the same conclusions can be drawn by directly using these equations.

11.5.1.3 Radiation-Hardening of CIS Pixels

As mentioned previously, in-pixel MOSFETs (all except the TG) can be radiation-hardened by using classical RHBD techniques. Unfortunately, RHBD for CMOS ICs cannot be used to mitigate the TID effects presented in this section and dedicated techniques have to be used. In APS and CIS based on conventional photodiodes (e.g., 3T pixels), the main issue is the depletion region extension on the photodiode perimeter and the buildup of interface states along this depleted interface. Several techniques have been proposed in the literature [7,9,10,70,139,144,159–161] and all are based on the recess of the isolation oxide to control the potential on the photodiode perimeter by using P+ implants or a polysilicon gate. One possible solution is the gated photodiode layout that is presented in Figure 11.12a in which a polysilicon gate is accumulated to control the potential on the photodiode perimeter. This type of solution generally brings a higher dark current before irradiation but delays the dark current increase with TID and prevents the loss of pixel functionality that can happen if the isolation oxide becomes inverted.

According to the degradation mechanisms presented in Figure 11.8, the main issues in PPD pixels come from the positive trapped charge in the PMD (including the TG spacer) and from the Si/PMD interface. These two degradation sources could possibly be mitigated by placing an electrode on top of the PPD (Figure 11.12b and c) to prevent or compensate the positive charge trapping in the PMD as proposed in [141]. Using an enclosed layout TG has also been proposed to improve the radiation hardness of PPD CIS [162,163]. This latter solution prevents the creation of the parasitic TG subthreshold leakage path and it should also reduce the TID-induced FD leakage. Unfortunately, this solution is inefficient [141] against the main degradation source discussed here: the PMD/spacer trapped charge.

Some radiation-hardening-by-process techniques have also been reported in the literature to improve the radiation hardness of 3T pixels [144], such as the use of a custom surface P+ pinning implant [13] (leading to a partially pinned photodiode

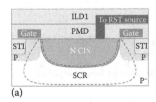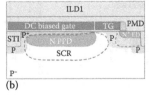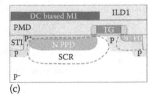

FIGURE 11.12 Photodiode RHBD technique examples. (a) One of the several proven solutions for conventional photodiodes: the gated photodiode layout. (b) A recently proposed solution for BSI PPD CIS (From V. Goiffon, M. Estribeau et al., *IEEE Trans. Nucl. Sci.*, vol. 61, no. 6, Dec. 2014.): the use of a DC biased polysilicon gate on top of the PPD to mitigate the PMD trapped charge. (c) Same solution but with a metallic gate instead of the polysilicon gate.

structure), but these custom process steps may not be available in the targeted manufacturing process. For PPD CIS, the hole-based PPD [158,164,165] (also called
PMOS PPD or P-channel PPD) seems to be a promising radiation-hardening-by-
process solution but very few data has been published [164] on this technology
behavior when exposed to ionizing radiation.

11.5.2 DISPLACEMENT DAMAGE EFFECTS

11.5.2.1 Overview

Nonionizing interactions create SRH centers (point defects or cluster of defects) in
the silicon bulk [24]. In APS and CIS, these defects are mainly active in the photodiode vicinity, as pictured in Figure 11.13a. They act as SRH generation centers in
depleted regions leading to a growth of the bulk generation dark current (Equation
11.2) discussed more in detail hereafter. These defects also act as recombination
centers that reduce the recombination lifetime of photogenerated minority carriers

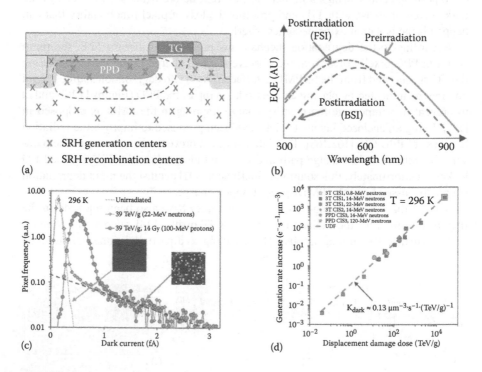

FIGURE 11.13 Overview of displacement damage in CIS. (a) Illustration of the main active
defects induced by displacement damage in a CIS pixel. (b) Illustration of the expected quantum efficiency degradations at high displacement damage dose. (c) Dark current distribution
measured on a 3T pixel CIS exposed to neutron and protons. (Modified from C. Virmontois,
V. Goiffon et al., *IEEE Trans. Nucl. Sci.* 57, 2010.) (d) Measured mean generation rate increase
versus D_d for several CIS design and technologies exposed to different particles. (Modified
from C. Virmontois, V. Goiffon et al., *IEEE Trans. Nucl. Sci.* 59, 2012, 927.)

[166,167]. When signal electrons are generated near the collecting depletion region, they have very little chance to recombine before being collected. On the other hand, electrons generated far from the PN junction must diffuse over a large distance before reaching the collecting well. Therefore, in FSI CIS, displacement damage mainly impacts the EQE at long wavelengths, whereas in the case of BSI CIS all the wavelengths can suffer from an EQE reduction due to displacement damage (with a higher effect at short wavelengths), as illustrated in Figure 11.13b. Indeed, in BSI devices, the majority of signal carriers are generated near the back interface (whatever the photon wavelength) far from the depletion region. Carrier lifetime reduction is expected for displacement damage dose beyond 40 TeV/g ($\approx 2 \times 10^{10}$ cm^{-2} 1-MeV neutron equivalent fluence and more than 1×10^{10} cm^{-2} of 50-MeV protons [167]). However such an effect is rarely reported in CIS, even after several PeV/g of absorbed dose, because the effective recombination lifetime is generally limited by the surface recombination or by the thin thickness of image sensors optimized for visible wavelengths (typically a few microns). On the other hand, charge collection efficiency drops are typically observed on MAPSs after 200 TeV/g (1×10^{11} cm^{-2} 1-MeV equivalent fluence) [9,121,168,169] because their sensitive layer is often thicker than in CISs. At even higher D_d, change of effective doping concentration may occur (as suggested in [170] to justify the FWC reduction*) because of carrier removal and type inversion can be [171], but the particle fluence necessary to see these effects on APS/CIS are rarely reached.

Finally, these SRH centers can also act as bulk traps able to capture a signal carrier and to release it later. But contrary to CCDs [58], D_d-induced CTE degradation due to bulk trapping in CIS has never been reported. This is mainly due to the limited number of charge transfers necessary to read the signal charge (none in 3T pixels and one in PPD CISs) compared to CCDs. Capture and delayed emission could possibly occur in the quasi-neutral silicon bulk far from the junction, but such phenomena are not likely to be visible in classical slowly varying illumination conditions used to characterize CISs.

11.5.2.2 Dark Current, Dark Current Nonuniformity, and RTS

As for TID, the main D_d effect is the dark current augmentation. Figure 11.13c shows a typical dark current distribution after exposure to nonionizing radiation. Unlike TID, D_d leads to strong nonuniformities that appear as a hot pixel tail.[†] For a given dose (and for NIEL > 10^{-4} MeV cm^2/g), the mean dark current increase induced by nonionizing radiation in a depleted silicon volume V_{dep} can be determined by using the Srour universal damage factor (UDF) [172]:

$$\Delta I_{dark} = K_{dark} \times V_{dep} \times D_d \qquad (11.5)$$

---

* However, this FWC reduction could be due to the dark current increase, as discussed in Section 11.5.1.2.

† It is interesting to notice in Figure 11.13c that charged particle irradiation (such as protons) leads to both TID and D_d effects compared to neutron that mainly induce displacement damage effects.

The validity of this equation has been verified several times in modern CISs (as illustrated in Figure 11.13d) where the optimized manufacturing process limits the high electric field regions [94,138,140,173,174]. This is generally not the case in APSs [175] and MAPSs [169] where EFE [176] appears in parasitic high electric field regions because of the use of high-voltage or nonoptimized doping profiles. In this case, the measured value is supposed to be greater than the value predicted by the UDF (because of the EFE). In a CIS with negligible EFE, the UDF is a very useful tool to anticipate the average displacement damage degradation or to estimate the depletion volume of an irradiated sensor.

Several modeling approaches have also been proposed to describe the whole D_d-induced dark current nonuniformity (DCNU) distributions.* Most are based on physical modeling starting from the nonionizing interaction (usually computed using Monte Carlo simulation tools such as GEANT4†) to the calculation of displacement damage dose deposited in each microvolume (i.e., pixel-depleted volume) [173,177–182]. When these models manage to reproduce the observed distribution, they bring very valuable information on the physical process underlying the displacement-damage-induced dark current distribution. However, most of recently reported D_d-induced dark current distributions in CIS (with no EFE) exhibit a clear exponential hot pixel tail that appears difficult to match with physical models.

Some empirical approaches have also been proposed [137,174,183]. Such models are generally used to perform quick analysis, interpolation, or prediction of DCNU histograms (without the use of Monte Carlo simulations). One of these approaches [174] suggests that the exponential distribution tail created by the nonionizing interactions may also be a function of the D_d only, whatever the particle type (as the mean dark current). First, based on experimental observation, it is assumed that the probability density function (PDF) of the dark current increase created by a nonionizing interaction in a sensitive volume follows this exponential function:

$$f(\Delta I_{\text{dark}}) = \frac{1}{\upsilon_{\text{dark}}} \exp\left(\frac{-\Delta I_{\text{dark}}}{\upsilon_{\text{dark}}}\right) \tag{11.6}$$

with υ_{dark} a factor that gives the average dark current increase generated by a nonionizing interaction. In [174], a second factor γ_{dark} is used to scale the PDF to the measured DCNU histogram for a given D_d. Similar to the UDF, υ_{dark} and γ_{dark} seemed independent of particle type, sensor design, or technology in the experimental conditions tested in [174]. If this elementary empirical model is validated with broader experimental conditions, it will represent a very convenient tool to predict the D_d-induced DCNU in any CIS (without EFE) but it will raise new questions about the physics underlying this common exponential behavior (that could be answered by physical modeling approaches).

As for TID-induced interface generation centers, some of the generation centers created by nonionizing interactions exhibit an RTS behavior in irradiated CCDs,

---

* That leads to dark signal nonuniformity (DSNU) on a dark frame captured with a fixed integration time.
† Tabulated values coming from Monte Carlo simulations are often used, when available, to avoid performing a new Monte Carlo computation each time a DCNU histogram is calculated.

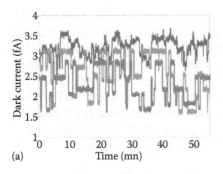

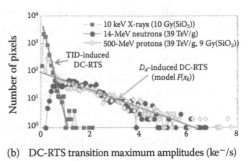

(a)　Time (mn)　　(b)　DC-RTS transition maximum amplitudes (ke⁻/s)

FIGURE 11.14　Radiation-induced DC-RTS at 296K. (a) Typical multilevel DC-RTS created by displacement damage. (b) DC-RTS maximum transition amplitude distributions measured on 3T pixel CIS sensors exposed to X-rays, neutrons, and protons. (Modified from C. Virmontois, V. Goiffon et al., *IEEE Trans. Nucl. Sci.* 58, 2011, 3085.) The TID-induced DC-RTS and the D_d-induced DC-RTS populations exhibit different exponential slopes (typically 100 e⁻/s for TID DC-RTS and 1200 e⁻/s for D_d DC-RTS at 296K).

MAPSs, and CISs [85–94,96,97,100], as presented in Figure 11.14a. Their amplitudes are generally much larger than the TID DC-RTS amplitudes and they are most of the time multilevel (more frequently than the probability to have several bilevel DC-RTS centers in a single pixel). EFE is not likely to be the cause of these very high amplitudes [94] since they are also observed in modern CISs where EFE is very infrequent. These high amplitudes and multilevel behaviors are probably related to defect clusters as discussed in [184].

D_d-induced DC-RTS maximum transition amplitude histograms are also exponentially distributed (as shown in Figure 11.14b) and they can be represented by the following discrete function [94,97]:

$$F(x_k) = \frac{BN_{pix}D_d V_{dep} K_{RTS}}{A_{RTS}} \exp\left(\frac{-x_k}{A_{RTS}}\right)$$

(11.7)

with $x_k = k \times B$ and $k \in N$

A_{RTS} is the average maximum transition amplitude, B is the histogram bin width (same unit as A_{RTS}), N_{pix} is the total number of pixels in the tested sensor, and K_{RTS} is a damage factor that gives the number of generated DC-RTS centers per unit of depleted volume and of D_d. The number of DC-RTS pixels can be estimated by the product $N_{pix} \times D_d \times V_{dep} \times K_{RTS}$.* A_{RTS} and K_{RTS} factors also seem to be independent of design, technology, and particle type with the following typical values in CCDs and CISs [94,96,97,185] at room temperature (≈23°C): $A_{RTS} \approx 1200$ e⁻/s and $K_{RTS} \approx$ 30–35 centers·cm⁻³·(MeV/g)⁻¹. As oxide DC-RTS, bulk DC-RTS is temperature-activated (amplitude and time constants [85,87,88,92]) and the typical amplitude activation energy in modern CIS is about 0.63 eV [94,96], which corresponds to the expected value for a midgap SRH generation center in absence of EFE.

---

* If at a given dose $N_{pix} \times D_d \times V_{dep} \times K_{RTS} > N_{pix}$, it means that all the pixels exhibit a DC-RTS behavior.

Some small deviations to this pure exponential distribution have also been reported [185,186] and need to be further studied, but this simple model generally provides a good *a priori* prediction of the amplitude distributions. Figure 11.14b illustrates how this approach can be used to separate the TID and D_d-induced RTS pixels populations [96].

It is likely that other DC-RTS parameters (such as time constants, number of levels, etc.) may be used to model or investigate further the origin of DC-RTS, but there is still a long way to go to unravel all the RTS mysteries: notably the complex multilevel behaviors [91], the evolution during annealing [90,96], the erratic transient behavior of DC-RTS centers [187], and their intense switching amplitudes.

11.6 CONCLUSION

Whereas CIS peripheral circuits benefit from the intrinsically good radiation hardness of CMOS circuits, that can be extended well beyond the 10-kGy range by using RHBD techniques, CIS and APS pixels are still sensitive to cumulative radiation effects (ionizing and nonionizing) and SEEs. TID- and D_d-induced dark current increases remain the major issues and the existing techniques never manage to completely mitigate them. With the advent of the pinned photodiode in sensors meant to be used in radiation environments, radiation effects new to CISs are appearing, such as charge transfer efficiency and full-well-capacity variations. These can lead to pixel functionality loss at a lower dose than a conventional photodiode sensor. Since image sensors are made to collect photogenerated charges, pixel SETs will always be a source of image quality degradation in high-radiation flux applications, but here again, trade-off can be made at the design level to reduce the radiation sensitivity. SEEs outside the pixel array have not been a major issue in the past, but with the growing complexity of CISs and the increasing number of on-chip integrated functions, this topic may get more interest in the near future. Radiation-induced RTSs, both TID- and D_d-induced ones, will also have a growing importance with the constant improvement of CIS sensitivity (lower dark current, lower noise, and higher gain), making these devices sensitive to very few parasitic electrons.

In conclusion, compared to CCDs (and even charge-injection devices), CISs remain the technology of choice for imaging applications in harsh radiation environments but there is still much work to do to fully understand, model, and mitigate the radiation effects on these imagers to push the limits of their radiation hardness well beyond what they can stand today.

REFERENCES

1. E.R. Fossum, "Active pixel sensors: Are CCD's dinosaurs?," *Proc SPIE*, vol. 1900, pp. 1–14, 1993.
2. E.R. Fossum, "CMOS image sensors: Electronic camera-on-a-chip," *IEEE Trans. Electron Devices*, vol. 44, no. 10, pp. 1689–1698, Oct. 1997.
3. E.R. Fossum, "Camera-on-a-chip: Technology transfer from saturn to your cell phone," *Technol. Innov.*, vol. 15, no. 3, pp. 197–209, Dec. 2013.
4. S. Mendis, E.R. Fossum, *CMOS Active Pixel Image Sensor*, Jul. 1993.
5. Yole Développement, *Status of the CMOS Image Sensors Industry*, 2014.
6. IC Insights, *OSD Report*, 2014.

7. B.R. Hancock, G.A. Soli, "Total dose testing of a CMOS charged particle spectrometer," *IEEE Trans. Nucl. Sci.* vol. 44, no. 6, pp. 1957–1964, Dec. 1997.

8. E.-S. Eid, T.Y. Chan et al., "Design and characterization of ionizing radiation-tolerant CMOS APS image sensors up to 30 Mrd (Si) total dose," *IEEE Trans. Nucl. Sci.* vol. 48, no. 6, pp. 1796–1806, Dec. 2001.

9. W. Dulinski, G. Deptuch et al., "Radiation hardness study of an APS CMOS particle tracker," in *IEEE Nucl. Sci. Symp. Conf. Rec.*, vol. 1, pp. 100–103, 2001.

10. B.R. Hancock, T.J. Cunningham et al., "Multi-megarad (Si) radiation-tolerant integrated CMOS imager," in *Proc. SPIE*, vol. 4306, pp. 147–155, 2001.

11. M. Cohen, J.-P. David, "Radiation-induced dark current in CMOS active pixel sensors," *IEEE Trans. Nucl. Sci.* vol. 47, no. 6, pp. 2485–2491, Dec. 2000.

12. G.R. Hopkinson, "Radiation effects in a CMOS active pixel sensor," *IEEE Trans. Nucl. Sci.* vol. 47, no. 6, pp. 2480–2484, Dec. 2000.

13. J. Bogaerts, B. Dierickx et al., "Total dose and displacement damage effects in a radiation-hardened CMOS APS," *IEEE Trans. Electron Devices* vol. 47, no. 6, pp. 2480–2484, Dec. 2000.

14. R. Turchetta, J.D. Berst et al., "A monolithic active pixel sensor for charged particle tracking and imaging using standard VLSI CMOS technology," *Nucl. Instr. Meth. A* vol. 458, no. 3, pp. 677–689, 2001.

15. G. Deptuch, J.D. Berst et al., "Design and testing of monolithic active pixel sensors for charged particle tracking," in *Nucl. Sci. Symp. Conf. Rec.*, vol. 1, pp. 3/103–3/110, 2000.

16. M. Baze, "Single Event Effects in Digital and Linear ICs," in *IEEE NSREC Short Course*, 2011.

17. T.P. Ma, P.V. Dressendorfer, *Ionizing Radiation Effects in MOS Devices and Circuits*, Wiley-Interscience, New York, 1989.

18. T.R. Oldham, F.B. McLean, "Total ionizing dose effects in MOS oxides and devices," *IEEE Trans. Nucl. Sci.* vol. 50, pp. 483–499, Jun. 2003.

19. J.R. Schwank, M.R. Shaneyfelt et al., "Radiation effects in MOS oxides," *IEEE Trans. Nucl. Sci.* vol. 55, no. 4, pp. 1833–1853, Aug. 2008.

20. P.E. Dodd, M.R. Shaneyfelt et al., "Current and future challenges in radiation effects on CMOS electronics," *IEEE Trans. Nucl. Sci.* vol. 57, no. 4, pp. 1747–1763, Aug. 2010.

21. D.M. Fleetwood, "Total ionizing dose effects in MOS and low-dose-rate-sensitive linear-bipolar devices," *IEEE Trans. Nucl. Sci.* vol. 60, no. 3, pp. 1706–1730, Jun. 2013.

22. H.J. Barnaby, "Total-ionizing-dose effects in modern CMOS technologies," *IEEE Trans. Nucl. Sci.* vol. 53, no. 6, pp. 3103–3121, Dec. 2006.

23. G.C. Messenger, "A summary review of displacement damage from high energy radiation in silicon semiconductors and semiconductor devices," *IEEE Trans. Nucl. Sci.* vol. 39, no. 3, pp. 468–473, Jun. 1992.

24. J.R. Srour, C.J. Marshall et al., "Review of displacement damage effects in silicon devices," *IEEE Trans. Nucl. Sci.* vol. 50, no. 3, pp. 653–670, Jun. 2003.

25. J.R. Srour, J.W. Palko, "A framework for understanding displacement damage mechanisms in irradiated silicon devices," *Nucl. Sci. IEEE Trans. On* vol. 53, no. 6, pp. 3610–3620, Dec. 2006.

26. J.R. Srour, J.W. Palko, "Displacement damage effects in irradiated semiconductor devices," *IEEE Trans. Nucl. Sci.* vol. 60, no. 3, pp. 1740–1766, Jun. 2013.

27. D. Durini, D. Arutinov, "Operational principles of silicon image sensors," in *High Performance Silicon Imaging: Fundamentals and Applications of CMOS and CCD Sensors*, D. Durini, ed., Woodhead Publishing, Cambridge, UK, pp. 25–77, 2014.

28. A.J.P. Theuwissen, "CMOS image sensors: State-of-the-art," *Solid-State Electron* vol. 52, no. 9, pp. 1401–1406, Sept. 2008.

29. O. Yadid-Pecht, R. Etienne-Cummings, eds., *CMOS Imagers: From Phototransduction to Image Processing*, Springer Science & Business Media, New York, 2004.

30. J. Ohta, *Smart CMOS Image Sensors and Applications*, CRC Press, Boca Raton, FL, 2007.
31. A. El Gamal, H. Eltoukhy, "CMOS image sensors," *IEEE Circuits Devices Mag.* vol. 21, no. 3, pp. 6–20, May 2005.
32. R.J. Gove, "Complementary metal-oxide-semiconductor (CMOS) image sensors for mobile devices," in *High Performance Silicon Imaging*, pp. 191–234, 2014.
33. B. Choubey, W. Mughal et al., "5—Circuits for high performance complementary metal-oxide-semiconductor (CMOS) image sensors," in *High Performance Silicon Imaging: Fundamentals and Applications of CMOS and CCD Sensors*, D. Durini, ed., Woodhead Publishing, Cambridge, UK, pp. 124–164, 2014.
34. J.A. Leñero-Bardallo, J. Fernández-Berni et al., "Review of ADCs for imaging," vol. 9022, p. 90220I–90220I-6, 2014.
35. G. Casse, "Recent developments on silicon detectors," *Nucl. Instrum. Methods Phys. Res. Sect. Accel. Spectrometers Detect. Assoc. Equip.* vol. 732, pp. 16–20, Dec. 2013.
36. R. Fontaine, "The evolution of pixel structures for consumer-grade image sensors," *IEEE Trans. Semicond. Manuf.* vol. 26, no. 1, pp. 11–16, Feb. 2013.
37. R. Fontaine, "A review of the 1.4 μm pixel generation," in *Proc. Int. Image Sens. Workshop IISW*, 2011.
38. R. Fontaine, "Trends in consumer CMOS image sensor manufacturing," in *Proc. Int. Image Sens. Workshop IISW*, 2009.
39. A. Lahav, A. Fenigstein et al., "Backside illuminated (BSI) complementary metal-oxide-semiconductor (CMOS) image sensors," in *High Performance Silicon Imaging: Fundamentals and Applications of CMOS and CCD Sensors*, D. Durini, ed., Woodhead Publishing, Cambridge, UK, pp. 98–123, 2014.
40. S. Wolf, *Silicon Processing for the VLSI Era, Volume 4: Deep-Submicron Process Technology*, vol. 4. Lattice Press, California, 2002.
41. J. Gambino, B. Leidy et al., "CMOS imager with copper wiring and lightpipe," in *IEDM Tech. Dig.*, pp. 1–4, 2006.
42. G. Agranov, R. Mauritzson et al., "Pixel continues to shrink, pixel development for novel CMOS image sensors," in *Proc. 2009 Int. Image Sens. Workshop*, pp. 58–61, 2009.
43. B.J. Park, J. Jung et al., "Deep trench isolation for crosstalk suppression in active pixel sensors with 1.7 μm pixel pitch," *Jpn. J. Appl. Phys.* vol. 46, no. 4S, p. 2454, Apr. 2007.
44. B. Pain, "Backside illumination technology for SOI-CMOS image sensors," in *Symp. Backside Illum. Solid-State Image Sens.*, 2009.
45. S. Wuu, "BSI technology with bulk Si wafer," in *Symp. Backside Illum. Solid-State Image Sens.*, 2009.
46. H. Rhodes, "Mass production of BSI image sensors: Performance results," in *Symp. Backside Illum. Solid-State Image Sens.*, 2009.
47. N. Teranishi, A. Kohno et al., "An interline CCD image sensor with reduced image lag," *IEEE Trans. Electron Devices* vol. 31, no. 12, pp. 1829–1833, Dec. 1984.
48. B.C. Burkey, W.C. Chang et al., "The pinned photodiode for an interline-transfer CCD image sensor," in *IEDM Tech. Dig.*, pp. 28–31, 1984.
49. P. Lee, R. Gee et al., "An active pixel sensor fabricated using CMOS/CCD process technology," in *Proc IEEE Workshop CCDs Adv. Image Sens.*, pp. 115–119, 1995.
50. E.R. Fossum, D.B. Hondongwa, "A review of the pinned photodiode for CCD and CMOS image sensors," *IEEE J. Electron Devices Soc.* 2014.
51. J.A. Ballin, J.P. Crooks et al., "Monolithic active pixel sensors (MAPS) in a quadruple well technology for nearly 100% fill factor and full CMOS pixels," *Sensors* vol. 8, no. 9, pp. 5336–5351, Sep. 2008.
52. S. Zucca, L. Ratti et al., "A quadruple well CMOS MAPS prototype for the layer0 of the SuperB SVT," *Nucl. Instrum. Methods Phys. Res. Sect. Accel. Spectrometers Detect. Assoc. Equip.* vol. 718, pp. 380–382, Aug. 2013.

53. B. Fowler, C. Liu et al., "A 5.5 Mpixel 100 frames/sec wide dynamic range low noise CMOS image sensor for scientific applications," in *Proc. SPIE*, p. 753607, 2010.
54. J. Ohta, *Smart CMOS image sensors and applications*, CRC Press, 2007.
55. A. Moini, *Vision Chips*, Kluwer Academic Publishers, 2000.
56. A. El Gamal, D.X.D. Yang et al., "Pixel-level processing: Why, what, and how?," in *Proc. SPIE*, vol. 3650, pp. 2–13, 1999.
57. G.R. Hopkinson, T.M. Goodman et al., *A Guide to the Use and Calibration of Detector Array Equipment*, SPIE, 2004.
58. J.R. Janesick, *Scientific Charge-Coupled Devices*, SPIE, Bellingham, 2001.
59. A.J.P. Theuwissen, *Solid-State Imaging with Charge-Coupled Devices*, Kluwer Academic, 1995.
60. V. Goiffon, M. Estribeau et al., "Pixel level characterization of pinned photodiode and transfer gate physical parameters in CMOS image sensors," *IEEE J. Electron. Devices Soc.* vol. 2, no. 4, pp. 65–76, Jul. 2014.
61. A. Krymski, N. Bock et al., "Estimates for scaling of pinned photodiodes," in *IEEE Workshop CCD Adv. Image Sens.*, 2005.
62. J. Tan, B. Buttgen et al., "Analyzing the radiation degradation of 4-transistor deep submicron technology CMOS image sensors," *IEEE Sens. J.* vol. 12, no. 6, pp. 2278–2286, June 2012.
63. A. Pelamatti, V. Goiffon et al., "Estimation and modeling of the full well capacity in pinned photodiode CMOS image sensors," *IEEE Electron. Device Lett.* vol. 34, no. 7, pp. 900–902, Jun. 2013.
64. A. Pelamatti, J.-M. Belloir et al., "Temperature dependence and dynamic behaviour of full well capacity in pinned phototiode CMOS image sensors," *IEEE Trans. Electron Devices* 2015.
65. G. Meynants, "Global shutter pixels with correlated double sampling for CMOS image sensors," *Adv. Opt. Technol.* vol. 2, no. 2, pp. 177–187, 2013.
66. S.M. Sze, *Physics of Semiconductor Devices*, 2nd ed., Wiley, New York, 1981.
67. W. Shockley, W.T. Read, "Statistics of the recombination of holes and electrons," *Phys. Rev.* vol. 87, pp. 835–842, 1952.
68. R.N. Hall, "Electron-hole recombination in germanium," *Phys. Rev. B* vol. 87, Jul. 1952.
69. A.S. Grove, *Physics and Technology of Semiconductor Devices*, Wiley International, 1967.
70. V. Goiffon, P. Cervantes et al., "Generic radiation hardened photodiode layouts for deep submicron CMOS image sensor processes," *IEEE Trans. Nucl. Sci.* vol. 58, no. 6, pp. 3076–3084, Dec. 2011.
71. S. Place, J.-P. Carrere et al., "Radiation effects on CMOS image sensors with sub-2um pinned photodiodes," *IEEE Trans. Nucl. Sci.* vol. 59, no. 4, pp. 909–917, Aug. 2012.
72. V. Goiffon, C. Virmontois et al., "Identification of Radiation induced dark current sources in pinned photodiode CMOS image sensors," *IEEE Trans. Nucl. Sci.* vol. 59, no. 4, pp. 918–926, Aug. 2012.
73. S.W. Benson, *The Foundations of Chemical Kinetics*, McGraw-Hill Education, 1960.
74. F. Domengie, J.L. Regolini et al., "Study of metal contamination in CMOS image sensors by dark-current and deep-level transient spectroscopies," *J. Electron. Mater.* vol. 39, no. 6, pp. 625–629, Jun. 2010.
75. G. Meynants, W. Diels et al., "Emission microscopy analysis of hot cluster defects of imagers processed," presented at the *International Image Sensor Workshop* (IISW), 2013.
76. A. Papoulis, S.U. Pillai, *Probability, Random Variables, and Stochastic Processes*, 4th ed., McGraw Hill, 2002.
77. M.J. Kirton, M.J. Uren, "Noise in solid-state microstructures: A new perspective on individual defects, interface states and low-frequency (1/f) noise," *Adv. Phys.* vol. 38, no. 4, pp. 367–468, 1989.

78. Y. Mori, K. Takeda et al., "Random telegraph noise of junction leakage current in submicron devices," *J. Appl. Phys.* vol. 107, no. 1, p. 014509, Jan. 2010.
79. D.S. Yaney, C.Y. Lu et al., "A meta-stable leakage phenomenon in DRAM charge storage—Variable hold time," in *IEDM Tech. Dig.*, pp. 336–339, 1987.
80. P.J. Restle, J.W. Park et al., "DRAM variable retention time," in *IEDM Tech. Dig.*, pp. 807–810, 1992.
81. Y. Mori, K. Ohyu et al., "The origin of variable retention time in DRAM," in *IEDM Tech. Dig.*, IEEE, pp. 1034–1037, 2005.
82. J.R. Srour, R.A. Hartmann et al., "Permanent damage produced by single proton interactions in silicon devices," *IEEE Trans. Nucl. Sci.* vol. 33, no. 6, pp. 1597–1604, Dec. 1986.
83. P.W. Marshall, C.J. Dale et al., "Displacement damage extremes in silicon depletion regions," *IEEE Trans. Nucl. Sci.* vol. 36, no. 6, pp. 1831–1839, Dec. 1989.
84. G.R. Hopkinson, "Cobalt60 and proton radiation effects on large format, 2-D, CCD arrays for an Earth imaging application," *IEEE Trans. Nucl. Sci.* vol. 39, no. 6, pp. 2018–2025, Dec. 1992.
85. I.H. Hopkins, G.R. Hopkinson, "Random telegraph signals from proton-irradiated CCDs," *IEEE Trans. Nucl. Sci.* vol. 40, no. 6, pp. 1567–1574, Dec. 1993.
86. I.H. Hopkins, G.R. Hopkinson, "Further measurements of random telegraph signals in proton-irradiated CCDs," *IEEE Trans. Nucl. Sci.* vol. 42, no. 6, pp. 2074–2081, 1995.
87. A.M. Chugg, R. Jones et al., "Single particle dark current spikes induced in CCDs by high energy neutrons," *IEEE Trans. Nucl. Sci.* vol. 50, no. 6, pp. 2011–2017, 2003.
88. D.R. Smith, A.D. Holland et al., "Random telegraph signals in charge coupled devices," *Nucl. Instr. Meth. A* vol. 530, no. 3, pp. 521–535, Sep. 2004.
89. T. Nuns, G. Quadri et al., "Measurements of random telegraph signal in CCDs irradiated with protons and neutrons," *IEEE Trans. Nucl. Sci.* vol. 53, no. 4, pp. 1764–1771, Aug. 2006.
90. T. Nuns, G. Quadri et al., "Annealing of proton-induced random telegraph signal in CCDs," *IEEE Trans. Nucl. Sci.* vol. 54, no. 4, pp. 1120–1128, 2007.
91. G.R. Hopkinson, V. Goiffon et al., "Random telegraph signals in proton irradiated CCDs and APS," *IEEE Trans. Nucl. Sci.* vol. 55, no. 4, Aug. 2008.
92. J. Bogaerts, B. Dierickx et al., "Random telegraph signals in a radiation-hardened CMOS active pixel sensors," *IEEE Trans. Nucl. Sci.* vol. 49, pp. 249–257, 2002.
93. M. Deveaux, S. Amar-Youcef et al., "Random telegraph signal in monolithic active pixel sensors," in *Nucl. Sci. Symp. Conf. Rec., IEEE*, pp. 3098–3105, 2008.
94. V. Goiffon, G.R. Hopkinson et al., "Multilevel RTS in proton irradiated CMOS image sensors manufactured in a deep submicron technology," *IEEE Trans. Nucl. Sci.* vol. 56, no. 4, pp. 2132–2141, Aug. 2009.
95. V. Goiffon, P. Magnan et al., "Evidence of a novel source of random telegraph signal in CMOS image sensors," *IEEE Electron. Device Lett.* vol. 32, no. 6, pp. 773–775, Jun. 2011.
96. C. Virmontois, V. Goiffon et al., "Total ionizing dose versus displacement damage dose induced dark current random telegraph signals in CMOS image sensors," *IEEE Trans. Nucl. Sci.* vol. 58, no. 6, pp. 3085–3094, Dec. 2011.
97. C. Virmontois, V. Goiffon et al., "Dark current random telegraph signals in solid-state image sensors," *IEEE Trans. Nucl. Sci.* vol. 60, no. 6, Dec. 2013.
98. V. Goiffon, C. Virmontois et al., "Investigation of dark current random telegraph signal in pinned PhotoDiode CMOS image sensors," in *IEDM Tech. Dig.*, pp. 8.4.1–8.4.4, 2011.
99. V. Goiffon, M. Estribeau et al., "Radiation effects in pinned photodiode CMOS image sensors: Pixel performance degradation due to total ionizing dose," *IEEE Trans. Nucl. Sci.* vol. 59, no. 6, pp. 2878–2887, Dec. 2012.
100. E. Martin, T. Nuns et al., "Proton and -rays irradiation-induced dark current random telegraph signal in a 0.18-CMOS image sensor," *IEEE Trans. Nucl. Sci.* vol. 60, no. 4, pp. 2503–2510, Aug. 2013.

101. C. Leyris, F. Martinez et al., "Impact of random telegraph signal in CMOS image sensors for low-light levels," in *Proc ESSCIRC*, pp. 376–379, 2006.
102. X. Wang, P.R. Rao et al., "Random telegraph signal in CMOS image sensor pixels," in *IEDM Tech. Dig.*, pp. 1–4, 2006.
103. R.-V. Wang, Y.-H. Lee et al., "Shallow trench isolation edge effect on random telegraph signal noise and implications for flash memory," *IEEE Trans. Electron Devices* vol. 56, no. 9, pp. 2107–2113, Sep. 2009.
104. P. Vu, B. Fowler et al., "Evaluation of 10MeV proton irradiation on 5.5 Mpixel scientific CMOS image sensor," in *Proc SPIE*, vol. 7826, 2010.
105. V. Goiffon, S. Girard et al., "Vulnerability of CMOS image sensors in megajoule class laser harsh environment," *Opt. Express* vol. 20, no. 18, pp. 20028–20042, Aug. 2012.
106. V. Lalucaa, V. Goiffon et al., "Single event effects in 4T pinned photodiode image sensors," *IEEE Trans. Nucl. Sci.* vol. 60, no. 6, Dec. 2013.
107. V. Lalucaa, V. Goiffon et al., "Single-event effects in CMOS image sensors," *IEEE Trans. Nucl. Sci.* vol. 60, no. 4, pp. 2494–2502, Aug. 2013.
108. M. Beaumel, D. Hervé et al., "Proton, electron, and heavy ion single event effects on the HAS2 CMOS image sensor," *IEEE Trans. Nucl. Sci.* vol. 61, no. 4, pp. 1909–1917, Aug. 2014.
109. C. Virmontois, A. Toulemont et al., "Radiation-induced dose and single event effects on digital CMOS image sensors," *IEEE Trans. Nucl. Sci.* vol. 61, no. 6, Dec. 2014.
110. L. Gomez Rojas, M. Chang et al., "Radiation effects in the LUPA4000 CMOS image sensor for space applications," in *Proc RADECS*, pp. 800–805, 2011.
111. F. Kastensmidt, "SEE mitigation strategies for digital circuit design applicable to ASIC and FPGAs, in *IEEE NSREC Short Course*, 2007.
112. R.C. Lacoe, "Improving integrated circuit performance through the application of hardness-by-design methodology," *IEEE Trans. Nucl. Sci.* vol. 55, no. 4, pp. 1903–1925, Aug. 2008.
113. H.L. Hughes, J.M. Benedetto, "Radiation effects and hardening of MOS technology: devices and circuits," *IEEE Trans. Nucl. Sci.* vol. 50, pp. 500–501, Jun. 2003.
114. F. Faccio, "Design hardening methodologies for ASICs," in *Radiat. Eff. Embed. Syst.*, Springer Netherlands, pp. 143–160, 2007.
115. T.S. Lomheim, R.M. Shima et al., "Imaging charge-coupled device (CCD) transient response to 17 and 50 MeV proton and heavy-ion irradiation," *IEEE Trans. Nucl. Sci.* vol. 37, no. 6, pp. 1876–1885, Dec. 1990.
116. C.J. Marshall, K.. LaBel et al., "Heavy ion transient characterization of a hardened-by-design active pixel sensor array," in *2002 IEEE Radiat. Eff. Data Workshop*, pp. 187–193, 2002.
117. J. Baggio, M. Martinez et al., "Analysis of transient effects induced by neutrons on a CCD image sensor," in *Proc. SPIE*, vol. 4547, pp. 105–115, 2002.
118. J.C. Pickel, R.A. Reed et al., "Radiation-induced charge collection in infrared detector arrays," *IEEE Trans. Nucl. Sci.* vol. 49, no. 6, pp. 2822–2829, Dec. 2002.
119. M. Raine, V. Goiffon et al., "Modeling approach for the prediction of transient and permanent degradations of image sensors in complex radiation environments," *IEEE Trans. Nucl. Sci.* vol. 60, no. 6, Dec. 2013.
120. G. Rolland, L. Pinheiro da Silva et al., "STARDUST: A code for the simulation of particle tracks on arrays of sensitive volumes with substrate diffusion currents," *IEEE Trans. Nucl. Sci.* vol. 55, no. 4, pp. 2070–2078, Aug. 2008.
121. L. Ratti, L. Gaioni et al., "Modeling charge loss in CMOS MAPS exposed to non-ionizing radiation," *IEEE Trans. Nucl. Sci.* vol. 60, no. 4, pp. 2574–2582, Aug. 2013.
122. G.J. Yates, B.T. Turko, "Circumvention of radiation-induced noise in CCD and CID imagers," *IEEE Trans. Nucl. Sci.* vol. 36, no. 6, pp. 2214–2222, Dec. 1989.

123. V. Goiffon, S. Girard et al., "Mitigation technique for use of CMOS image sensors in megajoule class laser radiative environment," *Electron. Lett.* vol. 48, no. 21, p. 1338, 2012.

124. M. Gaillardin, S. Girard et al., "Investigations on the vulnerability of advanced CMOS technologies to MGy dose environments," *IEEE Trans. Nucl. Sci.* vol. 60, no. 4, pp. 2590–2597, Aug. 2013.

125. M.R. Shaneyfelt, P.E. Dodd et al., "Challenges in hardening technologies using shallow-trench isolation," *IEEE Trans. Nucl. Sci.* vol. 45, no. 6, pp. 2584–2592, Dec. 1998.

126. F. Faccio, G. Cervelli, "Radiation-induced edge effects in deep submicron CMOS transistors," *IEEE Trans. Nucl. Sci.* vol. 52, no. 6, pp. 2413–2420, Dec. 2005.

127. M. Gaillardin, V. Goiffon et al., "Enhanced radiation-induced narrow channel effects in commercial 0.18 μm bulk technology," *IEEE Trans. Nucl. Sci.* vol. 58, no. 6, pp. 2807–2815, Dec. 2011.

128. D. Contarato, P. Denes et al., "High speed, radiation hard CMOS pixel sensors for transmission electron microscopy," *Phys. Procedia* vol. 37, pp. 1504–1510, 2012.

129. D. Contarato, P. Denes et al., "A 2.5 μm pitch CMOS active pixel sensor in 65 nm technology for Electron Microscopy," in *2012 IEEE Nucl. Sci. Symp. Med. Imaging Conf.*, NSSMIC, pp. 2036–2040, 2012.

130. K. Kruckmeyer, J.S. Prater et al., "Analysis of low dose rate effects on parasitic bipolar structures in CMOS processes for mixed-signal integrated circuits," *IEEE Trans. Nucl. Sci.* vol. 58, no. 3, pp. 1023–1031, Jun. 2011.

131. J. Verbeeck, Y. Cao et al., "A MGy radiation-hardened sensor instrumentation SoC in a commercial CMOS technology," *IEEE Trans. Nucl. Sci.* vol. 61, no. 6, Dec. 2014.

132. G.R. Hopkinson, M.D. Skipper et al., "A radiation tolerant video camera for high total dose environments," in *2002 IEEE Radiat. Eff. Data Workshop*, pp. 18–23, 2002.

133. G.R. Hopkinson, A. Mohammadzadeh et al., "Radiation effects on a radiation-tolerant CMOS active pixel sensor," *IEEE Trans. Nucl. Sci.* vol. 51, no. 5, pp. 2753–2761, Oct. 2004.

134. H.N. Becker, M.D. Dolphin et al., *Commercial Sensor Survey Fiscal Year 2008 Compendium Radiation Test Report*, 2008.

135. H.N. Becker, J.W. Alexander et al., *Commercial Sensor Survey Fiscal Year 2009 Master Compendium Radiation Test Report*, Jet Propulsion Laboratory, 2009.

136. B. Dryer, A. Holland et al., "Gamma radiation damage study of 0.18 m process CMOS image sensors," in *Proc SPIE*, vol. 7742, 2010.

137. M. Beaumel, D. Herve et al., "Cobalt-60, proton and electron irradiation of a radiation-hardened active pixel sensor," *IEEE Trans. Nucl. Sci.* vol. 57, no. 4, pp. 2056–2065, Aug. 2010.

138. E. Martin, T. Nuns et al., "Gamma and proton-induced dark current degradation of 5T CMOS pinned photodiode 0.18μm CMOS image sensors," *IEEE Trans. Nucl. Sci.* vol. 61, no. 1, pp. 636–645, Feb. 2014.

139. V. Goiffon, P. Magnan et al., "Total dose evaluation of deep submicron CMOS imaging technology through elementary device and pixel array behavior analysis," *IEEE Trans. Nucl. Sci.* vol. 55, no. 6, pp. 3494–3501, Dec. 2008.

140. V. Goiffon, M. Estribeau et al., "Overview of ionizing radiation effects in image sensors fabricated in a deep-submicrometer CMOS imaging technology," *IEEE Trans. Electron Devices* vol. 56, no. 11, pp. 2594–2601, Nov. 2009.

141. V. Goiffon, M. Estribeau et al., "Influence of transfer gate design and bias on the radiation hardness of pinned photodiode CMOS image sensors," *IEEE Trans. Nucl. Sci.* vol. 61, no. 6, Dec. 2014.

142. P.R. Rao, X. Wang et al., "Degradation of CMOS image sensors in deep-submicron technology due to γ-irradiation," *Solid-State Electron.* vol. 52, no. 9, pp. 1407–1413, Sep. 2008.

143. A. BenMoussa, S. Gissot et al., "Irradiation damage tests on backside-illuminated CMOS APS prototypes for the extreme ultraviolet imager on-board solar orbiter," *IEEE Trans. Nucl. Sci.* vol. 60, no. 5, pp. 3907–3914, Oct. 2013.

144. B. Pain, B.R. Hancock et al., "Hardening CMOS imagers: Radhard-by-design or rad-hard-by-foundry," in *Proc SPIE*, San Diego, CA, vol. 5167, pp. 101–110, 2004.
145. V. Goiffon, C. Virmontois et al., "Analysis of total dose induced dark current in CMOS image sensors from interface state and trapped charge density measurements," *IEEE Trans. Nucl. Sci.* vol. 57, no. 6, pp. 3087–3094, Dec. 2010.
146. L. Ratti, C. Andreoli et al., "TID effects in deep N-Well CMOS monolithic active pixel sensors," *IEEE Trans. Nucl. Sci.* vol. 56, no. 4, pp. 2124–2131, Aug. 2009.
147. V. Goiffon, C. Virmontois et al., "Radiation damages in CMOS image sensors: Testing and hardening challenges brought by deep sub-micrometer CIS processes," in 2010, vol. 7826, p. 78261S–78261S–12.
148. L. Ratti, M. Dellagiovanna et al., "Front-end performance and charge collection properties of heavily irradiated DNW MAPS," in *Proc RADECS*, pp. 33–40, 2009.
149. C. Virmontois, C. Codreanu et al., "Space environment effects on CMOS micro-lenses and color filters," presented at the *Image Sensors Optical Interfaces Workshop*, Toulouse, France, Nov. 27, 2015.
150. V. Re, M. Manghisoni et al., "Impact of lateral isolation oxides on radiation-induced noise degradation in CMOS technologies in the 100-nm Regime," *IEEE Trans. Nucl. Sci.* vol. 54, no. 6, pp. 2218–2226, Dec. 2007.
151. P. Martin-Gonthier, V. Goiffon et al., "In-pixel source follower transistor RTS noise behavior under ionizing radiation in CMOS image sensors," *IEEE Trans. Electron Devices* vol. 59, no. 6, pp. 1686–1692, Jun. 2012.
152. Y. Chen, J. Tan et al., "X-ray radiation effect on CMOS imagers with in-pixel buried-channel source follower," in *Proc ESSDERC*, pp. 155–158, 2011.
153. E.R. Fossum, "Charge transfer noise and lag in CMOS active pixel sensors," in *IEEE Workshop Charge-Coupled Devices Adv. Image Sens.*, Elmau, 2003.
154. B. Fowler, X. Liu, "Charge transfer noise in image sensors," in *IEEE Workshop Charge-Coupled Devices Adv. Image Sens.*, 2007.
155. J. Janesick, J. Pinter et al., in A.D. Holland, D.A. Dorn (Eds.), "Fundamental performance differences between CMOS and CCD imagers, part IV," pp. 7740B–77420B–30, 2010.
156. J.P. Carrere, J.P. Oddou et al., "New mechanism of plasma induced damage on CMOS image sensor: Analysis and process optimization," in *Proc ESSDERC*, pp. 106–109, 2010.
157. I. Inoue, N. Tanaka et al., "Low-leakage-current and low-operating-voltage buried photodiode for a CMOS imager," *IEEE Trans. Electron Devices* vol. 50, no. 1, pp. 43–47, Jan. 2003.
158. J.R. Janesick, T. Elliott et al., "Fundamental performance differences of CMOS and CCD imagers: Part V," in *Proc SPIE*, pp. 865902–865902, 2013.
159. W. Dulinski, A. Besson et al., "Optimization of tracking performance of CMOS monolithic active pixel sensors," *IEEE Trans. Nucl. Sci.* vol. 54, no. 1, pp. 284–289, Feb. 2007.
160. M.A. Szelezniak, A. Besson et al., "Small-scale readout system prototype for the STAR PIXEL detector," *IEEE Trans. Nucl. Sci.* vol. 55, no. 6, pp. 3665–3672, Dec. 2008.
161. M. Battaglia, D. Contarato et al., "A rad-hard CMOS active pixel sensor for electron microscopy," *Nucl. Instr. Meth. A* vol. 598, no. 2, pp. 642–649, Jan. 2009.
162. M. Innocent, "A radiation tolerant 4T pixel for space applications," in *Proc IISW*, 2009.
163. M. Innocent, "A Radiation Tolerant 4T pixel for Space Applications: Layout and Process Optimization," in *Proc IISW*, 2013.
164. S. Place, J.-P. Carrere et al., "Rad tolerant CMOS image sensor based on hole collection 4T pixel pinned photodiode," *IEEE Trans. Nucl. Sci.* vol. 59, no. 6, pp. 2888–2893, Dec. 2012.
165. E. Stevens, H. Komori et al., "Low-crosstalk and low-dark-current CMOS image-sensor technology using a hole-based detector," in *ISSCC Tech. Dig.*, pp. 60–595, 2008.
166. D.K. Schroder, "Carrier lifetimes in silicon," *IEEE Trans. Electron Devices* vol. 44, no. 1, pp. 160–170, Jan. 1997.

167. A. Johnston, "Optoelectronic devices with complex failure modes," in *IEEE NSREC Short Course*, 2000.
168. M. Deveaux, G. Claus et al., "Neutron radiation hardness of monolithic active pixel sensors for charged particle tracking," *Nucl. Instr. Meth. A* vol. 512, pp. 71–76, 2003.
169. S. Zucca, L. Ratti et al., "Characterization of bulk damage in CMOS MAPS with deep N-Well collecting electrode," *IEEE Trans. Nucl. Sci.* vol. 59, no. 4, pp. 900–908, Aug. 2012.
170. C. Virmontois, V. Goiffon et al., "Displacement damage effects in pinned photodiode CMOS image sensors," *IEEE Trans. Nucl. Sci.* vol. 59, no. 6, pp. 2872–2877, Dec. 2012.
171. J.R. Srour, "Displacement damage effects in devices," in *IEEE NSREC Short Course*, 2013.
172. J.R. Srour, D.H. Lo, "Universal damage factor for radiation induced dark current in silicon devices," *IEEE Trans. Nucl. Sci.* vol. 47, no. 6, pp. 2451–2459, Dec. 2000.
173. C. Virmontois, V. Goiffon et al., "Displacement damage effects due to neutron and proton irradiations on CMOS image sensors manufactured in deep sub-micron technology," *IEEE Trans. Nucl. Sci.* vol. 57, no. 6, Dec. 2010.
174. C. Virmontois, V. Goiffon et al., "Similarities between proton and neutron induced dark current distribution in CMOS image sensors," *IEEE Trans. Nucl. Sci.* vol. 59, no. 4, pp. 927–936, Aug. 2012.
175. J. Bogaerts, B. Dierickx et al., "Enhanced dark current generation in proton-irradiated CMOS active pixel sensors," *IEEE Trans. Nucl. Sci.* vol. 49, no. 3, pp. 1513–1521, Jun. 2002.
176. J.R. Srour, R.A. Hartmann, "Enhanced displacement damage effectiveness in irradiated silicon devices," *IEEE Trans. Nucl. Sci.* vol. 36, no. 6, pp. 1825–1830, Dec. 1989.
177. C.J. Dale, P.W. Marshall et al., "The generation lifetime damage factor and its variance in silicon," *IEEE Trans. Nucl. Sci.* vol. 36, no. 6, pp. 1872–1881, Dec. 1989.
178. P.W. Marshall, C.J. Dale et al., "Proton-induced displacement damage distributions and extremes in silicon microvolumes charge injection device," *IEEE Trans. Nucl. Sci.* On, vol. 37, no. 6, pp. 1776–1783, Dec. 1990.
179. C.J. Dale, P.W. Marshall et al., "Particle-induced spatial dark current fluctuations in focal plane arrays," *IEEE Trans. Nucl. Sci.* vol. 37, no. 6, pp. 1784–1791, Dec. 1990.
180. C.J. Dale, L. Chen et al., "A comparison of Monte Carlo and analytical treatments of displacement damage in Si microvolumes," *IEEE Trans. Nucl. Sci.* vol. 41, pp. 1974–1983, Dec. 1994.
181. M. Robbins, "High-energy proton-induced dark signal in silicon charge coupled devices," *IEEE Trans. Nucl. Sci.* vol. 47, no. 6, pp. 2473–2479, Dec. 2000.
182. C. Inguimbert, T. Nuns et al., "Monte Carlo based DSNU prediction after proton irradiation," presented at the *Radiation Effects on Components and Systems* (RADECS), Biarritz, France, 2012.
183. O. Gilard, M. Boutillier et al., "New approach for the prediction of CCD dark current distribution in a space radiation environment," *IEEE Trans. Nucl. Sci.* vol. 55, no. 6, pp. 3626–3632, Dec. 2008.
184. J.W. Palko, J.R. Srour, "Amorphous inclusions in irradiated silicon and their effects on material and device properties," in *IEEE Nucl. Space Radiat. Eff. Conf.*, 2008.
185. M.S. Robbins, L. Gomez Rojas, "An assessment of the bias dependence of displacement damage effects and annealing in silicon charge coupled devices," *IEEE Trans. Nucl. Sci.* vol. 60, no. 6, pp. 4332–4340, Dec. 2013.
186. O. Gilard, E. Martin et al., "Statistical analysis of random telegraph signal maximum transition amplitudes in an irradiated CMOS image sensor," *IEEE Trans. Nucl. Sci.* vol. 61, no. 2, pp. 939–947, Apr. 2014.
187. M. Raine, V. Goiffon et al., "Exploring the kinetics of formation and annealing of single particle displacement damage in microvolumes of silicon," *IEEE Trans. Nucl. Sci.* vol. 6, Dec. 2014.

12 Natural Radiation Effects in CCD Devices

Tarek Saad Saoud, Soilihi Moindjie,
Daniela Munteanu, and Jean-Luc Autran

CONTENTS

12.1 INTRODUCTION

Charge-coupled devices (CCDs) and other solid-state image sensors [1,2] are widely recognized to be sensitive to cosmic rays and cosmic-ray-induced secondaries that produce numerous artifacts in their electrical response (images). In astronomical CCD imagers, for example [3], a long dark exposure generally shows a collection of straight lines, worms, and spots that can be attributed to the interactions of different types of particles with the sensor active pixelated region [4,5]. Considering this sensitivity to the ambient radiation, CCD imagers can provide an excellent solution for the detection and investigation of interactions between natural radiation and electronics. Indeed, because the target material is mainly silicon, as in any integrated circuit, and the architecture of a CCD device is similar to that of a memory chip, a link between radiation-induced events observed in CCDs with phenomena occurring in circuits can be envisaged. The CCD can also "image" with a pixel size resolution (a few μm^2) the charge deposited by ionizing particles, and its high sensitivity of a few tens of electrons at the pixel level [2] is ideal for the detection of lightly ionizing particles, such as protons and muons at ground level (especially low-energy particles), which are susceptible to contribute to single event effects (SEEs) in modern nano-electronics circuits [6,7].

This chapter is dedicated to the particular effects of natural terrestrial radiation in CCD devices at atmospheric and ground levels. The text is divided into two main parts. In the first part, we shortly review recent works [4,5,8–24], typically published during the last decade in literature, that demonstrated the capability of such CCDs or other solid-state image sensors in general to investigate, with a high spatial resolution (in the range of order of the pixel size, typically a few microns in these studies) and a high sensitivity (of a few tens of electrons collected at pixel level), the interactions between atmospheric particles and silicon in the natural radiation background or using artificial sources of radiation. In the second part of the chapter, we will detail a recent study [24] in which a simple commercial CCD camera has been used as an atmospheric radiation detector. New experimental evidence concerning the following points will be reviewed: (i) the predominant role of charged particles in the CCD outdoor response, (ii) the impact of contamination by alpha particle emitters from underground experiments, (iii) the influence of CCD orientation and the characterization of particle flux anisotropy at ground level. Experimental conclusions are supported by the results of numerical simulation that considers a simplified 3-D architecture of the device and takes into account the impact of both alpha contamination and atmospheric particles on the CCD response for the different radiation environments.

12.2   SINGLE EVENT EFFECTS IN CCD DEVICES

This part briefly summarizes the main types of radiation effects susceptible to impact CCD devices; it proposes a survey of the recent literature concerning the effects of natural radiation (terrestrial cosmic rays) on the CCD electrical response and the use of CCDs to specifically detect radiation at ground or atmospheric levels.

12.2.1   RADIATION EFFECTS IN CCD DEVICES

As most solid-state devices, CCDs are vulnerable to total ionizing dose effects, displacement damages, and/or transient effects when exposed to natural or artificial irradiation [10]. The exact nature of the observed radiation effects depends of course on (i) the characteristics of the radiation field in which devices are immerged and (ii) the conditions (temperature) and duration of exposition. Total ionizing dose (TID) and displacement damage effects permanently degrade the performances of the CCDs. Schematically, TID produces threshold voltage shifts on the CCD gates and displacement damage reduces the charge transfer efficiency (CTE), increases the dark current, produces dark current nonuniformities, and creates random telegraph noise in individual pixels [14]. These two failure mechanisms are characterized by cumulative effects induced by the interaction of large quantities of energetic particles with the circuit materials: At low doses, they do not directly cause the malfunctioning of devices, but after a certain dose received, they may result in the alteration or the interruption of the device operation. Single event effects are different in nature: They are induced by individual energetic particles and can result in the so-called "soft" or "hard" errors. Soft errors cause a temporary malfunctioning of the device whereas hard errors result in permanent damage to the whole device or localized parts of it. We detail, in the following, these three main classes of radiation effects in CCDs.

- Total ionizing dose effects: Because CCDs are based on a metal–insulator–semiconductor vertical structure for photodetection and readout operations, these devices are susceptible to ionization damage within the gate oxide and other insulator layers. Silicon dioxide (SiO_2) is almost exclusively used as the insulator in CCDs. The main TID effects are the buildup of trapped charge in the oxide and the generation of interface (and border) traps at the silicon dioxide/silicon interface [14]. In such CMOS devices, this electrically active defect generation produces shifts in flat-band and threshold voltages (i.e., the effective bias voltages applied to the device are changed), increases in the surface dark current (i.e., the component of thermal dark current that is generated at the silicon dioxide/silicon interface), increased amplifier noise, and changes in linearity. These effects are relatively well understood in CCDs and can, in principle, be reduced by an appropriate choice of device architecture and oxide technology. Therefore, CCD performances in space are not generally limited by total ionizing dose effects because displacement effects are more often the limiting reliability mechanism.
- Displacement damages: These are produced by energetic particles, such as protons and neutrons, that collide with silicon atoms and displace them from their lattice sites [14]. As a result, many vacancy interstitial pairs are formed, most of which recombine. The vacancies that survive migrate in the lattice and form stable defects. These point defects degrade the CCD performances by decreasing the charge transfer efficiency, increasing the average dark current and dark current nonuniformities, by introducing individual pixels with very high dark currents (or "spikes") and by introducing random telegraph noise in pixels. In fact, bulk displacement damage effects often dominate the radiation response in state-of-the-art scientific imagers when operated in natural particle environments. The flat-band shifts and dark current increases that occur for ionizing dose levels below 10–20 krads (Si) are often not significant and can be overcome with minor changes in voltages and operating temperature. In contrast, significant displacement damage induced CTE losses are frequently observed for proton exposures of less than 1 krad (Si) [11]. Nevertheless, the degree of CTE loss that is tolerable is very application-dependent, and it is still possible for a device to ultimately fail as a result of either TID or displacement damage effects at higher exposure levels.
- Single event effects (SEEs): These have a different nature from TID and displacement damages and are due to the ionization-induced generation of electrical charge by individual energetic particles within the active region of the CCD. The effects are not permanent, and the spurious charge is swept out during readout, but this additional charge constitutes a significant source of noise (nonsense signal) in the image/video data. These aspects are detailed in the following section.

12.2.2   CCD-Based Radiation Detectors

CCDs are very sensitive devices that are able to give a digital pixelated image of charge deposition events occurring in the sensitive region of the imager. This is accomplished by reading out spurious charges collected from direct or indirect ionization processes in the CCD pixels as signal electrons [16]. Ionizations occur as charged particles or nuclear recoils travel through the silicon lattice and generate electron–hole pairs along their whole path. As a result of these charge deposition, transport, and collection processes, high-resolution images are produced showing the amount of charge collected in each pixel by proportionally increasing gray-level intensities in the CCD image/video frame output data. Using imaging processing techniques, it is thus possible to distinguish and characterize different types of interaction processes and possibly the particle type by measuring event features, including length, total collected charge, charge deposition rate, and charge moment [16].

Single event effects in CCDs can be generated by direct or indirect ionization [25].

- Direct ionization is achieved through Coulomb interactions between a charged particle and atoms of the device. The charged particle strips electrons of atoms as it passes through the device, thereby causing ionizations. Heavy ions (including charged nuclear recoils), low-energy protons, and muons directly ionize matter. Figure 12.1 illustrates the case of direct ionization events induced by the passage of charged particles through the CCD. As a function of the incidence angle of the particle with respect to the CCD plan, different signatures (spots, straight lines) can be obtained.
- Indirect ionization (Figure 12.2) is of concern for atmospheric neutrons and high energy protons (>100 MeV), which are able to ionize by collision with

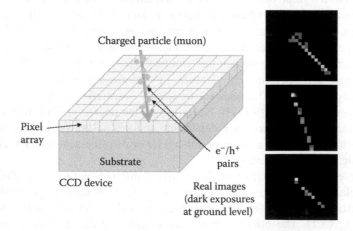

FIGURE 12.1   Schematic of a direct ionization event resulting from the passage of a charged particle within the CCD structure and corresponding real images recorded with the CCD camera described in Section 12.3.

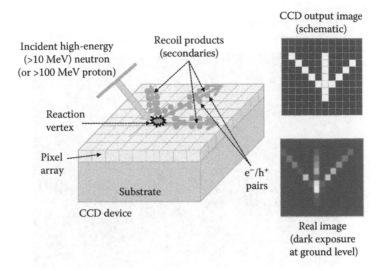

FIGURE 12.2 Schematic of a neutron–silicon interaction (indirect ionization) within the CCD device and corresponding real image recorded with the CCD camera described in Section 12.3.

the target nuclei [25]. Neutrons ionize indirectly; they do not interact via the Coulomb force, so they can travel through several centimeters of material without interacting with other particles and can remain undetected with CCD. Indirect ionization is accomplished through two mechanisms: elastic and inelastic scattering. Elastic scattering occurs when a neutron knocks out a target nucleus from its lattice but the nucleus remains in the same energy state. During inelastic scattering, the striking neutron interacts with the target nucleus such that the nucleus captures the neutron and thereby the nucleus becomes an isotope. The isotope then de-excites by the emission of secondary charged and uncharged radiation. The residual nucleus and the evaporation products may be highly ionizing and are able to deposit significant amounts of charge at various locations in a device and cause SEEs. Similarly to Figure 12.1, Figure 12.2 illustrates the case of an indirect ionization event induced by a neutron–silicon interaction within the volume of the CCD. As a function of the number and the momentum direction of the reaction products, the signature of the event can be more complex than a simple spot or straight line.

As a result of all these possible interactions between energetic particles and a CCD device, a long dark exposure in the natural radiation background (even at ground level) will generally show a zoo of particle signatures on the CCD output image. This point is illustrated in Figure 12.3 for dark exposures obtained in the framework of recent experiments described in the second part (Section 12.3) of this chapter. As already reported and analyzed in previous works [4,5,15–17,20,21], pixelated events have different topology shapes as a function of the type and/or the number of ionizing

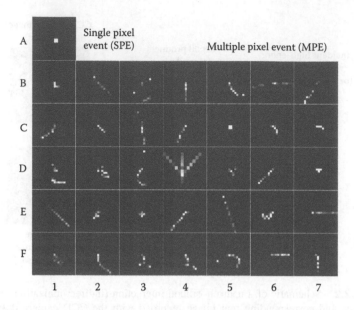

FIGURE 12.3   Panel of single pixel events (SPEs) and multiple pixel events (MPEs) detected by the CCD under complete darkness. Each individual event image is stored into a database with the corresponding values of different characteristic metrics (event date, position, topology, area, length, deposited charge, etc.). Note that the composition of this panel is not representative of the event statistics; in particular, MPEs corresponding to multibranch-shaped forms are rare.

particles involved in the underlying interaction mechanisms: The cluster of adjacent pixels may be a spot, a straight segment, a worm, or a multibranch-shaped form. For example, the D4 image of Figure 12.3 shows a typical event that may correspond to the signature of a neutron-induced reaction with three ejected evaporation particles that travel in the plane of the CCD. A high-energy proton could also have caused it.

"Straight line" events can be attributed to atmospheric muons [4,5] or alpha particles resulting from the disintegration of radioisotopes present as ultratraces in the circuit materials (principally from the uranium and thorium decay chains) [24]. Because more than 95% of sea-level cosmic rays are muons with a mean energy of 4 GeV, for a horizontal detector, the rate of muon-related events is relatively high and estimated to 0.8–1.0 $cm^{-2}min^{-1}$ [5]. Finally, the worms and spots are almost certainly recoil electrons from Compton scattering of environmental gamma rays. These gamma rays can be emitted from ^{40}K decay to ^{40}Ar, plus the U and Th decay chains, some being degraded by multiple Compton scattering. The rates are about twice that of cosmic ray muons. A simple lead shielding (1 cm) can generally reduce the Compton to below that of cosmic rays [5].

12.3   NATURAL RADIATION EFFECTS IN CCDs: A CASE STUDY

This second part reports a recent study that used a simple commercial CCD camera as an atmospheric radiation detector. It perfectly illustrates the capability of a standard CCD device to be used as a natural radiation detector able to "image" different types of particle interactions with silicon and to monitor radiation background at both ground and avionic levels.

12.3.1 EXPERIMENTAL SETUP

An experimental setup based on a commercial USB2.0 CCD monochrome camera (The Imaging Source, model DMK 41BU02) taking 3.75 frames/s in the complete darkness has been developed. The camera was used without any optical system in front of the sensor. The square image sensor (reference Sony ICX205AL) is an interline CCD [25] with a diagonal of 8 mm. It is composed of an array (1.45 × 10^6 effective pixels) of square pixels (4.65 × 4.65 μm^2) with a vertical antiblooming structure. Progressive scan allows all pixel signals to be output independently within approximately 143 ms. Pixels have a charge saturation capacity of 13,000 electrons. The setup also includes a control PC and a sophisticated image-processing software developed under Visual Basic and linked with MATLAB® routines. This later performs real-time frame cleaning and analysis, event extraction under the form of small images (with adjusted dimensions to capture each event) and the storage of these "event images" and related information into a database implemented on the same PC. This database also contains information about all damaged or instable pixels, subjected, for example, to random telegraph signal (RTS [26,27]) noise or other characteristic electrical instabilities [2].

When a frame is captured by the CCD, a preliminary frame cleaning is performed by subtracting the current frame with the previous one; events are then detected by applying to this new image a series of mathematical treatments that isolate pixels or group(s) of connected pixels with an electrical charge clearly above the background and verifying certain threshold criteria. This operation consists of identifying pixels above a first threshold value and examining if the neighboring pixels (up to the second neighbors) are also above a second value (inferior to the first threshold but superior to the image background). As a function of the test result, the detected events have been classified into two types (see Figure 12.3 in Section 12.2.2): single pixel events (SPEs) that correspond to isolated pixels and multiple pixel events (MPEs) that correspond to a group of adjacent or neighboring pixels (i.e., which present pixel connectivity). Finally, the software calculates different event rates during the experiment: a single pixel event rate (SPER) and a multiple pixel event rate (MPER), typically expressed per hour. The deposited charge corresponding to each detected event is estimated by summing the values (i.e., the readings) of the event pixels, considering a linear dependence between a pixel value and its electrical charge up to the charge saturation. The Sony ICX205AL being characterized by a low smear and a vertical antiblooming structure, if a pixel (or a group of pixels) reaches the saturation, the evaluation of the deposited charge may be under evaluated in this case.

12.3.2 EXPERIMENTAL RESULTS

Different experimental campaigns of measurements were performed in 2012, 2013, and 2014 using the same CCD camera transported in three different locations: at

TABLE 12.1

Averaged SPE and MPE Hourly Rates Measured for the Camera Located at LSM, Sea Level (Marseille), and at Aiguille du Midi (AM) during the Measurement Campaigns

Hourly Rate (h⁻¹)	LSM (Underground)	Marseille (Sea Level)	Aiguille du Midi (AM)
Single pixel event (SPE)	9.4	10.7	20.1
Multiple pixel event (MPE)	4.1	4.5	8.1
Total (SPE + MPE)	13.5	15.2	28.2

sea level in Marseille (altitude of 120 m), in high altitude at the Aiguille du Midi (3780 m) to increase the atmospheric radiation, and underground at LSM laboratory (–1700 m under the surface) to completely screen this radiation. Four other DMK 41BU02 cameras have been also tested in Marseille to verify their long-term radiation response; no significant difference from one camera to another has been observed in the event rates reported in the following.

As indicated, Figure 12.3 shows a panel of typical SPE and MPE events detected by the CCD under complete darkness. A collection of several thousands of such images has been obtained from the different measurement campaigns. From the date and time stamping of all the detected events, the image control and processing software evaluates in real time the hourly SPERs and MPERs, reported in Table 12.1 for the three test locations. These values have been averaged over several weeks for each location. On one hand, rather close results have been obtained for underground and sea-level tests (+20% in the global event rate with respect to the underground value), strongly suggesting for this commercial sensor the dominant role of alpha particle emitters present under the form of traces of contaminants in CCD and/or packaging materials. On the other hand, high-altitude measurements show a clear increase (by a factor ×2) of both SPERs and MPERs with respect to those of sea level, attesting to the additional detection of atmospheric particles but with an increase well below the ratio of the site acceleration factors usually considered for neutron flux [28]. The distribution of event size and charge will be discussed later in this chapter.

In order to identify the nature of the atmospheric particles responsible for the CCD response at ground level, we performed correlation studies with the responses of two additional instruments available in our laboratory in Marseille at sea level:

- The CCD signal was first compared with the signal of the TERRAMU [29] neutron monitor (Figure 12.4a), which is the clone of the Plateau de Bure neutron monitor installed on the ASTEP platform [30,31]. Figure 12.4b shows the comparison of these two signals: no frank correlation between the event pixel rate and the neutron monitor signal has been obtained,

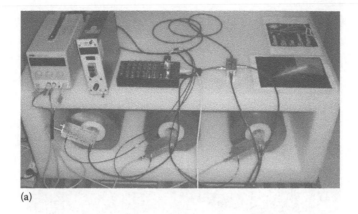

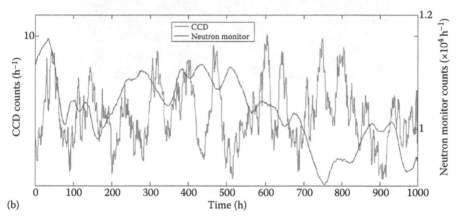

(a)

(b)

FIGURE 12.4 (a) Front view of the TERRAMU neutron monitor installed in Marseille for the continuous monitoring of the atmospheric neutron flux. The instrument is composed of three high-pressure (2280 Torr) cylindrical He^3 detectors surrounded by lead rings and a polyethylene box (see [31] for details). (b) Comparison between the neutron monitor and CCD signals recorded during a period of 1000 h.

suggesting a negligible contribution of atmospheric neutrons in the CCD pixel event rate.

- A second comparison study was also conducted, considering this time the "round" signal of an ultralow background alpha particle counter (XIA, model UltraLo-1800 [32], see Figure 12.5a) also installed on the TERRAMU platform in the vicinity of the neutron monitor and CCD setups. "Round" events are the signature of charged atmospheric particles that passes through the argon ionization chamber of the counter as distinctly demonstrated in recent experimental and simulation works [33,34]. In contrast with the neutron monitor, a clear correlation (Pearson product-moment correlation coefficient $r = 0.6532$) was observed between the variations of the CCD pixel event rate (around its averaged and rather constant value at

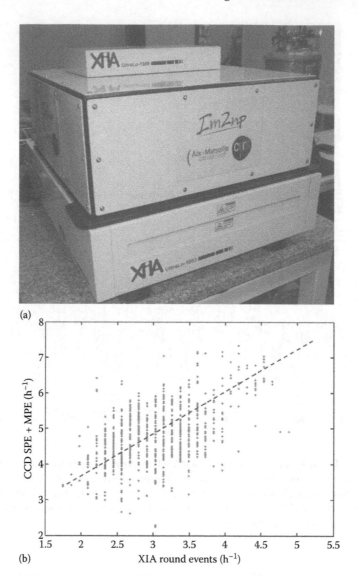

(a)

(b)

FIGURE 12.5 (a) General view of the XIA ultralow background alpha particle counter (model UltraLo-1800 [32]) installed on the TERRAMU platform in Marseille. (b) Correlation analysis between the total CCD signal (SPE + MPE hourly rates) and the "round" event rate detected by the XIA UltraLo-1800 counter.

sea level) and the variations of the "round" signal delivered by the ionization chamber. This correlation, analyzed in Figure 12.5b, suggests that the CCD signal variations may be roughly due to the interactions of charged particles with the CCD.

Finally, we investigated the impact of the CCD orientation on the event rates detected at sea level by measuring the SPE and MPE hourly rates for both vertical and horizontal orientations of the CCD plane (Figure 12.6). Even if the response of the CCD seems to be essentially dominated by the contribution of the alpha particle emitters at sea level as previously reported in the analysis of Table 12.1 (this contribution represents about 80% of the pixel event rate), a small but detectable and reproducible variation of the signals was evidenced in Figure 12.6 (around their average values) when switching the CCD from horizontal to vertical position (and vice versa): SPER decreases while MPER increases, which is consistent with an anisotropy of the atmospheric particle flux mainly responsible for the detected events. Indeed, although these events contribute to less than 20% of the total CCD response (SPE + MPE), they differently impact SPEs and MPEs as a function of the sensor orientation as experimentally demonstrated by results shown in Figure 12.6. Assuming a preferential CCD sensitivity to charged particles (see Section 12.3.3, notably Figure 12.10), Figure 12.6 and the corresponding event analysis shown in Figure 12.7 suggest that these particles predominantly arrive in the vertical

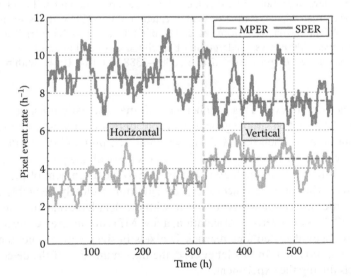

FIGURE 12.6 Experimental SPE and MPE rates for horizontal and vertical positions of the CCD (Marseille location). Data obtained with an integration time of 1 h and averaged over 24 h using a simple moving average. (Reprinted from Saad Saoud et al., Use of CCD to Detect Terrestrial Cosmic Rays at Ground Level: Altitude vs. Underground Experiments, Modeling and Numerical Monte Carlo Simulation, *IEEE Trans. Nucl. Sci.*, 61, 3380–3388. © [2014] IEEE. With permission.)

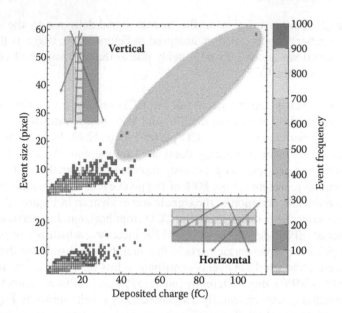

FIGURE 12.7 Distribution of event size (in number of pixels) versus deposited charge for measurements performed in both horizontal and vertical positions of the CCD (Marseille condition). The ellipse is only for eye: It indicates a domain of event size versus deposited charge where events are only detected when the CCD plan is vertical. The schematics illustrate how atmospheric charged particles intersect the active pixel area; the probability of SPE (resp. MPE) is lower (resp. higher) for the vertical orientation of the CCD with respect to the horizontal one. (Adapted from Saad Saoud et al., Use of CCD to Detect Terrestrial Cosmic Rays at Ground Level: Altitude vs. Underground Experiments, Modeling and Numerical Monte Carlo Simulation, *IEEE Trans. Nucl. Sci.*, 61, 3380–3388. © [2014] IEEE. With permission.)

direction at ground level. This is coherent with the schematic illustration inserted in Figure 12.7 based on simple geometrical considerations (intersections of particle tracks with pixels): when the CCD is horizontal, the probability of SPE is higher than the one of MPE; inversely, when the CCD is vertical, the probability to capture events with straight line topology is higher, due to the alignment of the pixel plan with the incoming particle trajectories; this situation is clearly favorable to MPEs. This first-order interpretation is consistent with the detailed event analysis reported in Figure 12.7: With a vertical orientation, a few MPE events greater than 20 pixels in size have been detected (one reaches 57 pixels in size), which is not the case for the horizontal position. In this latter case, the maximum size of the detected event is 18 pixels during the experiment.

12.3.3 MODELING AND SIMULATION

In a second part of this work, we performed the complete modeling and numerical simulation of the CCD device subjected to the terrestrial natural radiation environment. Our objective is to clearly reproduce the experimental results to explain the underlying mechanisms of the CCD response, notably in terms of particle sensitivity, image signatures, and both SPE and MPE probability distributions.

The first step of this work was to determine the internal architecture of the CCD device to correctly model this complex stack of materials. A sacrificed device was investigated using focused ion beam (FIB), electron microscopy (SEM, TEM), and chemical microanalysis (Electron Energy Loss Spectroscopy, EELS). From these data, we constructed a simplified but realistic 3-D model of the complete pixel array with exact dimensions in the three directions and exact chemical composition of the different layers. Unfortunately, with the detailed information about the pixel structure (internal architecture, doping profiles, etc.) being proprietary and difficult for us to analyze in the present approach, the accuracy of the model may be limited, especially at the silicon level. To simplify the device geometry modeling for the back-end-of-line (BEOL, which contains contacts, insulating layers, and metal levels above silicon), the material stack was replaced by a single layer of equivalent thickness and containing a mixture of all elements of the real BEOL (primarily W, SiO_2, Al, and Si_3N_4 with the following respective percentages: 10.2%, 18.9%, 54.3%, and 16.7%).

A dedicated C++ code based on computational libraries developed at IM2NP-CNRS has been specially developed to perform numerical simulations. Figure 12.8 shows the flowchart of this code. The particle generator considers atmospheric particle energy distributions given by the JEDEC [28] (neutrons) and the PARMA/

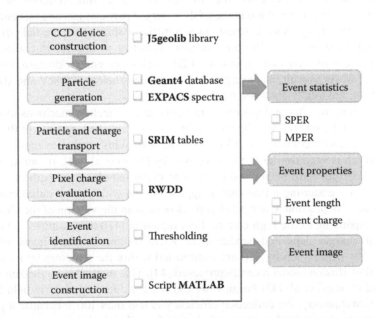

FIGURE 12.8 Flowchart of the numerical simulation code developed in this work. (Reprinted from Saad Saoud et al., Use of CCD to Detect Terrestrial Cosmic Rays at Ground Level: Altitude vs. Underground Experiments, Modeling and Numerical Monte Carlo Simulation, *IEEE Trans. Nucl. Sci.*, 61, 3380–3388. © [2014] IEEE. With permission.)

EXPACS [35] or QARM [36] (protons, muons) models for Marseille and Aiguille du Midi test locations; the particles are incident on the circuit with a given $\cos(\theta)^n$ distribution for the zenithal angle as discussed and used in [31]. Alternatively, alpha particles generated from randomly distributed impurities (uranium and thorium decay chains) in the circuit materials can be considered. Nuclear event databases, separately computed using Geant4 [37], have been considered for the description of neutron and high-energy proton interactions with the circuit materials. A numerical modeling of the SRIM tables has been also included to treat the transport of all charged particles within the device, including muons and low-energy protons [38]. Values for muons have been estimated by applying mass scaling to proton transport tables as specified in [39]. The code transports particles through the device, evaluates the electrical charge deposited and collected within each pixel, and computes the corresponding image and characteristics (size, charge, etc.) of each event, applying the same discrimination threshold criteria as considered in the experiment by the software.

Special attention has been paid to the impact of all material layers above the front-end-of-line (FEOL, i.e., silicon-level) of the CCD that can modify the characteristics of the incoming atmospheric particle flux. Indeed, the active region of the CCD is not directly exposed to the atmospheric radiation: overlayer stacks at the level of the BEOL and of the packaging (the protective glass above the CCD chip) can induce energy loss mechanisms for primary charged particles, that is, the incoming atmospheric muons and protons. Dedicated simulations (not shown here) have been performed to quantify the "filtering effects" of this additional structure (BEOL layer + protective glass) in the CCD. Charged particles passing through these layers undergo energy loss. It results in a significant shift of the particle distributions toward low energies. Protons are more sensitive to this effect compared to muons because of their more important LET in this low energy domain, typically below 2 MeV. As a result, incoming protons and muons below 400 keV and 200 keV, respectively, are totally absorbed by the overlayer stack.

Concerning the charge deposition, transport, and collection mechanisms occurring in the CCD, several specific models have been published, notably in the solid-state image sensor community [15,16,20,21,40–43]. With respect to these models, our approach is very similar to that proposed by Pickel et al. [41]. In our approach, electrical charges have been considered to be deposited along the path of the ionizing particles assuming a line charge approximation without any radial extension at the time of deposition. Charges directly deposited in the volume of the CCD pixels (corresponding to the high electric field region in [41]) are assumed to be fully collected; charges deposited outside these active volumes in the silicon substrate (i.e., the low field region in [41]) are transported within the structure by a random-walk drift-diffusion model recently proposed [44]. This method is equivalent to that proposed by Ratti et al. [43] because we did not consider any electric field in this region. Consequently, the collection efficiency is less than 100% because a part of the charges is recombined or transported away from the pixels.

The model contains a certain number of adjustable parameters, and, as indicated by Picker et al. in [41], it is important to carefully calibrate these parameters. To perform such a calibration, the main geometrical dimensions have been extracted from

SEM and TEM observations as previously mentioned. Typical values used for low-doped p-type silicon have been chosen for carrier mobility, diffusion coefficient, and minority carrier lifetime related to the substrate region. Note that there are no fitting or unphysical parameters in the adopted model. In absence of direct measurement or estimation, only two parameters have been finely tuned to reproduce experimental data: the alpha particle emitter concentration in the CCD material (see further in this section) and the thickness of the "active layer" fixed to 10 µm. Finally, note that the vertical antiblooming structure has not been modeled in this first work, which could constitute a certain limitation of our approach. This point will be investigated in a forthcoming work.

Monte Carlo numerical simulations have been analyzed in terms of frequency (i.e., occurrence probability) distributions of event size (expressed in number of pixels) versus event charge for the different atmospheric particle sources (neutrons, protons, and muons). In order to reach sufficient event statistics, 500 millions of atmospheric neutrons and 1 million muons and protons incident on the CCD surface have been considered. Only 0.33% of the incident atmospheric neutrons have been found to interact with the CCD materials and to induce detectable events, widely ranging in terms of collected charges and sizes. With respect to protons and muons, neutrons are responsible of the largest (but rare) events, characterized by a size typically above 80 pixels. In addition, the probability of multiple pixel events is larger for neutrons than for muons and protons. Around 95% of protons induce events via direct ionization, and protons were found to deposit more charge compared to muons. The contribution of the latter is comparable to that of protons because of their abundance at ground level.

Concerning the contribution of alpha particles, we made the hypothesis that these impurities are uniformly distributed as traces in the CCD packaging and chip materials. Because alpha particles emitted from the two decay chains have a maximum range of about 60 µm in silicon (56 µm for the 8.78 MeV alpha particle emitted by the ^{212}Po isotope in the thorium decay chain), they can induce MPEs with a size limited to 19 pixels in this particular device (with respect to pixel dimensions). The relatively high experimental pixel event rate measured underground (12.5 h^{-1}) has been numerically reproduced by forcing the uranium (and thorium) concentration to the value of 7 ppm in the device. This is an extremely high value [45] that certainly reflects a residual contamination problem in the chip and/or in the packaging. The cover protective glass, just in front of the sensor, should be at the origin of this problem because such a contamination with alpha radioactive impurities has been already observed in such materials [46,47]. Due to the extremely reduced surface of this piece of material, it was not possible to investigate this problem, for example, with a direct emissivity measurement using an ultralow background alpha particle counter. The origin of the alpha contamination of the CCD remains thus not completely explained at this level of our investigations.

Figure 12.9 shows the comparison of the simulated pixel event rates with experimental measurements for the three different test locations. This graph confirms the role of alpha particle emitters in the CCD response at sea level. Neutron contribution is found negligible for this device at ground level whereas muons and protons contribute to about 15% in the pixel event rate at sea level (Marseille) and about 45%

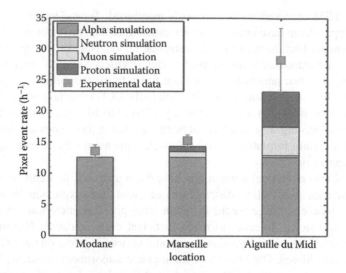

FIGURE 12.9 Comparison between experimental and simulated pixel event rates for sea level (Marseille), mountain altitude (Aiguille du Midi), and underground (Modane) test locations. (Reprinted from Saad Saoud et al., Use of CCD to Detect Terrestrial Cosmic Rays at Ground Level: Altitude vs. Underground Experiments, Modeling and Numerical Monte Carlo Simulation, *IEEE Trans. Nucl. Sci.*, 61, 3380–3388. © [2014] IEEE. With permission.)

at mountain altitude (AM). When cumulating the different contributions for alphas, neutrons, muons, and protons, the estimated pixel event rates for the three locations correctly match the experiment values. This quantitative result is strengthened by the very satisfactory agreement between simulations and measurements also obtained for the distribution of the MPE event size shown in Figure 12.10. This result supports the validity of our modeling approach that is capable of qualitatively reproducing the event images as illustrated in [24], showing that the interaction signatures obtained from simulation and experiments are very similar.

12.3.4 MODEL VALIDATION AT AVIONICS ALTITUDES

We took the opportunity of a long-haul flight at the end of this work to perform in-flight measurements (in the aircraft cabin) at avionics altitude (10,800 m) along a flight route characterized by important variations of the cosmic ray cutoff rigidity. This validation took a particular importance in view of results of Figure 12.9 that demonstrate a limited contribution in the CCD response of atmospheric particles at ground level, even at high mountain altitude. Figure 12.11 shows the flight route considered for this experiment; it corresponds to two commercial flights on May 29, 2014, connecting Frankfurt to Los Angeles and Los Angeles to Hawaii. Seven locations have been defined along the flight route; they are approximately equidistant in time-of-flight (roughly 1 hour). The duration of the experiment was limited to 7 hours due to the limited capacity of the power battery. Table 12.2 indicates the cosmic ray cutoff rigidity for these locations and the total flux above 1 MeV for protons, muons, and neutrons at the flight altitude (PARMA/EXPACS model, "aircraft

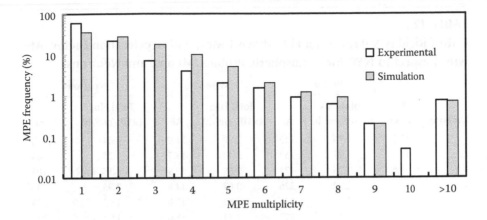

FIGURE 12.10 Experimental and simulated histograms of detected event sizes. The last point (>10) represents the proportion of all events with a size greater than 10 pixels. (Reprinted from Saad Saoud et al., Use of CCD to Detect Terrestrial Cosmic Rays at Ground Level: Altitude vs. Underground Experiments, Modeling and Numerical Monte Carlo Simulation, *IEEE Trans. Nucl. Sci.*, 61, 3380–3388. © [2014] IEEE. With permission.)

FIGURE 12.11 Definition of locations #1 to #7 along the flight route used for in-flight measurements using the CCD camera. (Reprinted from Saad Saoud et al., Use of CCD to Detect Terrestrial Cosmic Rays at Ground Level: Altitude vs. Underground Experiments, Modeling and Numerical Monte Carlo Simulation, *IEEE Trans. Nucl. Sci.*, 61, 3380–3388. © [2014] IEEE. With permission.)

cabin" environment). The corresponding "acceleration factors" are given by the ratio of these flux values with respect to the reference values at sea level. Data of Table 12.2 show that important variations (factor 4×) characterize the proton and neutron fluxes between locations #1 and #7. This is primarily due to the dramatic decrease of the cutoff rigidity along this route that impacts the intensity of terrestrial cosmic rays. Concerning the muon flux, this latter is less impacted (only by a factor 2×). These acceleration factors have been used to directly determine the contributions to the total pixel event rate of each particle type at avionics altitude from the simulated values at sea level. Table 12.3 shows the details of these calculations for protons, muons, and neutrons. Note that there is no fitting parameter or data adjustment; the

TABLE 12.2

Cutoff Rigidity Values, Total Flux above 1 MeV, and Acceleration Factor (AF) with Respect to NYC for Atmospheric Protons, Muons, and Neutrons

Location	Rc (GV)	Protons Total Flux (×10²/cm²/h)	AF	Muons Total Flux (×10²/cm²/h)	AF	Neutrons Total Flux (×10³/cm²/h)	AF
1	12.2	1.74	135	2.28	11.0	4.64	78
2	11	1.97	153	2.44	11.7	5.25	88
3	9.5	2.25	174	2.69	12.9	6.15	103
4	7.4	2.92	226	3.07	14.8	7.95	134
5	3.8	4.45	345	3.65	17.6	13.3	224
6	2.7	5.13	397	3.74	18.0	15.3	258
7	1.7	6.09	472	3.78	18.2	16.8	283

TABLE 12.3

Summary of Measured and Simulated Pixel Event Rates for Ground Level and In-Flight Locations

Location	Rc (GV)	Total Pixel Event Rate (h⁻¹) Protons	Muons	Neutrons	Alphas	Model	Experiment
LSM	4.9	0	0	0	12.5	12.5	13.5
Marseille	5.6	0.18	0.71	0.02	12.5	13.4	15.2
AM	4.8	5.6	4.6	0.3	12.5	23.0	28.2
1	12.2	24.4	7.8	1.5	12.5	46.2	49
2	11	27.7	8.3	1.7	12.5	50.2	62
3	9.5	31.6	9.2	2.0	12.5	55.2	52
4	7.4	41.0	10.4	2.6	12.5	66.5	74
5	3.8	62.5	12.4	4.3	12.5	91.7	96
6	2.7	72.0	12.7	5.0	12.5	102.3	97
7	1.7	85.5	12.9	5.4	12.5	116.3	94

pixel rate at avionics altitude is directly deduced from the simulated value at sea level for each component of the atmospheric radiation background multiplied by the acceleration factor. This supposes that the different spectra at sea level and at avionics altitudes have similar shapes, which is, of course, a first-order approximation. Also note that the alpha particle contribution to the pixel event rate has been taken as equal to the value measured from the underground test. Figure 12.12 shows the comparison between measurements and simulations. A good agreement is found, within the experimental and simulation uncertainties, roughly estimated to about 20%, notably due to the approximate estimations of the aircraft positions, the averaging of the pixel event rate on 1-hour period and the evaluation of the different acceleration factors per particle type previously discussed. Figure 12.12 demonstrates the capability

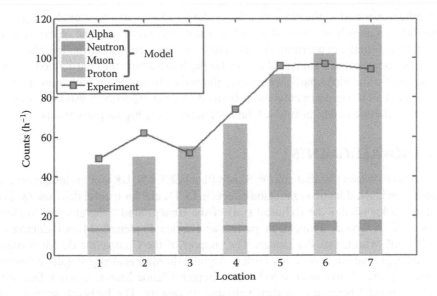

FIGURE 12.12 Comparison between in-flight measurements and simulation results for the different aircraft locations defined in Tables 12.2 and 12.3. The counting rates correspond to the sum of SPE and MPE signals. (Reprinted from Saad Saoud et al., Use of CCD to Detect Terrestrial Cosmic Rays at Ground Level: Altitude vs. Underground Experiments, Modeling and Numerical Monte Carlo Simulation, *IEEE Trans. Nucl. Sci.*, 61, 3380–3388. © [2014] IEEE. With permission.)

of such a simple CCD device to very correctly monitor the radiation ambient in the aircraft cabin. Moreover, simulation results highlight the respective contributions of the different particle species to the CCD signal. Neutrons are found to marginally contribute to the counting rate, even at these altitudes, whereas muons and primarily protons dominate the CCD detection response.

12.4 CONCLUSION

In conclusion, this work surveyed the effects of natural radiation on the response of CCD devices at terrestrial level. After introducing the main radiation effects in such solid-state imagers, we briefly described single event effects in CCD devices on the basis of the few dedicated works principally published in the last two decades in literature. In the following, we focused on our recent investigations demonstrating the use of a standard commercial CCD device to detect terrestrial cosmic rays at ground level and at avionics altitude. Underground experiments evidenced, for the particular device used, a surprisingly high contribution of the alpha contamination in the CCD response, largely contributing to screen the influence of atmospheric radiation at sea level. Nevertheless, the pixel event rate has been found to increase with altitude and to finely depend on the orientation of the device at ground level, suggesting an anisotropy of the particle flux to which the device is sensitive. A comparison with data obtained with other radiation detectors (neutron monitor, argon ionization chamber) operated in the close vicinity of the CCD experiment strongly suggested the role of atmospheric charged particles (muons and protons) in the observed pixel event rate. Finally, concerning our modeling and simulation contribution, the complete development of a numerical model, taking into account the contribution of both atmospheric and alpha particle emitters, allowed us to reproduce with a satisfactory agreement all the experimental data obtained. A final experiment was conducted at avionics altitude to definitively validate our approach at higher particle flux.

ACKNOWLEDGMENTS

The authors would like to thank Dr. Simon Platt (UCLAN, UK) for his long-term collaboration in the domain of radiation effects in CCDs and for fruitful discussions. The authors acknowledge the technical staff of the Undeground Laboratory of Modane (LSM, CEA-CNRS, France), in particular Fabrice Piquemal, Michel Zampaolo, Guillaume Warot, and Pia Loaiza. The support of the Compagnie du Mont-Blanc for the experimental part of this work at Aiguille du Midi is also gratefully acknowledged. Special thanks are due to Laurent Berger, Claude Marin, Emerick Desvaux, and Christophe Bochatay for their help and hospitality. The logistical support and the reconnaissance work performed by José Autran are also sincerely acknowledged.

REFERENCES

1. J. Ohta, *Smart CMOS Image Sensors and Applications*, CRC Press, 2007.
2. D. Durini (Editor), *High Performance Silicon Imaging: Fundamentals and Applications of CMOS and CCD sensors*, Woodhead Publishing Series in Electronic and Optical Materials, Elsevier, 2014.
3. S. B. Howell, *Handbook of CCD Astronomy*, Cambridge University Press, 2006.
4. A. R. Smith, R. J. McDonald, D. C. Hurley, S. E. Holland, D. E. Groom, W. E. Brown, D. K. Gilmore, R. J. Stover, M. Wei, "Radiation events in astronomical CCD images," Proc. SPIE 4669, Sensors and Camera Systems for Scientific, Industrial, and Digital Photography Applications III, pp. 172–183, 2002.
5. D. Groom, "Cosmic Rays and Other Nonsense in Astronomical CCD Imagers," in *Scientific Detectors for Astronomy*, Astrophysics and Space Science Library Vol. 300, pp. 81–94, 2004. Available at: http://www.astronomy.ohio-state.edu/MDM/OSMOS /CCD_CosmicRays_groom.pdf.
6. L. W. Massengill, B. L. Bhuva, W. T. Holman, M. L. Alles, T. D. Loveless, "Technology Scaling and Soft Error Reliability," 2012 IEEE Intl. Reliability Physics Symposium (IRPS), pp. 3C.1.1–3C.1.7, 2012.
7. P. Roche, J. L. Autran, G. Gasiot, D. Munteanu, "Technology Downscaling Worsening Radiation Effects in Bulk: SOI to the Rescue," International Electron Device Meeting (IEDM 2013), Washington, D.C., USA, December 9–11, 2013, pp. 31.1.1–31.1.4.
8. T. S. Lomheim, R. M. Shima, J. R. Angione, W. F. Woodward, D. J. Asman, R. A. Keller, and L. W. Schumann, "Imaging charge-coupled device (CCD) transient response to 17 and 50 MeV proton and heavy-ion irradiation," *IEEE Transactions on Nuclear Science*, Vol. 37, no. 6, pp. 1876–1885, Dec. 1990.
9. R. Bailey, C. J. S. Damerell, R. L. English, A. R. Gillman, A. L. Lintern, S. J. Watts, and F. J. Wickens, "First measurements of efficiency and precision of CCD detectors for high energy physics," *Nuclear Instruments and Methods in Physics Research*, Vol. 213, no. 2–3, pp. 201–215, Aug. 1983.

10. G. R. Hopkinson, "Radiation effects on solid state imaging devices," *Radiation Physics and Chemistry,* Vol. 43, no. 1/2, pp. 79–91, 1994.
11. G. R. Hopkinson, C. J. Dale, and P. W. Marshall, "Proton effects in charge-coupled devices," *IEEE Transactions on Nuclear Science,* Vol. 43, no. 2, pp. 614–627, Apr. 1996.
12. A. M. Chugg, R. Jones, P. Jones, P. Nieminen, A. Mohammadzadeh, M. S. Robbins, and K. Lovell, "CCD miniature radiation monitor," *IEEE Trans. Nucl. Sci.,* Vol. 49, pp. 1327–1332, 2002.
13. A. M. Chugg, R. Jones, M. J. Moutrie, C. S. Dyer, K. A. Ryden, P. R. Truscott, J. R. Armstrong, D. B. S. King, "Analyses of CCD images of nucleon-silicon interaction events," *IEEE Trans. Nucl. Sci.,* Vol. 51, pp. 2851–2856, 2004.
14. G. R. Hopkinson, A. Mohammadzadeh. "Radiation Effects in Charge-Coupled Device (CCD) Imagers and CMOS Active Pixel Sensors," *Journal of High Speed Electronics and Systems,* Vol. 14, No. 2, pp. 419–443, 2004.
15. Z. Török and S. P. Platt, "Application of imaging systems to characterization of single-event effects in high-energy neutron environments," *IEEE Trans. Nucl. Sci.,* Vol. 53, pp. 3718–3725, 2006.
16. Z. Török, "Development of image processing systems for cosmic ray effect analysis," Ph.D. Thesis, University of Central Lancashire (UK), 2007.
17. S. P. Platt and Z. Török, "Analysis of SEE-inducing charge generation in the neutron beam at The Svedberg Laboratory," *IEEE Trans. Nucl. Sci.,* Vol. 54, pp. 1163–1169, 2007.
18. A. M. Chugg, A. J. Burnell, and R. Jones, "Webcam observations of SEE events at the Jungfraujoch research station," in Proc. 9th European Conference on Radiation Effects on Components and Systems (RADECS 2007), paper PD-1, 2007.
19. S. P. Platt, B. Cassels, and Z. Török, "Development and application of a neutron sensor for single event effects analysis," *J. Phys.: Conf. Ser.,* Vol. 15, pp. 172–176, 2005. [Online]. Available: http://stacks.iop.org/1742-6596/15/172.
20. X. X. Cai, S. P. Platt, W. Chen, "Modelling Neutron Interactions in the Imaging SEE Monitor," *IEEE Trans. Nucl. Sci.,* Vol. 56, pp. 2035–2041, 2009.
21. X. X. Cai and S. P. Platt, "Modeling Neutron Interactions and Charge Collection in the Imaging Single-Event Effects Monitor," *IEEE Trans. Nucl. Sci.,* Vol. 58, pp. 910–915, 2011.
22. X. X. Cai, S. P. Platt, S. D. Monk, "Design of a Detector for Characterizing Neutron Fields for Single-Event Effects Testing," *IEEE Trans. Nucl. Sci.,* Vol. 58, pp. 1123–1128, 2011.
23. G. Hubert, A. Cheminet, T. Nuns, and V. Lacoste, "Atmospheric Radiation Environment Analyses Based-on CCD Camera, Neutron Spectrometer and Multi-Physics Modeling," *IEEE Trans. Nucl. Sci.,* Vol. 60, pp. 4660–4667, 2013.
24. T. Saad Saoud, S. Moindjie, J. L. Autran, D. Munteanu, F. Wrobel, F. Saigne, P. Cocquerez, L. Dilillo, M. Glorieux, "Use of CCD to Detect Terrestrial Cosmic Rays at Ground Level: Altitude vs. Underground Experiments, Modeling and Numerical Monte Carlo Simulation," *IEEE Transactions on Nuclear Science,* Vol. 61, pp. 3380–3388, 2014.
25. J. L. Autran, D. Munteanu, *Soft Errors: From Particles to Circuits,* CRC Press, 2015.
26. S. Ochi, T. Lizuka, M. Hamasaki, Y. Sato, T. Narabu, H. Abe, Y. Kagawa, K. Kato, *Charge-Coupled Device Technology,* CRC Press, 1997.
27. D. R. Smith, A. D. Holland, I. B. Hutchinson, "Random telegraph signals in charge coupled devices," *Nuclear Instruments and Methods in Physics Research Section A: Accelerators, Spectrometers, Detectors and Associated Equipment,* Vol. 530, Issue 3, pp. 521–535, 2004.
28. JEDEC Standard Measurement and Reporting of Alpha Particles and Terrestrial Cosmic Ray-Induced Soft Errors in Semiconductor Devices, JESD89 Arlington, VA: JEDEC Solid State Technology Association [Online]. Available at: http://www.jedec.org/download/search/JESD89A.pdf.

29. TERRAMU Platform, Terrestrial Radiation Environment Characterization Platform Aix-Marseille University, France. Available at: http://www.natural-radiation.net.
30. ASTEP Platform, Altitude SEE Test European Platform, Dévoluy, France. Available at: http://www.astep.eu.
31. S. Semikh, S. Serre, J. L. Autran, D. Munteanu, S. Sauze, E. Yakushev, S. Rozov, "The Plateau de Bure Neutron Monitor: Design, Operation and Monte Carlo Simulation," *IEEE Transactions on Nuclear Science*, Vol. 59, no. 2, pp. 303–313, 2012.
32. XIA model UltraLo-1800, ultra-low background alpha particle counters [Online]. Available: http://www.xia.com/UltraLo/.
33. M. S. Gordon, K. P. Rodbell, H. H. K. Tang, E. Yashchin, E. W. Cascio, B. D. McNally, "Selected Topics in Ultra-Low Emissivity Alpha-Particle Detection," *IEEE Trans. Nucl. Sci.*, Vol. 60, pp. 4265–4274, 2013.
34. S. Moindjie, Master's Degree Thesis, Aix-Marseille University, 2013.
35. T. Sato, H. Yasuda, K. Niita, A. Endo, L. Sihver. "Development of PARMA: PHITS based Analytical Radiation Model in the Atmosphere," *Radiation Research,* Vol. 170, pp. 244–259, 2008.
36. Quotid Atmopsheric Radiation Model (QARM). [Online]. Available: http://82.24.196.225:8080/qarm.
37. S. Agostinelli et al., "Geant4—A simulation toolkit," *Nuclear Instruments and Methods in Physics Research Section A: Accelerators, Spectrometers, Detectors and Associated Equipment*, Vol. 506, pp. 250–303, 2003.
38. S. Martinie, T. Saad-Saoud, S. Moindjie, D. Munteanu, J. L. Autran, "Behavioral modeling of SRIM tables for numerical simulation," *Nuclear Instruments and Methods in Physics Research Section B: Beam Interactions with Materials and Atoms*, Vol. 322, pp. 2–6, 2014.
39. H. H. K. Tang, "SEMM-2: A new generation of single-event-effect modeling tools," *IBM Journal of Research and Development*, Vol. 52, pp. 233–244, 2008.
40. J. M. Pimbley, G. J. Michon, "Charge detection modeling in solid-state image sensors," *IEEE Trans. Electron Dev.*, Vol. 34, pp. 294–300, 1987.
41. J. C. Pickel, R. A. Reed, R. Ladbury, B. Rauscher, P. W. Marshall, T. M. Jordan, B. Fodness, G. Gee, "Radiation-Induced Charge Collection in Infrared Detector Arrays," *IEEE Trans. Nucl. Sci.*, Vol. 49, pp. 2822–2829, 2002.
42. G. Rolland, L. Pinheiro da Silva, C. Inguimbert, J. P. David, R. Ecoffet, M. Auvergne, "STARDUST: A Code for the Simulation of Particle Tracks on Arrays of Sensitive Volumes With Substrate Diffusion Currents," *IEEE Trans. Nucl. Sci.*, Vol. 55, pp. 2070–2078, 2008.
43. L. Ratti, L. Gaioni, G. Traversi, S. Zucca, S. Bettarini, F. Morsani, G. Rizzo, L. Bosisio, I. Rashevskaya, "Modeling Charge Loss in CMOS MAPS Exposed to Non-Ionizing Radiation," *IEEE Trans. Nucl. Sci.*, Vol. 60, pp. 2574–2582, Aug. 2013.
44. M. Glorieux, J. L. Autran, D. Munteanu, S. Clerc, G. Gasiot, P. Roche, "Random-Walk Drift-Diffusion Charge-Collection Model For Reverse-Biased Junctions Embedded in Circuits," presented at NSREC 2014 and submitted to *IEEE Trans. Nucl. Sci.*, 2014.
45. S. Kumar, S. Agarwal, J. P. Jung. "Soft error issue and importance of low alpha solders for microelectronics packaging," *Rev. Adv. Mat. Sci.*, Vol. 34, pp. 185–202, 2013.
46. Z. Török, S. P. Platt, X. X. Cai, "SEE-inducing effects of cosmic rays at the High-Altitude Research Station Jungfraujoch compared to accelerated test data," *RADECS 2007 Proc.*, pp. 1–6, 2007.
47. W. C. McColgin, C. Tivarus, C. C. Swanson, A. J. Filo, "Bright-Pixel Defects in Irradiated CCD Image Sensors," *MRS Proceedings*, Vol. 994, pp. 0994-F12-06, 2007.

13 Radiation Effects on Optical Fibers and Fiber-Based Sensors

Sylvain Girard, Aziz Boukenter,
Youcef Ouerdane, Nicolas Richard,
Claude Marcandella, Philippe Paillet,
Layla Martin-Samos, and Luigi Giacomazzi

CONTENTS

13.1   INTRODUCTION

The presence of ionizing or nonionizing radiation appears as a severe constraint for most of the electronic components and circuits when the total ionizing dose* (TID) exceeds 1 kGy(SiO$_2$) [1]. Such TID levels are encountered in natural environments, such as space, or in the artificial manmade environments associated with nuclear power plants and/or high-energy physics facilities [2]. Radiation induces a variety of transient or permanent effects that can alter the device functionality or sometimes, depending on the technology vulnerability level and irradiation conditions, lead to the complete loss of its functionality. Optical fibers and fiber-based devices have been shown since the '70s to be more radiation-tolerant than most of the electronic technologies, in addition to their other intrinsic advantages, such as their electromagnetic immunity, low weight, high multiplexing capacity, and high temperature resistance [3]. Then, this photonic technology was first routinely used for data transfer in radiation environments instead of copper cables, becoming later key parts of more complex systems or subsystems in physics facilities, such as the plasma diagnostics for fusion facilities [4]. More recently, optical fibers revealed exceptional advantages for the development of new classes of sensors [5]. These new optical fiber sensors (OFS) use the fiber material scattering properties as sensing phenomenon when the reflectometry technique offers high spatial resolution over large fiber distances. Today's Brillouin or Raman-based sensors allow monitoring strain and temperature changes over dozens of kilometers with resolution below 1 m along one fiber link [6] whereas Rayleigh-based sensors offer an exceptional resolution of less than 1 mm over 70 m of fiber [7]. More complex techniques are under development to increase even more the sensor performances in terms of resolution, sensitivity, and discrimination between the two measurands.

Despite their good tolerance, optical fibers are not immune to radiation. Section 13.2 introduces the basic mechanisms occurring at the microscopic scale into the amorphous (a-SiO$_2$) silica glass of the fiber core and cladding and leading to the generation of radiation-induced point defects. These defects are responsible for three main degradation mechanisms occurring at the macroscopic scale that are also presented in this chapter. Section 13.3 reviews the parameters, associated with the fiber itself or related to external factors, impacting the levels and kinetics of these macroscopic changes. In Section 13.4, a few resulting examples of challenges associated with recent applications are presented as the main achievements concerning the integration of radiation-tolerant fibers into these harsh environments. Fewer results exist today concerning the radiation vulnerability and hardening against radiation of OFS, but this topic is introduced and available results briefly reviewed as the perspectives for such sensors. Finally, Section 13.5 discusses the efforts done to model the complex response of an optical fiber submitted to radiation. Several pragmatic models (recently reviewed [4]) exist to extrapolate the fiber degradation in real environments from facility testing. A more ambitious multiscale approach is also under development, from ab initio modeling of microscopic radiation induced defects to the device

---

* The dose is the quantity of energy deposited by radiation into a given material. In this chapter, we consider as material the amorphous silica (a-SiO$_2$). The dose is expressed in Gray (SiO$_2$) or sometimes in rad (SiO$_2$) with the conversion factor 1 Gy = 100 rad.

macroscopic response [8–10]. Recent advances on this theme are presented to highlight the perspectives of this work.

13.2 MAIN RADIATION EFFECTS ON OPTICAL FIBERS

Most optical fibers are designed with a core and a cladding made of pure or doped amorphous silica (a-SiO_2) with an overlayer of acrylate or other types of coating ensuring their mechanical strength [11]. To permit the guiding of light into the fiber core, its refractive index must be larger than the cladding one. This implies to dope the various parts of the fiber with some chemical elements allowing the increase (*Germanium, Phosphorus, Nitrogen...*) or decrease (*Fluorine, Boron...*) of the a-SiO_2 refractive index [11]. Depending on the various possible manufacturing processes [11] and on the fiber optical specifications in terms of attenuation, dispersion or numerical aperture, very different refractive index profiles (*and then glass layer compositions*) are produced by the fiber manufacturers. As will be explained later in this chapter, the choices made by the fiber designer strongly impact its radiation response by changing the nature and properties of the microscopic defects created during irradiation. It should be noticed that other classes of optical fibers than those based on silica and total internal reflection phenomenon exist, such as hollow core optical fibers for which light is mainly guided into air [12] or lower cost polymer-based optical fibers [13]. The radiation response of these other fibers is less documented as some parameters still limit their interest for harsh environments, but preliminary data concerning their radiation response are available in [14–16].

13.2.1 RADIATION-INDUCED POINT DEFECTS AND STRUCTURAL CHANGES

Radiation generates point defects into amorphous silica glass by ionization processes and by displacement, depending on the nature of the incident particles. A complete description of the basic mechanisms associated with the defect creation and annihilation was given by D. L. Griscom in [17] and is adapted in Figure 13.1.

Generation mechanisms of these radiation-induced point defects are mostly identical to those occurring in the thin SiO_2 layer of MOSFETs. However, for optical fibers, one should note that no electric fields are applied throughout the glass. This absence of field has a huge impact on the possible migration mechanisms of generated charges into the glass and then on their recombination mechanisms. Furthermore in the fiber case, not only the electrically active point defects have to be considered to understand the device radiation response but all generated defects will be optically active with contribution to global losses varying with the operating wavelength. Figure 13.2a illustrates an a-SiO_2 cell of 108 atoms, showing the structure of glass before irradiation with the main element being a tetrahedron with a central silicon (Si) atom bonded to four oxygen (O) atoms occupying the corners of the tetrahedron. Figure 13.2b shows some structures of typical Si-related defects impacting the fiber radiation vulnerability. Point defects are associated with the appearance of new energy states into the 9 eV band-gap of silica (see Figure 13.2c), being able to trap electrons or holes and adding optical absorption (OA) bands in the fiber transmission spectrum. The OA amplitudes depend on the defect concentration and their

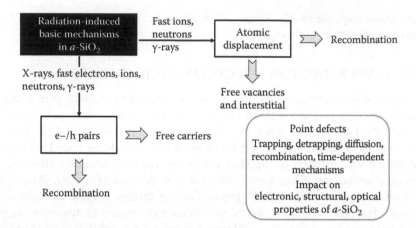

FIGURE 13.1 Illustration of the main mechanisms related to the radiation induced point defects creation and annihilation. (Adapted from D. L. Griscom, *Proc. of SPIE.* 541 "Radiation Effects in Optical Materials," pp. 38–59, 1985.)

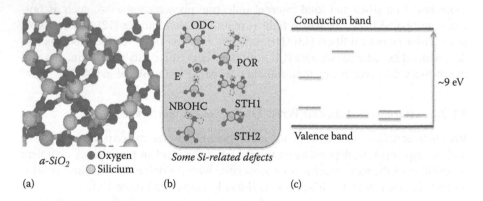

FIGURE 13.2 (a) Illustration of the structure of pure amorphous silica, (b) of some Si-related defects, and (c) of their consequences with the appearance of new energy levels in the bandgap of silica.

cross sections. These OA bands are responsible for the radiation-induced attenuation (RIA) described in Section 13.2.2. Some of these defects could also emit light and contribute to the radiation induced emission (RIE) detailed in Section 13.2.3. The study of these defects is mandatory to understand the origin of radiation-induced changes in optical fibers. As the fibers are based on various layers of differently doped silica glass, to fully explain the RIA measured for a given fiber, one should consider both the spatial distribution of the guided light and the distribution of defects in the fiber cross-section. The concentrations of the defects will be affected by the nature of dopants and impurities present in the layer but also by microscopic effects such as the residual strain along the fiber cross-section that can be related to the fiber manufacturing process.

13.2.2 Radiation-Induced Attenuation

Before irradiation, Telecom-grade single-mode optical fibers present very low attenuation in the Telecom windows, typically below 0.2 dB/km at 1550 nm. The attenuation level strongly depends on the operating wavelength chosen for the application and increases as the wavelength decreases.

Radiation alters the fiber transmission efficiency by causing an excess of attenuation, called radiation induced attenuation (RIA) that is mainly related to absorption by defects. The RIA level increases with TID, and its kinetics depend on the fiber type and its operating wavelength. RIA often decreases after the end of the irradiation highlighting the transient (or unstable) nature of some generated defects whereas others are stable at the operating temperature. As an example, an X-pulse of a few ns causes a RIA level into a Telecom grade germanosilicate fiber in the order of 2000 dB/km at 1550 nm after a 100 Gy dose, 99% of this RIA being recovered in less than 1 s [18].

Figure 13.3 illustrates the RIA levels in the visible to infrared domains and kinetics measured in two Telecom-grade Ge-doped single-mode and multimode optical fibers after such an exposure. In Figure 13.3 inset, the RIA recovery after irradiation at 1550 nm highlights the impact of the fiber profile of use with respect to irradiation chronology. It is remarkable that these excess losses also strongly depend on the wavelength of interest, being usually more intense in the ultraviolet and visible parts

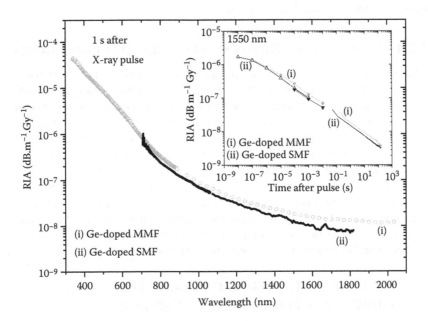

FIGURE 13.3 Spectral dependence of the radiation-induced attenuation (RIA) measured in two Telecom-grade Ge-doped single-mode fibers (SMFs) and multimode fibers (MMFs), 1 s after an X-ray pulse. In the inset is illustrated their RIA time dependence at 1550 nm after an X-ray pulse.

of the spectrum where more of the bands are peaking. No OA bands can be distinguished here, the RIA being related to overlapping between numerous OA bands. To design a radiation hardened fiber or to evaluate the vulnerability of commercial-off-the-shelf (COTS) fibers to a future environment (*e.g., that Earth facilities cannot yet reproduce*), the understanding of the nature of defects responsible for the RIA at the operating wavelength is mandatory. Improving the fundamental knowledge authorizes to imagine hardening-by-design strategies or to conceive predictive models. RIA is the main impacting and most studied effect for today's application of optical fibers.

13.2.3 Radiation-Induced Emission

It has been observed that light can also be generated under irradiation and guided to the detector. This is the so-called radiation induced emission (RIE) phenomenon. This parasitic light superposes to the signal and then strongly increases the signal-to-noise ratio, causing in worst cases the loss of information [19]. Different mechanisms can explain this excess of light such as Cerenkov light [20] or radiation-induced luminescence originating from the point defects [21]. For most of the applications, RIE is not as impacting as RIA but it should be noted that for radiation-tolerant or hardened fibers, it will be more probable to face RIE issues as this light is not as efficiently reabsorbed as in sensitive optical fibers. Furthermore, RIE has also been shown to be a useful way to monitor the nuclear power of a reactor core as its global intensity is directly related to this parameter [22].

13.2.4 Compaction and Radiation-Induced Refractive Index Changes

Radiation can lead to refractive-index changes. These changes can result from a contribution of the point defects and the RIA through the Kramers–Krönig relationships or to a density change via the Lorentz–Lorenz formula. Compaction was first observed by Primak in bulk silica glass exposed to high fluences ($>10^{18}$ n/cm²) of neutrons, leading to a 3% change of the refractive index [23]. These structural changes do not continuously increase with the fluence but tend to saturate when the amorphous silica reaches a new structure, called the metamict phase [24]. The radiation induced structural changes become of major importance when the optical fiber is used as the sensitive part of a sensor. In this case, the structural properties of the fibers are usually exploited to monitor external factors like strain or temperature. This is the case for example for the sensors based on Brillouin, Rayleigh, or Raman scattering of the fibers or those based on Fiber Bragg gratings. In fibers, these effects were recently discussed in [25].

13.3 INTRINSIC AND EXTRINSIC PARAMETERS IMPACTING FIBER RADIATION RESPONSE

One of the main reasons limiting the fiber integration in harsh environments is the diversity of the possible responses for COTS fibers and the lack of predicting tools allowing to evaluate a fiber vulnerability to given environments. This is due to the fact that a large number of parameters can affect the generation or bleaching processes

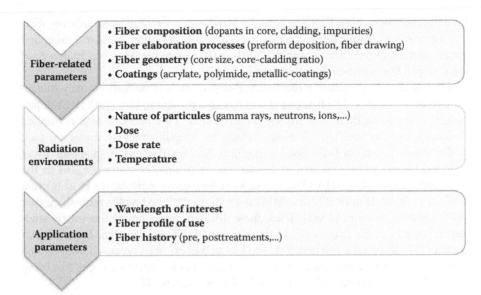

FIGURE 13.4 Review of the different intrinsic and extrinsic parameters affecting the fiber radiation vulnerability.

of radiation-induced point defects with an effect on the corresponding macroscopic changes. Figure 13.4 reviews the main parameters that have been shown to change the fiber radiation sensitivity.

13.3.1 Fiber-Related Parameters

First, the sensitivity of the fiber is mainly governed by the composition chosen for the different glass layers where the signal propagates. Into the core, the nature, concentrations of defects and associated RIA growth and decay kinetics are, at the first order, related to the dopants incorporated to increase the refractive index. In COTS fibers, Germanium is often used, and Ge-related defects are mainly responsible for observed RIA [26]. However, the presence of other codopants into the core can affect the Ge defect equilibrium and change the fiber response [27]. This is the case, for example, when F or Ce is added into the core [28,29]. Furthermore, as part of the light propagates into the cladding (up to 40% at 1550 nm for single-mode fibers [SMFs]), the defects created in this region can also totally change the response of fibers designed with similar cores [30].

For applications associated with moderate radiation constraints, low-cost Ge-doped fibers are usually acceptable; this is the case, for example, for Laser Mégajoule applications [31], large hadron collider [32], or space applications. For the other main dopants used in the core, such as phosphorus or nitrogen, defects related to these chemical species also explain the fiber response. Nitrogen-doped fibers are of interest for environments associated with moderate to high steady-state irradiation doses [33,34]. Moreover, it has been well established that phosphorus-doped optical fibers present very high RIA levels under steady-state irradiation prohibiting their use for nearly all applications needing radiation resistant fiber. A notable exception concerns the one needing to operate shortly after an irradiation pulse: In this case, the P-doped fibers present low transient RIA [35].

Despite their high radiation sensitivity, P-doped fibers remain still largely studied for two main reasons. The first one is that P is also present in rare earth (RE)-doped optical fibers to facilitate the RE ion incorporation. This research axis is implied to search ways to neutralize P-doped defects by adopting various strategies, such as Cerium-codoping or hydrogen-loading [36]. Second, the very high sensitivity of P-doped fibers can be exploited to develop online dosimetry systems based on the dose dependence of the RIA [37–39]. For high-dose environments, the most tolerant SMF or multimode optical fibers (MMFs) are those designed with a pure-silica core (PSC) or F-doped core [40,41]. Then, these different fibers have been widely studied, especially in the framework of the International Thermonuclear Experimental Reactor (ITER) project. For both PSC and F-doped fibers, RIA is caused by Si-related defects, whose concentration depends on the impurity content, these impurities being mainly hydroxyls groups OH groups and chlorine species [42].

Two main classes of pure-silica core fibers exist: the "wet" ones (*high OH content, low Cl content*) optimized for transmission in the ultraviolet and visible part of the spectrum and "dry" ones (*low OH content, high Cl content*) for the infrared domain. To further enhance the hardness of these PSC fibers, new waveguides have been developed with both low-OH and low-Cl contents to reduce all defect concentrations [43,44]. Then, "new" additional defects, such as the self-trapped holes, have been revealed even at room temperature, limiting their use for high-dose rate experiments due to transient RIA at the beginning of irradiation [45–47]. For such fibers, preirradiation can be a way to enhance the fiber radiation hardness [48]. Finally, the most promising results have been obtained with PSC fibers preloaded with gas such as H_2 or D_2 that can passivate the defects as soon as they are created [49]. Work is still in progress to validate the positive influence of such treatments during the whole lifetime of physics facilities.

In addition to the composition, preform deposition process parameters as well as its drawing conditions into fiber can impact the fiber radiation response [50–52]. All these fabrication steps determine the glass properties of the core and cladding and then their sensitivities to radiation. As an example, during the drawing of a preform into a fiber, some residual stress is more or less frozen into the waveguide, in particular at the core—cladding interface, generating strained regular Si–O–Si bonds that have been shown to act as precursor sites for the generation of non-bridging oxygen hole centers (NBOHCs) and E′ defects [50]. Furthermore, in some case, the opto-geometric parameters of the optical fiber can change its radiation response, like the size of the core or the core-cladding diameter ratio [53]. Such effects can be explained by diffusion-limited mechanisms or by changes in the overlap between the guided modes and the various layers constituting the waveguide [54].

13.3.2 EXTERNAL PARAMETERS

A large variety of harsh environments of interest for optical fibers exists. Those are associated with very different constraints in terms of dose, dose rate, nature of particles, and temperature. These differences explain that one fiber tolerant to one environment for a given application will probably not fulfill the requirements of another application under different irradiation conditions. Indeed, all these extrinsic parameters directly affect the nature, concentration, generation, and bleaching mechanisms or the radiation-induced point defects and consequently the amplitudes of the macroscopic changes (RIA, RIE, compaction...) measured under irradiation.

As discussed in Section 13.2.1, depending on the nature of the irradiation, different processes can lead to the generation of point defects. This is the case, for example, for neutrons that can induce different defects in silica glasses and cause major structural changes of the glass at very high fluences ($>10^{17}$ n/cm^2) [55]. The most complete study on this gamma/neutron irradiation has been done by CIEMAT in the framework of ITER project on bulk silica glasses and are described in [56–58].

Usually the RIA increases with the deposited dose as the defect concentration increases with the quantity of energy deposited into the glass [59,60]. However, especially under steady-state irradiation, this defect creation is counterbalanced by the instability of the generated centers absorbing at the wavelength of interest at the temperature of the irradiation. RIA evolution results from a competition between different processes implying generally several point defects. The RIA can then grow linearly with the dose or with a second-order law at moderate dose whereas RIA can saturate at doses larger than 1 kGy [61], or for some fibers RIA decreases after a dose threshold [62]. These competitive processes are also strongly influenced by the dose rate, the speed of energy deposition into the material. For nearly all optical fibers, it has been shown that increasing the dose rate leads to an increase of the RIA as for a given dose, the available time for defects to recover diminishes. In a few papers, an enhanced low-dose rate effect, comparable to the enhanced low dose rate sensitivity (ELDRS) in electronics has been observed for RE-doped optical fibers [63]. Such behavior is less probable, but it was demonstrated in [64] that it appears under some specific conditions. This dose rate dependence has a very strong impact on radiation testing procedures for space applications. For such applications, dose rates of interest are very low and the TID associated to a space mission of 20 years duration can only be achieved on Earth using a larger dose rate. Then, from these ground results, the vulnerability of the fiber to the real environment is usually extrapolated with simple models such as those presented in [65,66].

Finally, the temperature is another major parameter to consider when evaluating the fiber sensitivity. If a fiber is irradiated at a low temperature and then thermally treated at higher temperature, the RIA usually decreases. The thermal treatment eliminates the part of the defects created at low-T and unstable at higher temperatures. This positive temperature effect is less obvious when we consider irradiations occurring at different temperatures. Indeed, if higher temperature during irradiation increases the number of unstable defects and then accelerates their bleaching, it can also increase the defect generation efficiency and finally result in higher RIA levels. If this second effect is larger than the positive one, increasing the temperature of irradiation can increase the defects concentration and so the RIA. This is demonstrated in [67] and has still to be fully investigated.

13.4 MAIN APPLICATIONS AND CHALLENGES

13.4.1 FOR OPTICAL FIBERS

There are plenty of applications for optical fibers in harsh environments. In this paragraph, we discuss only a few of them that have been the subject of recent publications since 2000s. The first and second ones deal with the various optical fibers needed for fusion facilities, by magnetic confinement (ITER), and by inertial confinement (Laser Mégajoule—LMJ, National Ignition Facility—NIF), respectively. The third section consists of the identification and qualification of an optical fiber for the optical links of the large hadron collider (LHC). Finally, the last part concerns the hardening approaches applied to the RE-doped optical fibers for space applications.

13.4.1.1 International Thermonuclear Experimental Reactor (ITER) Studies

For ITER, the integration of radiation-tolerant optical fibers has been investigated for several applications. The most studied optical fibers concern the multimode ones intended for use as part of the plasma diagnostics [68,69]. These fibers will have to resist to high doses of radiations (a few MGy and fast neutron fluences 10^{16}–10^{18} n/cm^2). For this purpose, several research groups from Belgium (SCK-CEN), Russia (Fiber Optic Research Center), and Japan (Tohoku University) have investigated a large number of multimode optical fibers (see for example [70]). From their studies, it has been pointed out that pure silica core and F-doped optical fibers were the most promising candidates for transmission of the plasma light emission. Furthermore, with appropriate mitigation techniques, such as a hydrogen gas loading pretreatment, it is possible to further increase their radiation hardness [68]. Another very interesting study concerns the development of a fiber optic current sensor to monitor the plasma current and the magnetic field to control the plasma magnetic equilibrium [71]. The operation of this sensor is based on the Faraday effect: The magnetic field creates a birefringence into the fiber [68].

13.4.1.2 Mégajoule Class Laser Studies

For Mégajoule class lasers, applications for optical fibers located in the experimental halls of the French (LMJ) or US (NIF) facilities range from data transfer links to plasma or laser diagnostics [72]. These fibers will be submitted to very specific radiation constraints. The mixed radiation environment (gamma, 14 MeV neutrons, X-rays) is characterized by a very high dose rate and short irradiation duration: most of the dose is deposited in less than 300 ns [73]. For most of the applications, radiation-tolerant optical fibers have been identified or mitigation approaches applied to decrease the radiation constraints and increase the fiber-based system lifetimes. Particularly, optimizing the profile of use of the various fiber links is crucial as the RIA levels are strongly depending on the time between the irradiation pulse and the fiber operation (see Figure 13.3). The most sensitive systems are those that should operate during or just after the shots, such as the plasma or laser diagnostics. Moreover these diagnostics work in the ultraviolet and visible parts of the spectrum where the RIA values are the higher. Developments are then still in progress for this class of optical fibers to provide radiation-hardened systems.

13.4.1.3 Large Hadron Collider (LHC) Studies

A very complete study was done by LHC at CERN ("Conseil Européen pour la Recherche Nucléaire") to identify one single-mode fiber fulfilling the requirements of the 2500 km of optical links in the beam cleaning sections of this facility [74]. This SMF has to be tolerant to doses up to 100 kGy. Various SMFs from different manufacturers have been tested [75] at various doses, dose rates, and injected power levels under γ-rays and to a mixed environment representative of the application. From these tests, a COTS F-doped optical fiber from Fujikura was selected and fully characterized, exhibiting a RIA level below 5 dB/km up to 100 kGy (the allowed losses were below 6 dB/km). A hardness assurance program ensures the quality of all the preforms and fibers elaborated in this framework, revealing the impact of process parameters on the fiber radiation response [76].

13.4.1.4 Rare-Earth Doped Fibers and Amplifiers for Space Applications

Rare-earth (RE)-doped optical fibers with erbium (Er), ytterbium (Yb) or erbium/ytterbium (Er/Yb) are of great interest for space or military applications [77,78] as these fibers are key element of fiber optic gyroscopes, amplifiers or lasers. Usually only small lengths of these fibers are used, but the RE-fibers have been shown to be very sensitive to radiation compared to passive SMFs [79]. This high radiation sensitivity was rather explained by the nature of the additional codopants (Al or P) incorporated in their cores to enhance their optical and amplifying properties rather than by the RE ions themselves. The RIA at the pump and signal wavelengths mainly drive the more complex response of the amplifiers based on these fibers and simulation codes exist that can consider the radiation effects on such system behaviors [80,81]. Then, it is possible to optimize the response of RE-based amplifiers by acting at the system levels, optimizing the system with the smallest length of RE-doped fiber or by changing the pump wavelength from around 980 nm to 1480 nm [82]. Several approaches have been followed at the component level to improve its radiation hardness. Ce-codoping of Yb/Er-doped optical fibers has been shown to be very efficient to reduce the fiber and amplifier vulnerability [83] and was also confirmed after that for Er- and Yb-doped optical fibers [84,85]. Another successful approach to enhance the radiation tolerance of Er-doped optical fibers is based on an innovative fabrication process avoiding Al use [86]. Hydrogen loading of the RE-fiber is also very efficient to increase its radiation hardness and the one of associated amplifiers [36,87]. The issue of this approach was the difficulty to load permanently the fiber without degrading the optical performance; this is now achieved with the new hole-assisted carbon-coated Er-doped fibers (HACC fibers, [88]). Amplifiers made with such fibers present a degradation below 4% of their 31 dB gain after an irradiation dose exceeding 3 kGy at 2×10^{-3} Gy/s (<3% after 63 MeV proton fluence of 7.5×10^{11} p/cm² [89]), which corresponds to the most challenging future space missions to Jupiter moons.

13.4.2 FOR FIBER-BASED SENSORS

Various optical fiber-based sensors have been studied. They can be mainly divided into two classes. The first one uses a sensitive tip and the fiber only as a transport medium of the signal generated by the sensitive tip to the detector [90]. This type of sensors will not be discussed in this chapter. The second one, that will be further discussed, uses at the same time the fiber as sensitive element and transport support. There are plenty of sensors whose radiation vulnerability is of great interest. We present hereafter a nonexhaustive list introducing the Fiber Bragg Grating (FBG), the fiber as dosimetry system, the sensors based on Brillouin, Rayleigh, or Raman scattering in silica-based glasses.

13.4.2.1 The Fiber Bragg Gratings (FBGs)

SMFs can be functionalized by writing gratings in their inner core to filter a specific wavelength (Bragg wavelength, λ_B); the value of λ_B is fixed by the gratings characteristics, its refractive index properties, and its period of variation along the FBG [91]. This λ_B shifts with the temperature or strain allowing the monitoring of these parameters at the location of the FBG sensors [92]. As several FBGs can be inscribed into a unique SMF, FBGs permit the making of distributed sensors [93]. Different procedures can be used to inscribe FBGs into fibers using, for example, cw or pulsed lasers or more recently femtosecond lasers. Depending on the used procedure and on additional pretreatments, such as hydrogen-loading, FBGs can be written in various fiber types, from those photosensitive (Ge-doped) to those less sensitive (PSC, N-doped...). The radiation effects on FBGs have been recently reviewed in [92]. Radiation induces a Bragg wavelength shift, λ_B, that directly affects the temperature or strain evaluation, increasing the measurement errors. Radiation can also modify the shape and reduce the amplitude of the FBG peak up to its complete erasing. The amplitudes of these changes are extremely dependent on the writing conditions, treatments, and temperature of irradiation. The FBG radiation tolerance is not related to the one of the fiber used for its inscription [94]. Moreover, even if the FBG is radiation-tolerant, RIA has to be considered if the fiber length used for the system is sufficiently long to allow RIA to affect the measurement dynamic. FBG is a very promising technology for operation in radiation environment. By applying a specific writing procedure into radiation-hardened optical fibers, FBG can resist to doses up to 3 MGy and at operation temperatures exceeding 200°C, with a Bragg wavelength shift limited to less than 15 pm (*corresponding to a temperature error of 1.5°C*) [95].

13.4.2.2 Fibers as Dosimetry Systems

The vulnerability study of various classes of optical fibers reveals an extreme radiation sensitivity for some of them. This is of great interest for the development of active or passive dosimetry systems, using the advantages of optical fibers to reduce the system cost, weight, and increasing their performances. These dosimeters are based on the various radiation effects previously mentioned and presented.

RIA: Some of the studied systems use the RIA dose dependence to monitor the radiation. For such systems, highly sensitive fibers, such as the P-doped ones or those with RE ions, are the most promising candidates [96]. Combining their use with reflectometry technique, the feasibility to monitor the dose using a unique fiber with spatial resolution below one meter was demonstrated [97,98].

RIE: Radioluminescence from preexisting defect centers or from the ones induced by radiation can be a way to monitor the flux (or dose rate) in real time. For example, recent work showed that such monitoring is possible using RE-doped optical fibers [99] or MMFs loaded with O_2 [100]. For these examples, the luminescence used for monitoring is in the infrared range and is then less affected by the RIA. As previously mentioned, Cerenkov emission can also be used in reactors to monitor the neutron flux as shown in [22].

Thermoluminescence: A lot of papers have been published about the potential of optical fibers (Telecom-grade or specialty ones) to act as thermoluminescent dosimeters (TLDs) for ionizing or nonionizing radiations [101]. As an example, it was recently demonstrated that highly Ge-doped optical fibers possess more interesting dosimetry properties than commercial TLDs like the TLD500 based on other materials [102–104].

OSL: Optically stimulated luminescence is studied for the development of online dosimetry systems [105,106], for example for proton therapy purposes. In this case, the excitation of the fiber with a laser probe at the end of the irradiation gives rise to a luminescence of the generated defects during their desexcitation. From this measurement, a dose value can be extracted a few seconds after the end of the irradiation.

13.4.2.3 Distributed Sensors

The distributed sensors using the scattering of propagating light along the fiber offers exceptional sensing advantages with a resolution of less than 1 m long up to 50 km long optical fibers. The Brillouin and Rayleigh scattering properties are dependent on the strain and temperature applied to the silica and can then be used to monitor both types of measurands (*discrimination between them remains a major issue*) [107]. Raman-based sensors are sensitive only to temperature, not to strain [108]. Figure 13.5 summarizes today's knowledge concerning radiation effects on their performances.

Raman-sensors are based on the measurement of the intensities ratio between the Stokes (S) and anti-Stokes (AS) scattered light into multimode optical fibers.

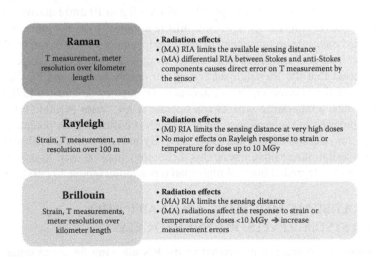

FIGURE 13.5 Review of the radiation effects on Raman, Rayleigh, and Brillouin sensors (MA means major effect and MI means minor effects).

The amplitude of the AS light changes with temperature whereas the S component remains unchanged allowing the monitoring of temperature changes along the fiber with a resolution of about 1 m for today's commercial systems. When submitted to radiation, the performances of Raman sensors are strongly degraded through two mechanisms: the RIA and the differential RIA, ΔRIA, between the S and AS wavelengths whereas the Raman scattering itself remains unchanged at least up to 10 MGy γ-ray dose [109]. RIA limits the range of sensing for the instrument and depending on the application, it implies the use of radiation-hardened optical fibers, such as PSC or F-doped ones to design systems with lengths over 100 m. Differential RIA between the S and AS lines causes large measurement errors on the temperature. ΔRIA is not directly linked to the RIA levels but to its spectral dependence (a similar effect is observed with H_2 loading for example). It leads to high-temperature reading errors by the single-ended instruments and without correction such Raman-based sensors cannot be used in harsh environments. If the RIA at the S and AS wavelengths are known, it is possible to correct these errors at the system level in some specific cases [110,111]. Work is still in progress, but Raman sensors are good candidates for the monitoring of nuclear waste storage facilities or nuclear power plants.

Brillouin sensors are based on the measurement of the shift in frequency of the Brillouin scattered light that linearly evolves with the temperature or with the strain. Like Raman sensors, these offer spatial resolution of less than 1 m over kilometer length and uses SMFs. Radiation induces also a shift of the Brillouin frequency that will lead to errors in the measurands evaluation. This radiation-induced shift remains limited in PSC and F-doped fibers [112] whereas it can be larger in Ge-doped optical fibers [113]. Once again, RIA is a main effect as it will strongly limit the range of the sensors, reducing it from kilometers to less than 200 m after a 10 MGy irradiation dose [112]. With adapted fibers, these Brillouin sensors will present acceptable performances for integration in extreme environments, such as those associated with the nuclear waste storages.

Rayleigh sensors offer the best spatial resolution (up to 10 µm) but over a limited fiber length (below 100 m) for today's available systems. Like for other classes of sensors based on scattering mechanism, their performances are affected by the RIA phenomenon and could degrade their ability to monitor strain and temperature. It is not expected that the Rayleigh scattering can deeply be affected by the γ-rays, maybe more under neutron exposure. Recent work done by A. Faustov [54] demonstrates that such temperature sensors can operate at low doses (<100 kGy) whereas work is in progress at higher doses (up to 10 MGy) confirming the potential of this technology for operation under radiation [114]. By using it as an OTDR associated with highly radiation sensitive fibers, this technology may be used to monitor dose deposition with a degraded but still high spatial resolution (about 15 cm) [54].

13.5 MULTISCALE MODELING FROM AB INITIO TO SYSTEM LEVELS: RECENT PROGRESS

As previously discussed [4], no predictive models allowing for an estimate of the vulnerability of one optical fiber in a future harsh environment exists, and nearly none is used as part of the different hardening approaches to enhance their radiation hardness. Due to the complexity of the consequences of radiation-induced point defects generation on the fiber macroscopic properties, such model should be based on multiscale tools from ab initio simulation of the amorphous glass and properties of point defects created in the different layers of the fibers to the impact of these defects on the guided light. The development of such approach was identified in 2008 as a major challenge for fusion facilities (inertial or magnetic confinement), implying the creation of strong collaboration between the different researchers and fiber providers [115].

It is obvious from previous sections that the sole experimental approach is not sufficient to have a complete view of the nature and properties of the various point defects and then of the origin of the fiber degradation. This is due to the intrinsic limitation of the main spectroscopic investigation tools: optical absorption (OA), optical and time-resolved luminescence (OL, TRL), and EPR (electron paramagnetic resonance). The RIA spectra, as the one illustrated in Figure 13.3 are usually very difficult to reproduce with the Gaussian bands associated with defects due to strong overlapping of the OA bands. Moreover, the origin of attenuation in the infrared, one of the most important spectral domains, remains unknown for most of the fiber types. Only a part of defects are emitting light, and EPR can only be used to investigate paramagnetic defects. Then, despite 50 years of research and great progress on this subject (see for examples reviews [116–118]), the exact structure of some defects as well as their properties are still debated and under study. To solve the open issues concerning defects, the simulation at ab initio level can be very well adapted, but until recently no clear procedure was elaborated allowing to investigate sufficiently large supercells of a-SiO$_2$ capable of reproducing the complex nature of the amorphous glass and of the subsequent statistical distribution of the point defects properties. LabHC and CEA group work on this subject and an approach has been developed and validated (see [8–10]) allowing the calculation of both optical properties and EPR signatures of point defects. This approach is detailed in Figure 13.6, which illustrates the main steps and used calculation commercial or homemade codes to obtain the distributions of structural, optical, and EPR signatures of microscopic defects. Recent progress has been published showing that this approach has been successful to obtain the optical absorption signature of some oxygen-deficient centers, the ODC(I) ones, in pure silica material and explains the lack of clear absorption signature for Ge-ODC(I) defects [119]. Furthermore, the different variants of Si or Ge related-centers have been theoretically investigated in terms of their optical and EPR signatures [120]. Moreover, promising results have been obtained concerning the modeling of structural properties of neutron-irradiated glasses, such as their Raman spectra [55]. Then, in the near future, these simulation tools, coupled with experimental investigations, will be able to help predicting a fiber radiation response to a given environment.

If the procedure given in Figure 13.6 could help to improve our knowledge concerning point defect generation mechanisms in dielectrics, additional tools to those ab initio will be needed to simulate up to the device or system scale. From the developed approach, the induced attenuation in one layer of the glass may be calculated, but such a work should be done for all the layers. Moreover, new codes like Monte Carlo ones are needed to consider mechanisms at the microscopic scale like diffusion

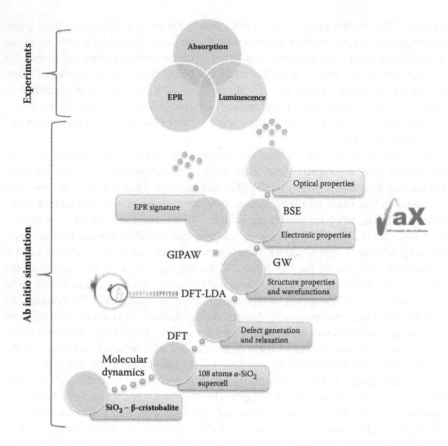

FIGURE 13.6   Ab initio part of the multiscale simulation approach used to calculate the optical and EPR properties of points defects in amorphous glasses (see [8–10] for more details).

of mobile species or the impact of the radially distributed stress into the fiber on the generation and bleaching mechanisms of radiation-induced point defects. After that, the interaction between the guided mode and the evolving material under irradiation should be considered to obtain the global response of the waveguide to the irradiation. This step includes the calculation of the guided modes into the various optical fibers but corresponds to well-known physics that can be solved with electromagnetic codes. However, here again, the impact of parameters appearing at this level will have to be taken into account at the ab initio level such as the photobleaching effect for example, needing the different scales to interact with each other's.

13.6   CONCLUSIONS

In this chapter, we briefly introduced the basics of radiation effects on silica-based optical fibers as well as the main recent achievements concerning this research axis. This research is mainly driven by industrial or space applications or by research labs involved in big facilities, such as ITER, LMJ-NIF, LHC. Today radiation-tolerant devices have been identified for most of the targeted applications, even for the RE-based devices that were one of the bigger issues. Their hardness or the ones of systems using them has been improved combining both hardening-by-component or hardening-by-system approaches, reducing the impact of radiation-induced point defects at the operating wavelengths. Today's challenge consists in functionalizing these silica-based fibers to design discrete or distributed sensors for extreme environments such as the reactors of next nuclear power plants. In addition to the point defect control, this implies to understand and limit the radiation induced structural changes occurring under exposure, particularly under non-ionizing radiation.

REFERENCES

1. J. R. Schwank, M. R. Shaneyfelt, D. M. Fleetwood, J. A. Felix, P. E. Dodd, P. Paillet and V. Ferlet-Cavrois, "Radiation Effects in MOS devices," *IEEE Transactions on Nuclear Science*, vol. 55, no. 4, pp. 1833–1853, 2008.
2. J. L. Barth, C. S. Dyer, E. G. Stassinopoulos, "Space, atmospheric, and terrestrial radiation environments," *IEEE Transactions on Nuclear Science*, vol. 50, no. 3, pp. 466–482, 2003.
3. F. Berghmans, B. Brichard, A. Fernandez, A. Gusarov, M. Van Uffelen, S. Girard, "An introduction to radiation effects on optical components and fiber optic sensors," *NATO Science for Peace and Security Series B: Physics and Biophysics*, Bock, W. J.; Gannot, I. & Tanev, S. (Eds.), (127–165), Springer Netherlands, ISBN 978-1-4020-6950, 2008.
4. S. Girard, J. Kuhnhenn, A. Gusarov, B. Brichard, M. Van Uffelen, Y. Ouerdane, A. Boukenter and C. Marcandella, "Radiation Effects on Silica-based Optical Fibers: Recent Advances and Future Challenges," *IEEE Transactions on Nuclear Science*, vol. 60, no. 3, pp. 2015–2036, 2013.
5. A. D. Kersey, "A Review of Recent Developments in Fiber Optic Sensor Technology," *Optical Fiber Technology*, vol. 2, no. 3, pp. 291–317, 1996.
6. X. Bao and L. Chen, "Recent Progress in Distributed Fiber Optic Sensors," Sensors, vol. 12, pp. 8601–8639, 2012.
7. S. Kreger, D. Gifford, M., Froggatt, A. Sang, R. Duncan, M. Wolfe, B. Soller, "High-Resolution Extended Distance Distributed Fiber-Optic Sensing Using Rayleigh Backscatter," Proc. SPIE 6530, Sensor Systems and Networks: Phenomena, Technology, and Applications for NDE and Health Monitoring, paper 65301R 2007.
8. S. Girard, Y. Ouerdane, G. Origlio, C. Marcandella, A. Boukenter, N. Richard, J. Baggio, P. Paillet, M. Cannas, J. Bisutti, J.-P. Meunier, and R. Boscaino, "Radiation Effects on Silica-Based Preforms and Optical Fibers—I: Experimental Study With Canonical Samples," *IEEE Transactions on Nuclear Science*, vol. 55, no. 6, pp. 3473–3482, 2008.
9. S. Girard, N. Richard, Y. Ouerdane, G. Origlio, A. Boukenter, L. Martin-Samos, P. Paillet, J.-P. Meunier, J. Baggio, M. Cannas, and R. Boscaino, "Radiation Effects on Silica-Based Preforms and Optical Fibers-II: Coupling Ab initio Simulations and Experiments," *IEEE Transactions on Nuclear Science*, vol. 55, no. 6, pp. 3508–3514, 2008.
10. N. Richard, S. Girard, L. Giacomazzi, L. Martin-Samos, D. Di Francesca, C. Marcandella, A. Alessi, P. Paillet, S. Agnello, A. Boukenter, Y. Ouerdane, M. Cannas, and R. Boscaino, "Coupled theoretical and experimental studies for the radiation hardening of silica-based optical fibers," *IEEE Transactions on Nuclear Science*, vol. 61, no. 4, pp. 1819–1825, 2014.
11. A. Mendez, T. Morse. *Specialty Optical Fibers Handbook*, Academic Press, Dec. 2006.

12. P. Russell, "Photonic Crystal Fibers," *Science*, vol. 299, pp. 358–362, 2003.
13. K. Peters, "Polymer optical fiber sensors—A review," *Smart Materials and Structures*, vol. 20, no. 1, 2011.
14. G. Cheymol, H. Long, J. F. Villard, and B. Brichard, "High level gamma and neutron irradiation of silica optical fibers in CEA OSIRIS nuclear reactor," *IEEE Transactions on Nuclear Science*, vol. 55, no. 4, pp. 2252–2258, 2008.
15. S. Girard, J. Baggio, and J.-L. Leray, "Radiation-induced effects in a new class of optical waveguides: The air guiding photonic crystal fibers," *IEEE Transactions on Nuclear Science*, vol. 52, no. 6, pp. 2683–2688, 2005.
16. S. O'Keeffe, A. Fernandez Fernandez, C. Fitzpatrick, B. Brichard and E. Lewis, "Real-time gamma dosimetry using PMMA optical fibres for applications in the sterilization industry," *Meas. Sci. Technol.*, vol. 18, pp. 3171–3176, 2007.
17. D. L. Griscom, "Nature of defects and defect generation in optical glasses," in Proc. of SPIE. 541 "Radiation Effects in Optical Materials," pp. 38–59, 1985.
18. S. Girard, J. Keurinck, Y. Ouerdane, J.-P. Meunier, and A. Boukenter, "Gamma-Rays and Pulsed X-Ray Radiation Responses of Germanosilicate Single-Mode Optical Fibers: Influence of Cladding Codopants," *J. Lightwave Technol.*, vol. 22, pp. 1915–1922, 2004.
19. A. Kh. Mukhsin, B. I. Maksudbek, G. M. Eldar, I. D. Jalil, N. Izzatillo, R. R. Igor, A. Mukhtor and S. Kakhramon, "Measurement Method of Radiation Induced Emission Spectra of Optical Fibers," *Jpn. J. Appl. Phys.*, vol. 47, p. 301, 2008.
20. K. W. Jang, W. J. Yoo, S. H. Shin, D. Shin, and B. Lee, "Fiber-optic Cerenkov radiation sensor for proton therapy dosimetry," *Opt. Express* 20, 13907–13914, 2012.
21. M. J. Marrone, "Radiation-induced luminescence in silica core optical fibers," *Appl. Phys. Lett.*, vol. 38, pp.115–117, 1981.
22. B. Brichard, A. F. Fernandez, H. Ooms et al., "Fibre-optic gamma-flux monitoring in a fission reactor by means of Cerenkov radiation," *Meas. Sci. and Technol.*, vol. 18, no. 10, p. 3257, 2007.
23. W. Primak, "Fast-neutron-induced changes in quartz and vitreous silica," *Phys. Rev. B*, vol. 110, no. 6, pp. 1240–1254, 1958.
24. E. Lell, N. J. Hensler, and J. R. Hensler, "Radiation Effects in Quartz, Silica and Glasses," *Prog. In Ceramic Sci.*, vol. 4, J. Burke, Pergamon Press, Oxford, NY, 3–93, 1966.
25. B. Brichard, O. V. Butov, K. M. Golant, and A. Fernandez Fernandez, "Gamma radiation-induced refractive index change in Ge- and N-doped silica," *J. Appl. Phys.* 103, 054905, 2008.
26. V. B. Neustruev, "Colour centres in germanosilicate glass and optical fibres," *J. Phys.: Condens. Matter*, vol. 6, pp. 6901–6936, 1994.
27. E. J. Friebele, C. G. Askins, C. M. Shaw, M. E. Gingerich, C. C. Harrington, D. L. Griscom, T.-E. Tsai, U.-C. Paek, and W. H. Schmidt, "Correlation of single-mode fiber radiation response and fabrication parameters," *Appl. Opt.*, vol. 30, pp. 1944–1957, 1991.
28. D. Di Francesca, A. Boukenter, S. Agnello, S. Girard, A. Alessi, P. Paillet, C. Marcandella, N. Richard and Y. Ouerdane, "X-ray irradiation effects on fluorine-doped germanosilicate optical fiber," *Optical Materials Express*, vol. 4, Issue 8, pp. 1683–1695, 2014.
29. D. Di Francesca, "Role of Dopants, Interstitial O2 and Temperature, in the Effects of Irradiation on Silica-based Optical Fibers", PhD Thesis, University of Saint-Etienne, 2015.
30. S. Girard, J. Keurinck, Y. Ouerdane, J.-P. Meunier, and A. Boukenter, "Gamma-Rays and Pulsed X-Ray Radiation Responses of Germanosilicate Single-Mode Optical Fibers: Influence of Cladding Codopants," *J. Lightwave Technol.*, vol. 22, pp. 1915–1922, 2004.
31. S. Girard, Y. Ouerdane, A. Boukenter, C. Marcandella, J. Bisutti, J. Baggio, and J.-P. Meunier, "Integration of Optical Fibers in Radiative Environments: Advantages and Limitations," *IEEE Transactions on Nuclear Science*, vol. 59 (4), pp. 1317–1322, 2012.

32. J. A. B. Arviddson, K. Dunn, D. Gong, T. Huffman, C. Issever, M. Jones, C. Kerridge, J. Kierstead, G. Kuyt, C. Liu, T. Liu, A. Povey, E. Regnier, N. C. Ryder, N. Tassie, T. Weidberg, A. C. Xiang, J. Ye, "The Radiation Tolerance of Specific Optical Fibers for the LHC Upgrades," *Physics Procedia*, vol. 37, pp. 1630–1643, 2012.

33. E. M. Dianov, K. M. Golant, R. R. Khrapko, A. S. Kurkov, A. L. Tomashuk, "Low-hydrogen silicon oxynitride optical fibers prepared by SPCVD," *J. Lightwave Technol.*, vol. 13 (7), pp. 1471–1474, 1995.

34. S. Girard, J. Keurinck, A. Boukenter, J.-P. Meunier, Y. Ouerdane, B. Azaïs, P. Charre and M. Vié, "Gamma-rays and pulsed X-ray radiation responses of nitrogen, germanium doped and pure silica core optical fibers," *Nucl. Instr. Methods in Phys. Res. B*, vol. 215, no. 1–2, pp. 187–195, 2004.

35. S. Girard, "Analyse de la réponse des fibres optiques soumises à divers environnements radiatifs," Thèse de Doctorat, Université de Saint-Etienne, 2003.

36. S. Girard, M. Vivona, A. Laurent, B. Cadier, C. Marcandella, T. Robin, E. Pinsard, A. Boukenter and Y. Ouerdane, "Radiation hardening techniques for Er/Yb doped optical fibers and amplifiers for space application," *Opt. Express* 20, 8457–8465, 2012.

37. S. Girard, Y. Ouerdane, C. Marcandella, A. Boukenter, S. Quenard, N. Authier, "Feasibility of radiation dosimetry with phosphorus-doped optical fibers in the ultraviolet and visible domain," *J. Non-Cryst. Solids*, vol. 357, pp. 1871–1874, 2011.

38. H. Henschel, M. Körfer, J. Kuhnhenn, U. Weinand, F. Wulf, "Fibre optic radiation sensor systems for particle accelerators," *Nucl. Instr. Methods in Phys. Res. A*, vol. 526 (3), pp. 537–550, 2004.

39. M. C. Paul, D. Bohra, A. Dhar, R. Sen, P. K. Bhatnagar, K. Dasgupta, "Radiation response behavior of high phosphorous doped step-index multimode optical fibers under low dose gamma irradiation," *J. Non-Cryst. Solids*, vol. 355 (28–30), pp. 1496–1507, 2009.

40. B. Brichard, and A. Fernandez-Fernandez, "Radiation effects in silica glass optical fibers," in RADECS 2005 Short Course, New challenges for Radiation Tolerance Assessment, Editor A. Fernandez Fernandez, 2005.

41. T. Kakuta, T. Shikama, T. Nishitani, B. Brichard, A. Krassilinikov, A. Tomashuk, S. Yamamoto, S. Kasai, "Round-robin irradiation test of radiation resistant optical fibers for ITER diagnostic application," *J. Nucl. Mater.*, vol. 307–11, pp. 1277–1281, 2002.

42. D. L. Griscom, "γ and fission-reactor radiation effects on the visible-range transparency of aluminum-jacketed, all-silica optical fibers," *J. Appl. Phys.*, vol. 80, pp. 2142–2155, 1996.

43. A. L. Tomashuk, E. M. Dianov, K. M. Golant, A. O. Rybaltovskii, "γ-radiation-induced absorption in pure-silica-core fibers in the visible spectral region: The effect of H2-loading," *IEEE Trans. Nucl. Sci.*, vol. 45 (3), pp. 1576–1579, 1998.

44. K. Nagasawa, M. Tanabe, and K. Yahagi, "Gamma-Ray-Induced Absorption Bands in Pure-Silica-Core Fibers," *Jpn. J. Appl. Phys.*, vol. 23, pp. 1608–1613, 1984.

45. S. Girard, D. L. Griscom, J. Baggio, B. Brichard, F. Berghmans, "Transient optical absorption in pulsed-X-ray-irradiated pure-silica-core optical fibers: Influence of self-trapped holes," *J. Non-Cryst. Solids*, vol. 352, no. 23–25, pp. 2637–2642, 2006.

46. D. L. Griscom, "γ-Ray-induced visible/infrared optical absorption bands in pure and F-doped silica-core fibers: Are they due to self-trapped holes?" *J. Non-Cryst. Solids*, vol. 349, pp. 139–147, 2004.

47. D. L. Griscom, "Self-trapped holes in pure-silica glass: A history of their discovery and characterization and an example of their critical significance to industry," *J. Non-Cryst. Solids*, vol. 352, pp. 2601–2617, 2006.

48. D. L. Griscom, "Radiation hardening of pure-silica-core optical fibers: Reduction of induced absorption bands associated with self-trapped holes," *Applied Physics Letters*, vol. 71, pp. 175–177, 1997.

49. B. Brichard, A. Fernandez Fernandez, H. Ooms, F. Berghmans, M. Decreton, A. Tomashuk, S. Klyamkin, M. Zabezhailov, I. Nikolin, V. Bogatyrjov, E. Hodgson, T. Kakuta, T. Shikama, T. Nishitani, A. Costley, G. Vayakis, "Radiation-hardening techniques of dedicated optical fibres used in plasma diagnostic systems in ITER," *J. Nucl. Mater*, vol. 329–333, pp. 1456–1460, 2004.

50. S. Girard, C. Marcandella, A. Alessi, A. Boukenter, Y. Ouerdane, N. Richard, P. Paillet, M. Gaillardin, and M. Raine, "Transient Radiation Responses of Optical Fibers: Influence of MCVD Process Parameters," *IEEE Transactions on Nuclear Science*, vol. 59, issue no. 6, pp. 2894–2901, 2012.

51. S. Girard, B. Vincent, J.-P. Meunier, Y. Ouerdane, A. Boukenter, and A. Boudrioua, "Spatial distribution of the red luminescence in pristine, gamma-rays and ultraviolet-irradiated multimode optical fibers," *Applied Physics Letters*, vol. 84, pp. 4215–4217, 2004.

52. S. Girard, Y. Ouerdane, A. Boukenter, and J.-P. Meunier, "Transient radiation responses of silica-based optical fibers: Influence of Modified Chemical Vapor Deposition process parameters," *Journal of Applied Physics*, vol. 99, pp. 023104, 2006.

53. J. Kuhnehnn, H. Henschel, U. Weinand, "Influence of coating material, cladding thickness, and core material on the radiation sensitivity of pure silica core step-index fibers," in Proc. RADECS 2005, 8th European Conference on Radiation and Its Effects on Components and Systems, 2005, paper A2.

54. A. Faustov, "Advanced fibre optics temperature and radiation sensing in harsh environments," PhD Thesis, Université Polytechnique de Mons, 2014.

55. M. León, L. Giacomazzi, S. Girard, N. Richard, P. Martin, L. Martin-Samos, A. Ibarra, A. Boukenter and Y. Ouerdane, "Neutron Irradiation Effects on the Structural Properties of KU1, KS-4V and I301 Silica Glasses," *IEEE Transactions on Nuclear Science*, vol. 61 (4), pp. 1522–1530, 2014.

56. M. León, P. Martín, D. Bravo, F. J. López, A. Ibarra, A. Rascón, F. Mota, "Thermal stability of neutron irradiation effects on KU1 fused silica," *J. Nucl. Mater.*, vol. 374, pp. 386–389, 2008.

57. J. C. Lagomacini, D. Bravo, M. León, P. Martín, Á. Ibarra, A. Martín, F. J. López, "EPR study of gamma and neutron irradiation effects on KU1, KS-4V and Infrasil 301 silica glasses," *Journal of Nuclear Materials*, vol. 417, Issues 1–3, pp. 802–805, 2011.

58. D. Bravo, J. C. Lagomacini, M. León, P. Martín, A. Martín, F. J. López, A. Ibarra, "Comparison of neutron and gamma irradiation effects on KU1 fused silica monitored by electron paramagnetic resonance," *Fusion Engineering and Design*, Volume 84, Issues 2–6, pp. 514–517, 2009.

59. M. Van Uffelen, "Modélisation de systèmes d'acquisition et de transmission à fibres optiques destinés à fonctionner en environnement nucléaire," PhD Thesis, Université de Paris, 2001.

60. E. J. Friebele, G. C. Askins, M. E. Gingerich, K. J. Long, "Optical fiber waveguides in radiation environments, II," *Nucl. Instr. Meth. Phys. Res. B*, Volume 1, Issue 2–3, pp. 355–369, 1984.

61. O. Deparis, "Etude physique et expérimentale de la tenue des fibres optiques aux radiations ionisantes par spectrométrie visible-infrarouge," PhD thesis, Faculté Polytechnique de Mons, 1997.

62. D. L. Griscom, "A Minireview of the Natures of Radiation-Induced Point Defects in Pure and Doped Silica Glasses and Their Visible/Near-IR Absorption Bands, with Emphasis on Self-Trapped Holes and How They Can Be Controlled," *Physics Research International*, vol. 2013, Article ID 379041, 2013.

63. O. Gilard, J. Thomas, L. Troussellier, M. Myara, P. Signoret, E. Burov, and M. Sotom, "Theoretical explanation of enhanced low dose rate sensitivity in erbium-doped optical fibers," *Appl. Opt.* vol. 51, pp. 2230–2235, 2012.

64. F. Mady, M. Benabdesselam, J.-B. Duchez, Y. Mebrouk, and S. Girard, "Global View on Dose Rate Effects in Silica-Based Fibers and Devices Damaged by Radiation-Induced Carrier Trapping," *IEEE Transactions on Nuclear Science*, vol. 60, no. 6, pp. 4241–4348, 2013.

65. D. L. Griscom, M. E. Gingerich, and E. J. Friebele, "Radiation-induced defects in glasses: Origin of power-law dependence of concentration on dose," *Phys. Rev. Lett.*, vol. 71, 1019, 1993.

66. O. Gilard, M. Caussanel, H. Duval, G. Quadri, and F. Reynaud, "New model for assessing dose, dose rate, and temperature sensitivity of radiation-induced absorption in glasses," *J. Appl. Phys.*, vol. 108, 093115, 2010.

67. S. Girard, C. Marcandella, A. Morana, J. Perisse, D. Di Francesca, P. Paillet, J.-R. Macé, A. Boukenter, M. Léon, M. Gaillardin, N. Richard, M. Raine, S. Agnello, M. Cannas and Y. Ouerdane, "Combined High Dose and Temperature Radiation Effects on Multimode Silica-based Optical Fibers," *IEEE Transactions on Nuclear Science*, vol. 60, no. 6, pp. 4305–4313, 2013.

68. B. Brichard, "Systèmes à fibres optiques pour infrastructures nucléaires: Du durcissement aux radiations à l'application," Thèse de doctorat, IES–Institut d'Electronique du Sud, Montpellier, 2008.

69. T. Shikama, T. Kakuta, N. Shamoto, T. Sagawa, M. Narui, "Behavior of developed radiation-resistant silica-core optical fibers under fission reactor irradiation," *Fusion Engineering and Design*, vol. 51–52, pp. 179–183, 2000.

70. B. Brichard, M. Van Uffelen, A. F. Fernandez, F. Berghmans, M. Décreton, E. Hogdson, T. Shikama, T. Kakuta, A. Tomashuk, K. Golant, A. Krasilnikov, "Round robin evaluation of optical fibres for plasma diagnostics," *Fusion Eng. and Design*, vol. 5–57, pp. 917–921, 2001.

71. M. Aerssens, A. Gusarov, B. Brichard, V. Massaut, P. Mégret, M. Wuilpart, "Faraday effect based optical fiber current sensor for Tokamaks," in Proceedings of ANIMMA-2011 Conference, Ghent, Belgium, 2011.

72. J.-L. Bourgade, R. Marmoret, S. Darbon, R. Rosch, P. Troussel, B. Villette, V. Glebov, W. Shmayda, J. C. Gommé, Y. Le Tonqueze, F. Aubard, J. Baggio, S. Bazzoli, F. Bonneau, J.-Y. Boutin, T. Caillaud, C. Chollet, P. Combis, L. Disdier, J. Gazave, S. Girard, D. Gontier, P. Jaanimagi, H.-P. Jacquet, J.-P. Jadaud, O. Landoas, J. Legendre, J.-L. Leray, R. Maroni, D. D. Meyerhofer, J.-L. Miquel, F. J. Marshall, I. Masclet-Gobin, G. Pien, J. Raimbourg, C. Reverdin, A. Richard, D. Rubins de Cervens, C. T. Sangster, J.-P. Seaux, G. Soullie, C. Stoeckl, I. Thfoin, L. Videau, and C. Zuber, "Present LMJ Diagnostics Developments Integrating its Harsh environment," *Review of Scientific Instruments*, vol. 79 no. 10, 10F301, 2008.

73. J.-L. Bourgade, V. Allouche, J. Baggio, C. Bayer, F. Bonneau, C. Chollet, S. Darbon, L. Disdier, D. Gontier, M. Houry, H.-P. Jacquet, J.-P. Jadaud, J.-L. Leray, I. Masclet-Gobin, J.-P. Negre, J. Raimbourg, B. Villette, I. Bertron, J.-M. Chevalier, J.-M. Favier, J. Gazave, J.-C. Gomme, F. Malaise, J.-P. Seaux, V. Y. Glebov, P. Jaanimagi, C. Stoeckl, T. C. Sangster, G. Pien, R. A. Lerche, and E. R. Hodgson, "New constraints for plasma diagnostics development due to the harsh environment of MJ class lasers," *Rev. Sci. Instrum.* 75 (10), 4204–4212 (2004).

74. B. Amacker, "CERN Main Optical Fibre Links," retrieved from http://ts-dep.web.cern.ch/ts-dep/groups/el/sections/OF/activities/Fo-cern-monitoring%20Model%20(1).pdf, 2010-06-26.

75. T. Wijnands, L. K. De Jonge, J. Kuhnhenn, S. K. Hoeffgen, U. Weinand, "Optical Absorption in Commercial Single Mode Optical Fibers in a High Energy Physics Radiation Field," *IEEE Trans. Nucl. Sci.*, vol. 55 (4), pp. 2216–2222, 2008.

76. T. Wijnands, K. Aikawa, J. Kuhnhenn, D. Ricci, U. Weinand, "Radiation Tolerant Optical Fibers: From Sample Testing to Large Series Production," *J. Lightwave Technol.*, vol. 29, no. 22, pp. 3393–3400, 2011.

77. M. Ott, "Radiation Effects Data on Commercially Available Optical Fiber: Database Summary," in IEEE NSREC 2002 Data workshop, pp. 24–31, 2002.

78. B. Singleton, "Radiation Effects on Ytterbium-doped Optical Fibers," PhD Thesis, Air Force University, Wright-Patterson Air Force Base, OH, June 2014.

79. H. Henschel, O. Kohn, H. U. Schmidt, J. Kirchof, S. Unger, "Radiation-induced loss of rare earth doped silica fibres," *IEEE Trans. Nucl. Sci.*, vol. 45, no. 3, pp. 1552–1557, 1998.

80. S. Girard, L. Mescia, M. Vivona, A. Laurent, Y. Ouerdane, C. Marcandella, F. Prudenzano, A. Boukenter, T. Robin, P. Paillet, V. Goiffon, M. Gaillardin, B. Cadier, E. Pinsard, M. Cannas, and R. Boscaino "Design of Radiation-Hardened Rare-Earth Doped Amplifiers Through a Coupled Experiment/Simulation Approach," *Journal of Lightwave Technology*, vol. 31, no. 8, pp. 1247–1254, 2013.

81. L. Mescia, S. Girard, P. Bia, T. Robin, A. Laurent, F. Prudenzano, A. Boukenter, Y. Ouerdane, "Optimization of the design of high power Er^{3+}/Yb^{3+}–codoped fiber amplifiers for space missions by means of particle swarm approach," *IEEE Journal of Selected Topics in Quantum Electronics*, vol. 20, no. 5, ID# 3100108, 2014.

82. K. V. Zotov, M. E. Likhachev, A. L. Tomashuk, A. F. Kosolapov, M. M. Bubnov, M. V. Yashkov, A. N. Guryanov, and E. M. Dianov, "Radiation resistant Er-doped fibers: optimization of pump wavelength," *IEEE Photon. Technol. Lett.*, vol. 20, no. 17, pp. 1476–1478, 2008.

83. M. Vivona, "Radiation hardening of rare-earth doped fiber amplifiers," PhD thesis, Université de Saint-Etienne, 2013.

84. Y. Sheng, L. Yang, H. Luan, Z. Liu, Y. Yu, Ji. Li "Improvement of radiation resistance by introducing CeO2 in Yb-doped silicate glasses," *J. Nucl Materials*, vol. 427, Issues 1–3, pp. 58–61, 2012.

85. R. Xing, Y. Sheng, Z. Liu, H. Li, Z. Jiang, J. Peng, L. Yang, J. Li, and N. Dai, "Investigation on radiation resistance of Er/Ce co-doped silicate glasses under 5 kGy gamma-ray irradiation," *Opt. Mater. Express* 2, 1329–1335, 2012.

86. J. Thomas, M. Myara, L. Troussellier, E. Burov, A. Pastouret, D. Boivin, G. Mélin, O. Gilard, M. Sotom, and P. Signoret, "Radiation-resistant erbium-doped-nanoparticles optical fiber for space applications," *Opt. Express*, vol. 20, pp. 2435–2444, 2012.

87. K. V. Zotov, M. E. Likhachev, A. L. Tomashuk, M. L. Bubnov, M. V. Yashkov, A. N. Guryanov, and S. N. Klyamkin, "Radiation-resistant erbium-doped fiber for spacecraft applications," *IEEE Trans. Nucl. Sci.*, vol. 55, no. 4, pp. 2213–2215, 2008.

88. S. Girard, A. Laurent, E. Pinsard, T. Robin, B. Cadier, M. Boutillier, C. Marcandella, A. Boukenter, and Y. Ouerdane, "Radiation-hard erbium optical fiber and fiber amplifier for both low and high dose space missions," *Optics Letters*, vol. 39, Issue 9, pp. 2541–2544, 2014.

89. S. Girard, A. Laurent, E. Pinsard, M. Raine, T. Robin, B. Cadier, D. Di Francesca, P. Paillet, M. Gaillardin, O. Duhamel, C. Marcandella, M. Boutillier, A. Ladaci, A. Boukenter and Y. Ouerdane, "Proton irradiation response of Hole-Assisted Carbon Coated Erbium-Doped Fiber Amplifiers," *IEEE Transactions on Nuclear Science*, vol. 61 (6), pp. 3309–3314, 2014.

90. P. Ferdinand, "Capteurs à fibres optiques et réseaux associés," Tec & Doc Eds, 1999.

91. A. Othonos and K. Kalli, "Fiber Bragg Gratings, Fundamentals and Applications in Telecommunications and Sensing," Norwood, MA, USA: Artech House, 1999.

92. A. Gusarov and S. Hoeffgen, "Radiation Effects on Fiber Gratings," *IEEE Transactions on Nuclear Science*, vol. 60, no. 3, pp. 2037–2053, 2013.

93. A. Cusano, A. Cutolo, and J. Albert, "Fiber Bragg Grating Sensors: Recent Advancements, Industrial Applications and Market Exploitation," Bentham Science, 2011.

94. H. Henschel, S. K. Hoeffgen, K. Krebber, J. Kuhnhenn, and U. Weinand, "Influence of Fiber Composition and Grating Fabrication on the Radiation Sensitivity of Fiber Bragg Gratings," *IEEE Trans. Nucl. Sci.* 55, 2235, 2008.

95. A. Morana, S. Girard, E. Marin, C. Marcandella, P. Paillet, J. Périsse, J.-R. Macé, A. Boukenter, M. Cannas, Y. Ouerdane, "Radiation tolerant Fiber Bragg Gratings for high temperature monitoring at MGy dose levels," *Optics Letters*, vol. 39, pp. 5313–5316, 2014.

96. P. Borgermans, "Spectral and Kinetic Analysis of Radiation Induced Optical Attenuation in Silica: Towards Intrinsic Fiber Optic Dosimetry?" Thèse de doctorat, Brussel, Vrije Universiteit, 2001.

97. A. Faustov, A. Gusarov, M. Wuilpart, A. A. Fotiadi, L. B. Liokumovich, I. Zolotovskii, A. L. Tomashuk, T. de Schoutheete, P. Mégret, "Comparison of Gamma-Radiation Induced Attenuation in Al-Doped, P-Doped and Ge-Doped fibres for Dosimetry," *IEEE Transactions on Nuclear Science*, vol. 60, no. 4, pp. 2511–2517, 2013.

98. H. Henschel, M. Körfer, J. Kuhnhenn, U. Weinand, F. Wulf, "Fibre optic radiation sensor systems for particle accelerators," *Nucl. Instr. Methods in Phys. Res. A*, vol. 526 (3), pp. 537–550, 2004.

99. A. Vedda, N. Chiodini, D. Di Martino, M. Fasoli, S. Keffer, A. Lauria, M. Martini, F. Moretti, G. Spinolo, M. Nikl, N. Solovieva, G. Brambilla, "Infrared luminescence for real time ionizing radiation detection," *Appl. Phys. Lett.* 85, 6356 (2004).

100. D. Di Francesca, S. Girard, S. Agnello, C. Marcandella, P. Paillet, A. Boukenter, F. M. Gelardi, and Y. Ouerdane, "Near infrared radio-luminescence of O2 loaded Rad-Hard silica optical fibers: A candidate dosimeter for harsh environments," *Applied Physics Letters*, vol. 105, 183508, 2014.

101. S. O'Keeffe, C. Fitzpatrick, E. Lewis, and A. I. Al-Shamma'a, "A review of optical fibre radiation dosimeters," *Sens. Rev.* 28, 136–142 (2008).

102. M. Benabdesselam, F. Mady, S. Girard, "Assessment of Ge-doped optical fibres as a TL-mode detector," *Journal of Non-Crystalline Solids*, vol. 360, pp. 9–12, 2013.

103. M. Benabdesselam, F. Mady, S. Girard, Y. Mebrouk, J.-B. Duchez, M. Gaillardin, P. Paillet, "Performance of Ge-doped Optical Fiber as a Thermoluminescent, Performance of Ge-doped Optical Fiber as a Thermoluminescent Dosimeter," *IEEE Transactions on Nuclear Science*, vol. 60, no. 6, pp. 4251–4256, 2013.

104. M. Benabdesselam, F. Mady, J.-B. Duchez, Y. Mebrouk, and S. Girard, "The Opposite Effects of the Heating Rate on the TSL Sensitivity of Ge-doped Fiber and TLD500 Dosimeters," *IEEE Transactions on Nuclear Science*, vol. 61 (6), pp. 3485–3490, 2014.

105. C. A. G. Kalnins, H. Ebendorff-Heidepriem, N. A. Spooner, and T. M. Monro, "Radiation dosimetry using optically stimulated luminescence in fluoride phosphate optical fibres," *Opt. Mater. Express* vol. 2, pp. 62–70, 2012.

106. A. L. Huston, B. L. Justus, P. L. Falkenstein, R. W. Miller, H. Ning, and R. Altemus, "Optically stimulated luminescent glass optical fibre dosemeter," *Radiat. Prot. Dosim.*, vol. 101, pp. 23–26, 2002.

107. D. Zhou, W. Li, L. Chen and X. Bao, "Distributed Temperature and Strain Discrimination with Stimulated Brillouin Scattering and Rayleigh Backscatter in an Optical Fiber," *Sensors*, vol. 13, 1836–1845, 2013.

108. G. Bolognini and A. Hartog, "Raman-based fibre sensors: Trends and applications," *Optical Fiber Technology*, vol. 19, Issue 6, Part B, pp. 678–688, 2013.

109. C. Cangialosi, S. Girard, A. Boukenter, M. Cannas, S. Delepine-Lesoille, J. Bertrand, P. Paillet, and Y. Ouerdane, "Effects of Radiation and Hydrogen-Loading on the

Performances of Raman Distributed Temperature Fiber Sensors," *IEEE/OSA Journal of Lightwave Technology*, vol. 33, Issue 12, pp. 2432–2438, 2015.

110. C. Cangialosi, Y. Ouerdane, S. Girard, A. Boukenter, S. Delepine-Lesoille, J. Bertrand, C. Marcandella, P. Paillet, M. Cannas, "Development of a Temperature Distributed Monitoring System Based On Raman Scattering in Harsh Environment," *IEEE Transactions on Nuclear Science*, vol. 61 (6), pp. 3315–3322, 2014.

111. A. Kimura, E. Takada, K. Fujita, M. Nakazawa, H. Takahashi, and S. Ichige, "Application of a Raman distributed temperature sensor to the experimental fast reactor JOYO with correction techniques," *Meas. Sci. Technol.* vol. 12, pp. 966–973, 2001.

112. X. Phéron, S. Girard, A. Boukenter, B. Brichard, S. Delepine-Lesoille, J. Bertrand, and Y. Ouerdane, "High γ-ray dose radiation effects on the performances of Brillouin scattering based optical fiber sensors," *Opt. Express* vol. 20, 26978–26985, 2012.

113. D. Alasia, A. F. Fernandez, L. Abrardi et al., "The effects of gamma-radiation on the properties of Brillouin scattering in standard Ge-doped optical fibres," *Meas. Sci. Technol.*, vol. 17, no. 5, pp. 1091–1094, 2006.

114. S. Rizzolo, A. Boukenter, E. Marin, M. Cannas, J. Perisse, S. Bauer, J.-R. Mace, Y. Ouerdane, S. Girard, "Vulnerability of OFDR-based distributed sensors to high γ-ray doses," accepted for publication in *Optics Express*, 2015.

115. J.-L. Bourgade, A. E. Costley, R. Reichle, E. R. Hodgson, W. Hsing, V. Glebov, M. Decreton, R. Leeper, J.-L. Leray, M. Dentan, T. Hutter, A. Moroño, D. Eder, W. Shmayda, B. Brichard, J. Baggio, L. Bertalot, G. Vayakis, M. Moran, T. C. Sangster, L. Vermeeren, C. Stoeckl, S. Girard, and G. Pien, "Diagnostic components in harsh radiation environments: Possible overlap in R&D requirements of inertial confinement and magnetic fusion systems," *Rev. Sci. Instrum.*, vol. 79, 10F304, 2008.

116. D. L. Griscom, "Intrinsic and extrinsic point defects in a-SiO2," in *The Physics and Technology of Amorphous SiO$_2$*, Editor R. A. B. Devine, Plenum Press, NY, pp. 125–134, 1988.

117. S. Agnello, "Gamma ray induced processes of point defect conversion in silica," PhD Thesis, Università di Palermo, 2000.

118. L. Skuja, M. Hirano, H. Hosono, K. Kajihara, "Defects in oxide glasses," *Phys. Stat. Sol.* (c), vol. 2 (1), pp. 15–24, 2005.

119. N. Richard, L. Martin-Samos, S. Girard, A. Ruini, A. Boukenter, Y. Ouerdane, "Oxygen deficient centers in silica: Optical properties within many-body perturbation theory," *Journal of Physics: Condensed Matter*, vol. 25, 335502, 2013.

120. L. Giacomazzi, L. Martin-Samos, A. Boukenter, Y. Ouerdane, S. Girard and N. Richard, "EPR parameters of E′ centers in v-SiO2 from first-principles calculations," *Physical Review B*, vol. 90, 014108, 2014.

Index

Page numbers followed by f, t and n indicate figures, tables and notes, respectively.